For more information, address

cactusbush@gmail.com

ISBN 979-8-234-08421-7

Preface

This book is part historic and part scientific in scope. It endeavors to entertain, not to be pedantic. Some simple subjects are outlined. Perhaps some practical, suggestions might be offered. The main goal though is to equip a furtive mind of the future, with a tool set of ideas.

This book contains hundreds of images; many more images than the number of pages in fact. Roughly one third of the approximate 820 images herein, are photographs or constructions by the author. But that means that the larger portion, the remainder are not. Those other images are humbly borrowed from public domain or Creative Commons type licensed works. The providence of each such image was examined and strictly vetted for appropriateness. This writer has the upmost respect for another individual's intellectual property.

Where feasible the source of a borrowed picture is included in a caption, in the nearby text or as a watermark within the image. While some publishers might eschew the practice, using old pictures makes a lot of sense, especially when one is exploring a historic topic like steam power. Here most of the excitement from research and development occurred centuries ago. The original interest in steam power has waned, is forgotten about or is practically dead today. Few in the present generation are contemplating, constructing or photographing - tantalizing new steam powered devices.

But tomorrow there might be urgent need for small, portable steam generators with engines. Should today's energy GRID collapse, the potential of small steam projects might quickly become self evident. But to realize this, the student needs to appreciate some past, forgotten machines.

To see what has gone before; to build upon what already works, to contemplate new possibilities.

A picture truly can be "worth a thousand words". Surely the long deceased creators who carved, drew or photographed some of these images - might have taken pride in knowing that their work can still be appreciated by future generations.

As many public libraries discard dusty old printed books to make room for computer terminals, posterity might suffer severely in the long run. Digitizing all images and prose is one thing, but afterwards depending solely upon the viability of digital archives could be a tragic mistake.

This volume reiterates a premise from the first book; that society might be ready to collapse at any moment. That a book dedicated to providing substance useful after such an event, might not go unappreciated.

> As of January 27, 2026, the "**Doomsday Clock**" is set at 85 seconds to midnight, the closest it has ever been. *"The Doomsday Clock is a symbolic representation of how close humanity is to global catastrophe, primarily due to nuclear risk, climate change, and disruptive technologies".*[1]

The first volume under this title (Dysfunctionally Civilized Vol 1) investigated the subjects of self orientation, dairy, food oils and party drugs. But this second volume unpretentiously wanders into the more immediate, more applicable subjects of manipulating fire, controlling fresh water, harnessing renewable steam, exploiting solar energy and finally some rudimentary construction. Short chapters with occasional solutions; projects fit for the contemplation and activity of isolated individuals, families or small communities.

1https://thebulletin.org/

As before the book's title, is defended by critically analyzing today's culture. It's not natural for any autonomous organism of a particular species to be so divorced from nature. But humans are. So hapless and alienated from nature in fact that most "civilized" individuals would perish in a matter of days if left unattended, no mater what the natural environment. They've no "horse sense", no survival skills whatsoever.

It is not natural to disfigure one's body with silicone implants or neurotoxic proteins. It's unnatural to be confused about one's own sex. It's not natural for complex organisms to breed promiscuously to the point of being irresponsible parents, and/or unnatural to exercise onanistic self-gratification or practice abortions to the point of never becoming parents to begin with. However this apolitical book does not mention or harp upon these issues or numerous other societal shortcomings.

Rather, this book attempts to assist worthy young individuals by offering information potentially conducive to their future survival. Survival potentially far outside and apart from the fragile and facile reality promulgated by today's 'nanny state'.

Tough times create strong men,
strong men create easy times.
Easy times create weak men,
weak men create tough times.

* A sentiment in passing. *(Origin: words widely attributed to G. Michael Hopf in his post-apocalyptic novel, Those Who Remain).*

But the notion is old. Consider biorhythms and the traditional Hindu cosmological notions of history being cyclic.

Table of Contents

let there be no Blank page

Chapter 1

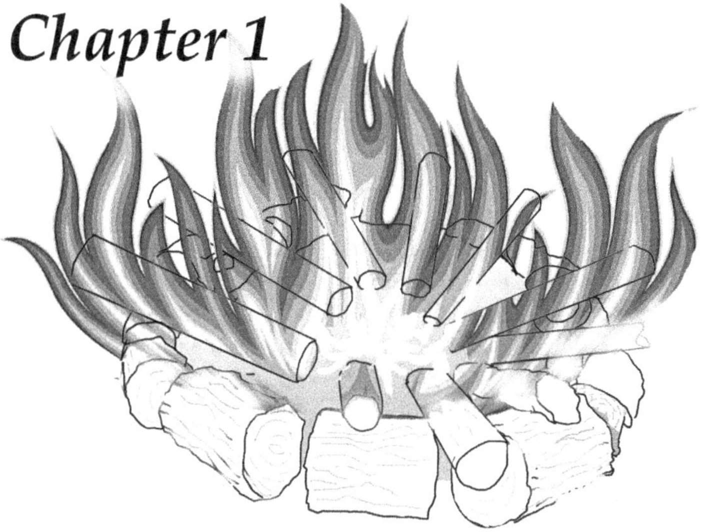

(wood stacked in preparation for a lime kiln perhaps)

Fire / Metallurgy / Ceramics

Controlling fire's concentration and increasing its heat were critical prerequisites for humanity's progress. Anthropological and archaeological categorization of humankind's progression, segregate our historic advancement into relatable periods or partitions. Three million years ago in what is considered the Old Stone Age, primitive humanoids did not know how to build a fire. Neanderthals lived and thrived and then disappeared as an independent species some 35,000 to 24,000 years ago; still within the Stone age.

Some academics advocate for the inclusion of an intermittent "Neolithic Age" (meaning "new stone"age); which begins about 10,000 years ago and runs up to about 4000 BC. During the Neolithic age, some humans domesticated plants and animals, ground and polished stone tools, lived in permanent housing (no longer nomadic), practiced weaving and crude pottery.

It was once the consensus of many anthropologist that humankind did not acquire reliable fire starting skills until somewhere around 9,000 years ago. In more recent decades though, that estimate was pushed back much further, given the arrival of early modern humans (*Homo sapiens / Cro-Magnons)*.

Homo Sapiens interacted and interbred with the indigenous Neanderthals *(Homo neanderthalensis)* of Europe and Western Asia, who went extinct about 35 millennia ago. Off and on, tribes of these people had fire. Fire can be found naturally. It is a struggle to **date fire-making** because fire leaves behind a poor record. Charcoal and ash dissolve or are blow away. So while scientists might place reliable human fire-making skills at around 50,000 years ago, this is still just an estimate. They have casually based their **"estimate"** solely upon a collection of factors like diet, brain growth, social life and migration patterns.

Recently (in late 2025), archaeologist in Barnham, England discovered what they assume to be proof that mankind's oldest technology dates back to an incredible 400,000 years ago. (This is a time before Cro-Magnons, Neanderthals and Denisovans (Asian pre-human hominids)). The soil at this site showed repeated heating at high temperatures, and presence of the mineral iron pyrite. Stones in the vicinity exhibited signs of thermal shock; which requires achieving high temperatures between 752 and 1,382 degrees F. (The heat of a normal campfire is about 600° F (315° C). Numerous fragments of iron pyrite were also found at the dig. (It makes sparks when struck with flint). Alien to this particular area, the iron pyrite had to have been brought there.

Just because it appears that someone was making fires at this location 400,000 years ago does not mean that the practice was widespread. *"Fire technology may have appeared, vanished, and reappeared across generations before becoming reliable and routine".*[2]

Ultimately it was the ability to better control heat, smelt metals and make better tools that elevated a society's development. Some early societies were isolated and so entered the "Bronze Age" at different times (say somewhere between 3000 BC and 1900 BC). But after that it took roughly another 3,000 years and an increase of about 500° Celsius in smelting temperatures, before humans ascended into the "Iron Age". Copper melts at 1,084° C. Iron melts at 1,538° C. That later temperature is impossible to achieve in an open fire pit. To smelt iron from ore you would at least need a "**bloomery**" (the earliest form of furnace).

As humans experimented with and improved the heat of fire,

2 https://www.nhm.ac.uk/press-office/press-releases/groundbreaking-discovery-shows-humans-were-making-fire-350-000-y.html

they even cooked rocks and dirt – and significantly thereby, created or discovered different metals, ceramics and glasses. Metals, ceramics and glass can be used to manufacture trade items. An activity which ranks right up there with other qualifying requirements (like plant and animal domestication, division of labor and written language) to define what civilization really is.

*(**Civilization** is characterized by having either cosmopolitan or squalid urban centers, then evidence of advanced technologies, layered social hierarchy, complex political and economic systems. **Societies are** groups of individuals living together,while sharing customs and culture; not all societies qualify as a civilizations. **Culture** represents inner qualities, the beliefs, values, and behaviors of a society).*

Furnaces allowed combustion air to be forced in at the base, making the fires much hotter. Human powered bellows supplied this compressed air at first. But humans tire quickly. Eventually, new bloomery furnaces and foundries were being built nearby streams or rivers, so that water power could be harnessed. Waterwheels were used to power the bellows that compressed air. Better air supply led to much larger furnaces and bigger metal yields. Increased heats that could now turn soft metals and even iron - into runny liquids.

* Waterwheels might be considered to be the **first engines**. They convert one form of energy into a mechanical force; rotational torque. (A mechanical device has parts that move when it is working). Waterwheels (engines) date back 2,500 years perhaps.

The Versatility of a Fire

*"One reason why people in industrial societies regard **open fire** as inefficient and unsustainable is because they simply don't know how their ancestors actually used it. If these days a fire is considered to be inefficient, it's because we only measure the efficiency of one of its functions, usually space heating. However, our ancestors did not only use the fire to warm themselves. They also **used it for cooking, illumination, food preservation, hot water production, clothes drying,** and **protection from predators and insects, among other things"**.*[3]

3 https://solar.lowtechmagazine.com/2019/12/too-much-combustion-too-little-fire/

Building a Fire

Before the invention of the phosphorous friction match only two short centuries ago, cultured society's best fire starting technologies were scarcely improvements upon, or less tedious to perform than those ancient methods used by wild aborigine contemporaries. We civilized moderns scoff at what we may consider to be the very simple technology behind 'strike anywhere kitchen matches' or butane cigarette lighters. But these were important inventions that took more than **90 - long, cold centuries to perfect**.

Very, very few comparatively sophisticated people living today could even build a fire in an emergency if they had to (without props like matches, cigarette lighters, or other devices).

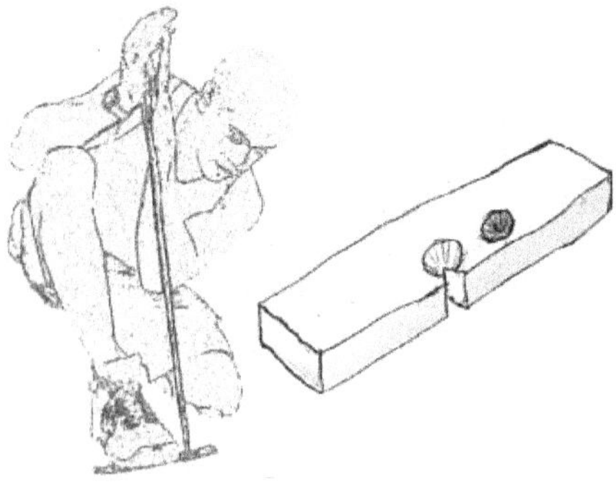

Starting from scratch. Possibly the earliest fire starting method was the **fire drill**. In this friction method the drill is a shaft of wood spun by hand pressure. **Dissimilar woods** are primarily chosen, usually a harder wood for the shaft and softer one for the plank. Continued friction causes powder or dust to separate from the softer wood and become heated. A **base plank** is specially prepared with a notch to allow the tiny precious glowing ember of hot dust to fall out onto some tinder. The person blows on the **ember**, creates a flame, and then adds more tinder and kindling – to make the flame grow.

The whole process appears easy enough, but in truth can be a significant chore. It's a question of experience, technique and good tinder. The world's record for getting a suitably hot coal with a hand-drill is 4.5 seconds.

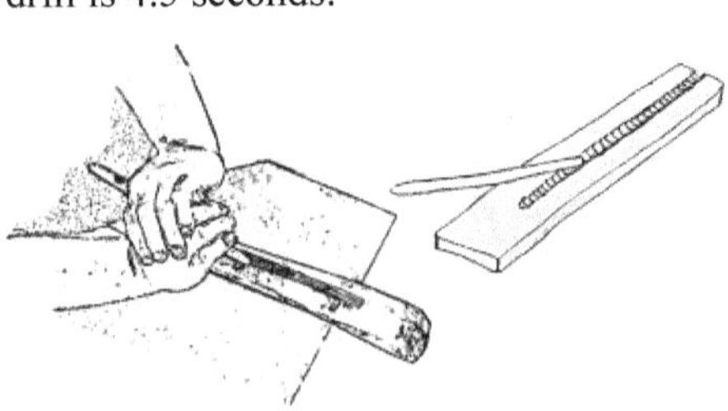

Another ancient friction fire starting method is the **wood plough**. Popular in Polynesian cultures, this method also requires careful selection of woods. The base is usually a small tree trunk or staff of soft wood with a grove worn into it. While the base is held immobile a plow of smaller diameter hardwood is draw back and fourth in the grove.

As with the fire drill, friction creates dust which turns into a hot ember, which is then dropped into some tinder. Like the fire drill, the fire plow requires experience and steadily applied pressure to work.

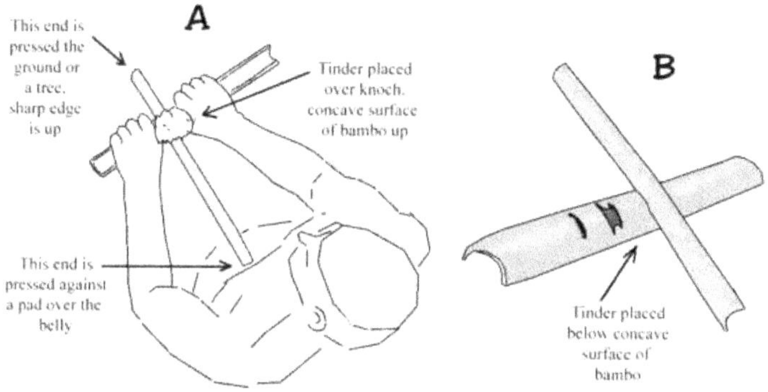

A method of primitive fire starting popular in Indonesia is the "**bamboo fire saw**". A short section of dry bamboo is split in half. On one half-section a small notch is started with a knife. The other half-section that is to be used for the sawing can be whittled down in size and one edge should be sharp. In method "A" (previous image), the saw is below the tinder and is held still by the body pressing it against a firm object. The ember is caught in the tinder above. In method "B" above, the position of the saw is reversed and it is held in the hand. The ember falls onto the tinder below.

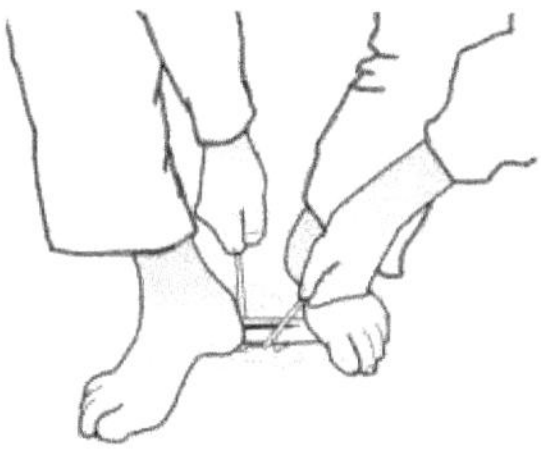

A New Guinea variation of the bamboo fire saw utilizes a thin strand of bamboo or tough local vine as **a rope saw**. The bamboo plank and tinder beneath are stood upon as the friction is applied.

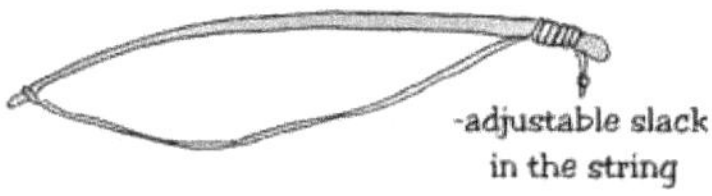

The **bow drill** is an adaptation of the fire drill / hand drill method. Also ancient. Egyptians were using this method while building their pyramids. Most other civilizations that used the bow for archery also probably learned to use it as a fire starting tool. This was the favored method of some American Indian tribes although they did not forget about the hand drill. An archery bow will work for fire starting, but a bow for such a chore does not need to be so big. In fact a small branch about 2 feet long, with a small curvature, is optimum.

The same kind of bottom plank or **fire-board** incorporated by the fire drill method is used. Other acceptable fire-boards are a pair of branches tied together, a branch that has a season split or a chunk of dead and dried tree fungus.

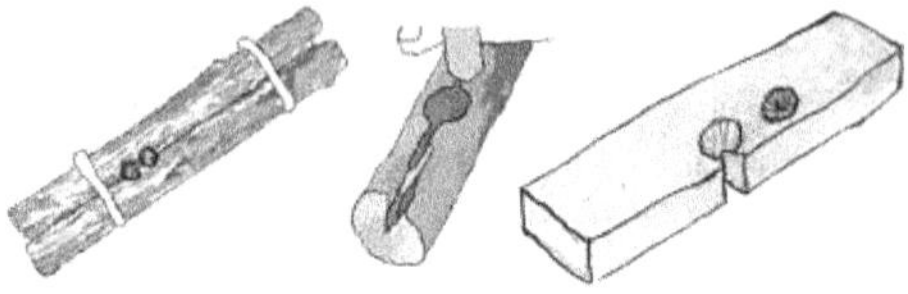

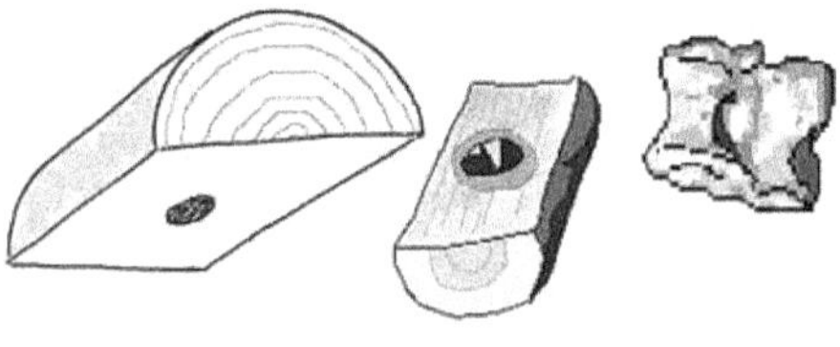

A **socket** of wood or bone knuckle is griped in one hand and applies downward pressure upon the spinning spindle. Since the friction and heat is wanted only at the fire-board end of the shaft, the top end which is held by hand will benefit from a socket with some form of lubrication or a hard insert to reduce friction. If the socket is wood then a metal bottle cap or a small concave stone insert will reduce friction while allowing more pressure to be gradually applied.

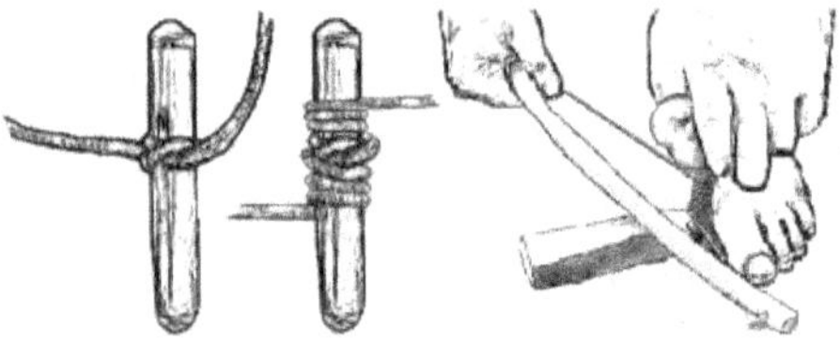

The types of wood used for spindle and fire-board make a big difference. Again, dissimilar woods work best. Given the choices at a random location – only experimentation will tell. Yucca and Elm rate highly but Maple and Pine do not. The **Egyptian bow drill** (above) used several millennia ago, was usually a tiny, refined kit of parts that worked efficiently and was easy to transport. One of its attributes was the fact that the spindle was attached to the string. Extra coils of string were wrapped around the spindle (wrapped both directions from center). This allowed better traction and control over the spindle. The spindle was fastened either by the string passing through a hole in the spindle or by the tying of a simple clove hitch knot. Someone skilled in its use can start a flame within 25 seconds, using a bow drill. The world record in the late 1930's for getting a flame with a bow drill was 7.5 seconds.

The ingenious "**fire piston**"has been around for some time. This device was first discovered by Europeans in the 1860's when visiting Indonesia. (* Indonesia incorporates Sumatra, Indo China, the Philippines, Borneo and about **17,508 islands** in between).

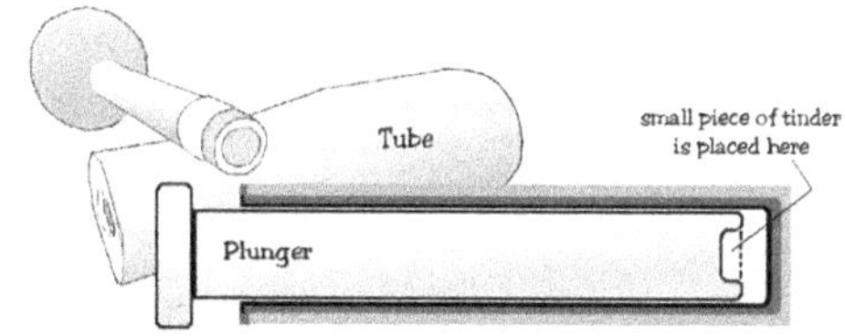

The very reliable fire piston is thought to be an ancient device because of its wide distribution. It may have been a co-development of the blow gun or blow tube. The fire piston works on the same principle as the Diesel engine, and tight tolerances are required. A hand sized tube is fitted with a close fitting rod (piston). For a tight seal the rod is fitted with a gasket of string or sinew and packed with animal fat or wax. A small piece of tinder is placed in the dimple of the plunger, the plunger inserted into the tube, and a pump or two of the piston generates enough heat (through pressure) to ignite the ember.

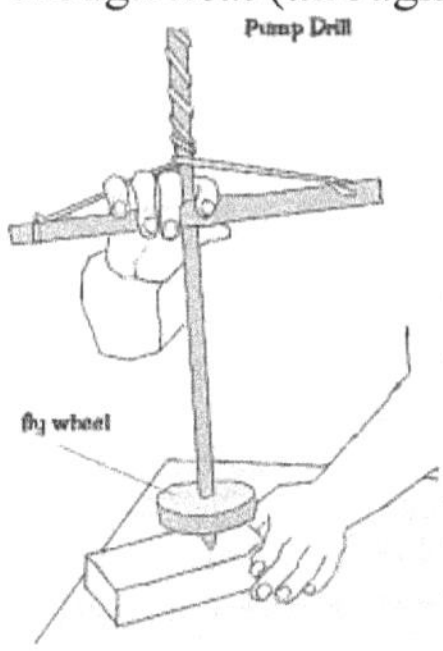

Another old, similar contraption to the bow drill is the **pump drill**. The pump drill would be only slightly more complicated to build than the bow drill. Used correctly, the spindle can be kept in continual motion by the centrifugal force or inertia of the flywheel and rhythmic motion of the pumping hand. Useful friction would only be exerted on the downward stroke however. So while this may or may not suffice to create a usable glowing – fire starting ember, the tool itself is useful for drilling holes. Has been for several millennia.

For eons, both **Pyrite** and **Marcasite** (white iron pyrite) have been used to start fires. Both minerals share the same chemical composition (iron sulfide / FeS_2) and at times might be confused, but they have different crystalline forms. Marcasite is lighter and more brittle than pyrite. Pyrite and a flint stone were carried by the 5,000 year old mummy known as "The Iceman"; who was found at high altitude along the Austria–Italy border in 1991. Iron pyrite, also known as "**fools gold**", was used to ignite gunpowder in some 16th and 17th century wheellock firearms.

Before the 19th century the most advanced means of initiating a fire in Europe was done with implements held or kept in a kit, called a **tinderbox**.

The tinderbox typically contained a sharp piece of flint-stone, a steel striker and some tinder. (*The term "tinderbox" has become almost synonymous with something so dry that it will ignite almost spontaneously).*

Tinderboxes usually had a carbon steel striker that was often bent to fit around the hand. These forged strikers date back as far as Roman times. (High carbon steels generate sparks easily, while softer iron doesn't).

A **flint-stone** was a piece of flint, chert or even iron pyrite. Like quartz, flint has a Moh's hardness of 7 or above. A rock with a cryprocrystalline (minute crystal) texture that when struck against metal, would cut tiny shavings off that would ignite. **It's the metal / not the rock – that ignites.**

Tinder is simply the initial fuel; some type of very combustible material. Tinder can be anything from **char cloth** (linen or cotton cloth that has been pre-burned in a low oxygen environment) to spider webs, various plant fibers, termite dust, grass, pitch wood, bird's nest, down, fungus, Spanish tree moss, paper from wasp or hornet's nest, oakum, cotton balls dipped in Vaseline or lint taken from a clothes dryer.

Tinder needs to be dry, fibrous, fluffy, and highly ignitable. Many materials can be masticated and crushed to make them more fibrous.

a short diversion

(explaining the hardness of flint:)

Geologist in the field, test and identify minerals by their physical properties. Those properties include luster, color, streak, crystal habit, cleavage, fracture and hardness. Hardness is where the informal **Mohs Hardness Scale** becomes useful. It's a simple, relative ranking of a material's ability to resist scratching (abrasion). Friedrich Moh created the scale (1812), which grades minerals from 1 (soft) to 10 (hard). But, the scale is not linear.

Mohs Hardness Scale

		substitute	*notes*
10	Diamond		
9	Corundum	ruby, sapphire	(a)
8	Topaz	emerald, aquamarine, beryl	
7	Quartz	agate, amethyst	(b)
6	Feldspar	rhodonite, orthoclase, glass	(c)
5	Apatite		(d)
4	Fluorite	malachite, rhodochrosite,	
3	Calcite	pearl, coral, penny	
2	Gypsum		
1	Talc		

Notes – on the previous table

(a) **Corundum** is a crystal formed by aluminum oxide and some transition metal *(like iron, chromium, t itanium, vanadium and 36 others)*. A ruby might look red because of the influence of chromium and a blue sapphire is colored by the presence of iron and titanium. Makes good sandpaper.

(b) **Topaz is** a gemstone and silicate of fluorine and aluminum. Masonry drill bit hardness is averaged somewhere around 8.5 (might be constructed with diamond dust, titanium or tungsten carbide). Synthetic zirconia and alexandrite (chrysoberyl) gemstones rate about 8.5. Minerals and gemstones like garnet, iolite and tourmaline rank 7.5 – harder than quartz.

(c) Zircon, andradite, zoisite, tanzanite, jade, pyrite, hematite and kunzite, **steel** and unglazed ceramic rank about 6.5 in hardness. **Feldspar** (#6) and or orthoclase make up about 60% of the earth's crust.

(d) Opal, lapis lazuli, turquoise, glass plate and a steel knife, all rank about 5.5 on the scale. Harder than apatite, which is a crystal often found in igneous and metamorphic rocks.

(+) Calcite is a form of calcium carbonate or limestone that can polymorph or crystallize into more than one crystal structure. A real copper penny (which before 1982 was an alloy of 95% copper and 5% zinc) is actually harder than calcite, with a rating near 3.5. A penny was still a degree of magnitude harder than silver, gold, a human fingernail, amber (fossilized tree resin) or serpentine (a hydrous magnesium silicate usually having a dull green color and mottled appearance) all rating about 2.5 in hardness.

(+) Gypsum which is a soft rock of calcium sulfate and talc (magnesium silicate) are rocks that can both be scratched with the human fingernail. Gypsum is used in drywall and plaster, while talc can be used to extend house paints, in cosmetics and as baby powder. Soapstone is metamorphic talc that can be sculpted with a knife or other bodkin.

(+) **Cardinal gems** are gemstones which have traditionally been considered precious above all others. Five traditional cardinal gems of the Old World were the purple amethyst, green emerald, red ruby, blue sapphire and clear diamond. A second rank of gemstones (**semiprecious**) includes topaz, aquamarine, morganite, and tourmaline (and now amethyst – demoted because it is no longer rare / large deposits were discovered in Brazil).

With gemstones, abrasion testing isn't done very often because it could lower the value of a gem. A diamond can easily cut or score a sapphire or ruby both (both are forms of corundum also) as well as all lesser gems. Diamond (rank 10) is 4-5 times harder than corundum (rank #9). A ruby would scratch an emerald.

* Diamond is no longer considered to be the hardest material known to man either. Scientist have created a few unnatural substances now, that are even harder.

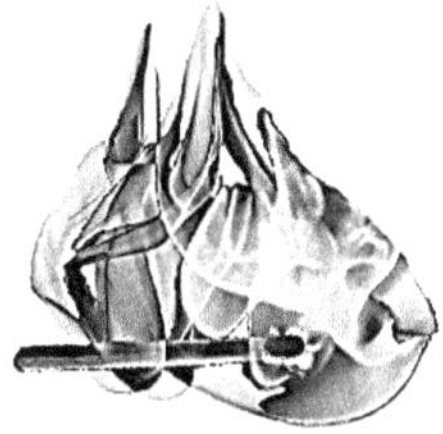

back to fire

In all these afore mentioned fire starting techniques the precious spark or glowing ember must be captured by the tinder, then skillfully manipulated. Blowing or fanning air provides extra oxygen to assist in the creation of a flame.

A more recent equivalent to the old flint and steel combination is the **ferrocerium rod**. Typically found in cigarette lighter "flints", wind-up toys that spark or in a welder's striker; ferrocerium is a man made mix of cerium and iron. (**Cerium (Ce**) atomic # 58, is a rare-earth metal).

In a toy or welder's striker a piece of ferrocerium is usually pushed by a spring, against an abrasive piece of steel, which moves; creating sparks. As with the old flint and steel method, friction ignites tiny shavings of metal. But in this situation however the iron in ferrocerium burns, not the harder abrasive steel. Cerium's low temperature pyrophoricity is responsible for the easy sparking. A modern survival kit might contain a single rod of ferrocerium as a fire starter. It's tough and waterproof.

Better yet, a survival kit might include a **magnesium fire starter**; basically the same ferrocerium rod, but now attached to its own tender. Shavings of magnesium are scraped off into a little pile (already atop paper or another tinder). The sparks are then scraped off the ferrocerium rod, down onto the magnesium flakes, which should then burst into a bright, but brief flame.

Some Match History

In 1669 Hamburg Germany, an alchemist was trying to convert some of "life's essence" into gold. He saved buckets of his own urine, let it rot, then boiled it down to a paste, then cooked it some more - "letting the vapors travel through water". What he got was a waxy substance that glowed in the dark (***ammonia sodium hydrogen phosphate***).

Two years later the Irishman physicist Robert Boyle (of-Boyle's Law) rubs this newfound **phosphorus** against some sulfur and creates a flame. Boyle did not exploit his opportunity to invent the friction match. Mankind was to wait yet another 1.5 centuries before finding an easier way to start a fire.

Along came an English apothecary and chemist in 1827. He invents a functional but impractical match called the "**Prometheus**". This was a wood splinter with a potassium chlorate head placed next to a tiny glass bead of sulfuric acid, then rolled in paper. A person used tweezers or a bite with the teeth to break the glass and set off the flame. More importantly, our apothecary later sticks a mixture of starch, gum Arabic, antimony sulfide and potassium chlorate onto a stick and lets it dry. This invention he calls a "**Congreve**", named after an officer who had introduced War Rockets to the British arsenal.

- Those **rockets** were large artillery projectiles (16' long) that could rise 9,000 feet in the sky and which sprouted great flames. Some were used against Fort McHenry- Baltimore harbor, in the **War of 1812**).

Our English chemist and friction match inventor did sell a few matches but he did not get rich. Another Englishman exploited the commercial market for these matches, and renamed them "**Lucifer s**". They became very popular with smokers, but stank.

In 1830 a French chemist created a match that did not stink, using white phosphorous which was highly reactive and toxic. But these worked very well.

1888/ public domain / early signs of Phossy Jaw

During the next 50 years, large match factories were created that mostly exploited the cheap labor of children, and young women.

"**Phossy Jaw**" was a famous ailment caused by inhalation of white and yellow phosphorus vapors in the match factories and often led to death.

* The English "**Suffragette** Movement" and a defining moment in trade union history started with women striking against conditions and hazards of the match factories, in the Bow district of London. Their first strike was in 1888.

In 1855 a Swede created the first **safety match**, using less dangerous red phosphorus and ignitable only on the box. In 1889 the first **matchbook matches** were invented and were called "*flexibles*".

By 1910 the Diamond Match Co. patented the first nonpoisonous match, using sequisulfide of phosphorous. Asked by President Taft to release their patent for the good of mankind, Diamond Match did so in 1911.

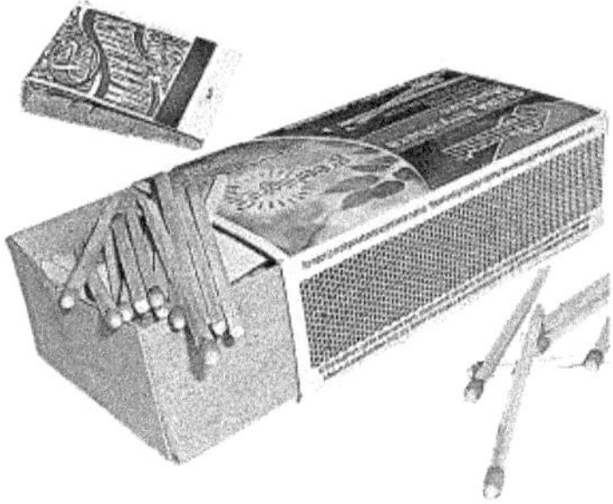

A century later the once common "**strike anywhere**" type of kitchen match has become **rare** in the US today. They are actually illegal in 7 states! *(Alaska, California, Minnesota, New Jersey, Ohio, Virginia and Wisconsin)*. Reasoning varies, from the higher rates shippers like UPS and FedEx charge for transporting "dangerous goods", to possibly fears that new upstarts imitating Ted Kaczynski (math prodigy and Unabomber) might use the chemicals to make pipe bombs.

Modern safety matches (the **strike on the box** kind) usually have antimony sulfide, potassium chlorate, and sulfur or charcoal in the heads, and red phosphorus in the striking surface. The non toxic **red phosphorous** is glued down so tightly to the cardboard of the box that it mostly defies being scraped off with a knife.

- (Otherwise **red phosphorus** is illegal or classified as a List I chemical by the DEA – because they are worried about it being used to manufacture methamphetamine.

The more useful (old fashioned, non-safety) **strike anywhere type** of matches usually have potassium chlorate and phosphorus sesquisulfide (P_4S_3) in the heads. Although the tip of the match may look like white phosphorous, it is not.

"Dangerous" white phosphorous and "safe" red phosphorous are but two of several **allotropes** of this same element (atomic # 15). White phosphorus, yellow phosphorus and tetraphosphorus are all the same thing, with four atoms (P_4) arranged in an unstable tetrahedral structure.

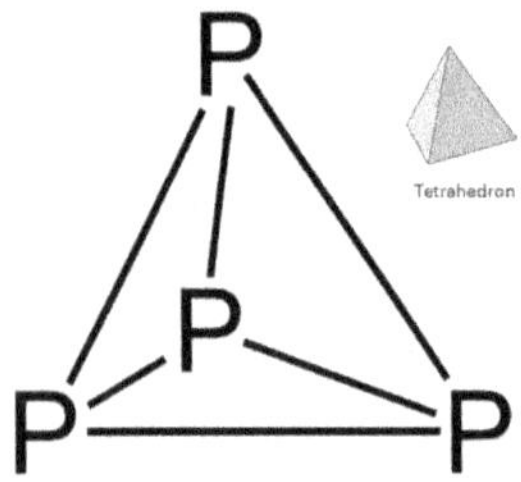

Tetrahedron

Pure **white phosphorus** can glow in the dark, which is a result of its vapors slowly being oxidized by the air. It is so thermodynamically unstable that it combusts in air. It is not commonly used in fireworks any longer but militarizes still use it in incendiary munitions. Red phosphorous has an amorphous structure and is more stable and less toxic. These two allotropes can be converted back and fourth from one to the other.

Alternative methods

Everyone probably knows that by using a magnifying glass, energy from sunlight can be concentrated enough to start a fire. Without a quality lens and the cooperation of good weather and direct sunlight however, this task is easier said than done.

A ray of light passing through the center of a thin lens keeps its original direction. A ray that strikes anywhere else is bent. The amount a light ray is bent increases with its distance from the center of the lens.

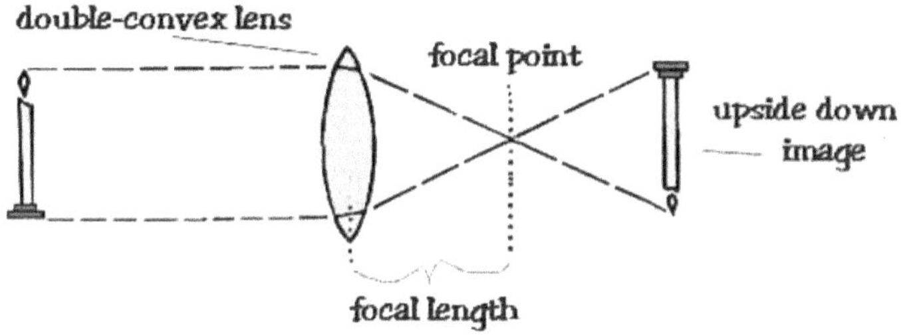

A magnifying glass is actually a **double-convex lens**. It can gather the energy from a broad area and concentrate it into a smaller area. The focal point or hot spot is where parallel light rays converge (cross) along the principle axis of the lens.

Dust and scratches or imperfections of the lens will considerably diffract the light and lessen the practicality of this fire starting method. Plastic or toy magnifying lenses generally diffract, diffuse or scatter so much sunlight that they are useless for starting fires. One is not likely to have or to run across a magnifying lens in an emergency situation either.

Other types of lenses might be available however, and might be drafted into making an improvised double-convex lens. Glass lenses can be salvaged from eye-wear, cameras, binoculars, and telescopes.

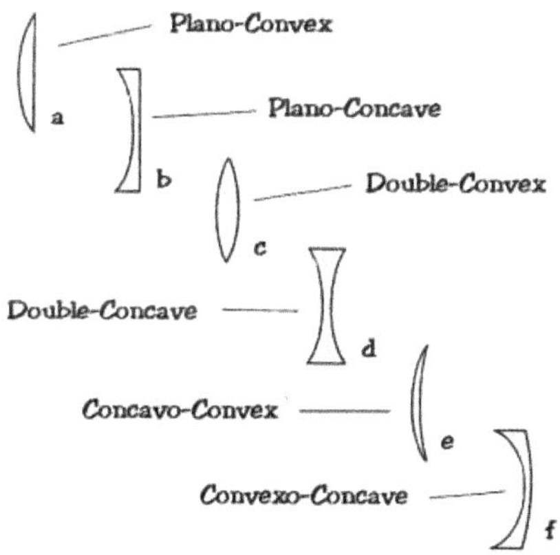

In this image, shape *a* & *b*, bend light in a way that is not beneficial to fire starting. A pair of lenses with the shape *a*, back to back however might adequately mimic a double-convex lens. The last two lens shapes *e* & *f* are the type normally found in eyeglasses.

A drop of water placed on the back or inside of lens type **e** or **f**, will produce a temporary double-convex shape. The surface tension of the water droplet should produce the opposing convex surface. It takes a very steady hand to find the optimum focal point and to hold eye-wear and water still; long enough to initiate a flame. Not all eye-wear is created equal so the vision prescription will actually be a factor in any success.

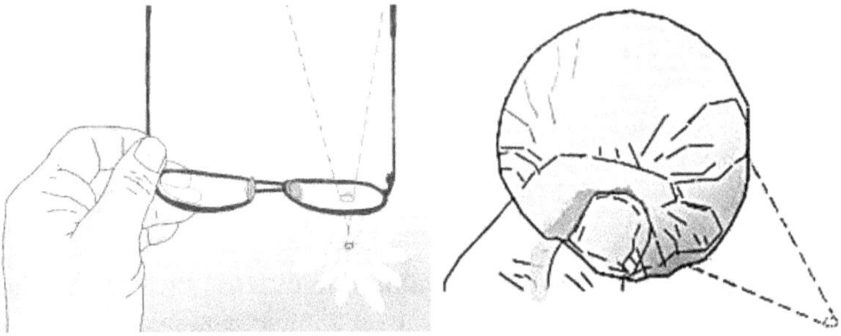

Another wannabe clever idea for making a fire in an emergency involves using a simple clear plastic sandwich baggie. The intention is that one fills the baggie with water, and then twists the contents into a sphere, thereby constructing a makeshift double convex lens. In real life, it is almost impossible to hold the water still enough for the concept to work.

A clear chunk of ice might also theoretically concentrate solar energy in one spot, long enough to ignite some tinder. Unlikely. Double-convex shaped ice cubes are hard to come by.

Notions of starting a fires with anything less than good clear lenses, are nifty but usually impractical.

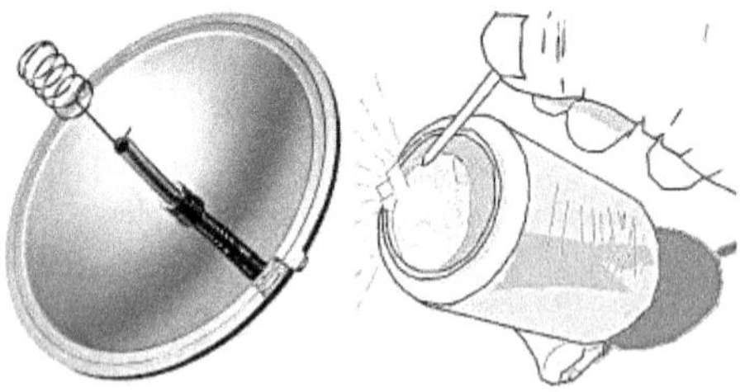

A **parabolic mirror** or highly polished parabolic surface can also capture heat from the sun. If held at the correct angle, the surface will concentrate the light into one small spot along the edge of the parabola. At left is a solar cigarette lighter. It has two springs; one to hold the cigarette at the proper focal length distance and the other to allow finger pressure to bend the holder around to exploit the hottest spot. It's not as easy as it looks to achieve ignition, everything needs to be held steady and still. At the right in the same image, is the **concave** bottom of a beer or soda can. Normally the surface of such a can would need to be polished as much as possible. If steel wool cannot be acquired, perhaps the diataneous silica inside the leaves of salt marsh plants or graphite, as from pencil lead might work. *(Graphite is commonly used as a lubricant but it can also perform as a mild abrasive)*. The polished parabolic surface is then propped up by rocks and pebbles until the sun is caught at a very small spot at the bottom edge of the can. Very small pieces of tinder are then dropped onto the hot spot and should become hot enough to ignite.

Flashlight batteries & steel wool are a handy way to start a fire. For many decades Boy Scout manuals have advocated this trick. One strips a small ribbon of wool to the proper length to reach both positive and negative terminals, and then shorts it out.

Two 1.5 volt cells in series provide 3 volts, which (if using "C" cells) is usually enough energy to make the steel wool glow red hot and then ignite. The procedure can work surprisingly well; but one is unlikely to have both steel wool and batteries in his pocket – in time of need.

It may not be appropriate to call someone a good **Boy Scout** because they can start a fire. The Boy Scout movement began 1908 in British Isles, and spread to America by 1910. The society attracted urban boys, but rarely their countrified counterparts, whom might have been able to teach the urban instructors a thing or two. For a century now, thirteen different editions of the Boy **Scout handbook** have traditionally placed more emphasis upon controlling fire rather than starting it. Never has skill in fire making been a requirement for scout advancement. In 1911 the only fire related merit badge was for Firemanship, which focused upon extinguishing fires, safety and avoiding panic. Today's equivalent merit badge for Fire Safety differs only slightly by including small requirements like that of igniting a camp stove prudently. Where modern Scout manuals might have more information concerned with outdoor skills, yesteryear's handbooks were more moralistically toned. Older editions focused more upon building leadership, character and integrity.

Yet another interesting proposal is that of using **firearm ammunition** to build a fire. The notion of dumping ½ of the gunpowder from a cartridge out and stuffing in a bit of rag cloth to replace the bullet is suggested. Ideally the propelled cloth should smolder long enough to ignite tinder and/or excess gunpowder. In practice however the cartridge primer usually just ejects the cloth and fails to ignite anything. The process might work for someone in a Hollywood movie, or under special conditions with the right firearm using a cartridge with the right powder. But mainly this is just a fanciful idea; that doesn't work.

> There is a wide variety of smokeless gunpowders (**propellants**) – and their burn rates differ widely . Pistol cartridges use the fastest, most volatile nitrocellulose powders. Shotgun powders use the equivalent of slower pistol powders. Rifle powders are generally the largest grained, slowest burning powders of all; and therefore should be the hardest to ignite. Firecracker or 'flash' power is faster than black powder and both deflagrate faster than nitrocellulose powders.

In any event, gunpowder from any firearm cartridge would compliment other tinder or help start a fire from an ember achieved by other means. Gunpowder can be unpredictable or sometimes troublesome to ignite. Many a fool has blackened his face or singed his eyebrows by patently holding a flame to the stuff.

Throughout history, flame and its heat have played an indispensable role in the development of human technology and civilization. Today, refined heat allows us to create sophisticated materials, it extends our lives by allowing us to survive cold weather, it helps generate our electrical power and kills diseases in our food.

Simple Metallurgy

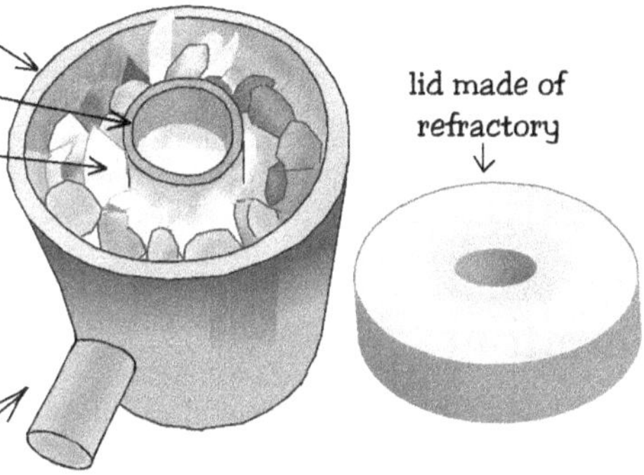

How hot is HOT?

While physicist agree upon an absolute lowest temperature (**absolute zero** – where even subatomic particles don't move) there is no consensus or formally defined limit for a maximum temperature. The best approximation of maximum temperature might be **Planck temperature (1.4168 × 10^32 Kelvin)**. That's about 100 million million million million million degrees in other words.

- We have no physical instruments like thermometers or thermocouples to physically measure these just mentioned temperatures; but instead must rely upon idealized **thermodynamic theory** to extrapolate these numbers.

The surface of the sun and the earth's inner core are both much cooler at about 5,778 K (5,505 °C) each.

The fission bomb "Little Boy" dropped on Hiroshima generated a heat of about 299,726 °C at its core.

Within a thermonuclear bomb a temperature of 50 million °C is needed to initiate the fusion of a deuterium and tritium tamper (both are isotopes of hydrogen; hydrogen-2 and radioactive hydrogen-3).

The temperature at the core of our sun is assumed to be about 15 million °C.

House fires and forest fires can sometimes reach impressive heats;

far hotter than a little campfire. At their flame front, wildfires can heat the surrounding air to 1,470 °F (800 °C). If fed by wind, the internal temperature of a wildfire might surpass 2,192°F. That's a temperature high enough to substantially soften steel or to liquefy several other types of metal. Heats able to warp and bend metal I-beams. For this reason Uniform Building Codes for new domiciles stipulate that any load bearing steel beams if present, must be wrapped and fireproofed to provide egress time for the residents.

Masonry like rocks in a chimney, can sometimes explode like bombs during a house fire too. The pressure comes from steam buildup created by moisture entrapped within the rocks.

Smelting

Smelting is the separation of metal from its ore. The reduction of aluminum using electrolysis instead of heat, can also be called smelting. Smelting with heat is often assisted by adding a reducing agent and a **flux.**

When smelting iron, coke or charcoal are added to the crushed ore within a traditional blast furnace and act as a **reducer** in the redox (reduction-oxidation) reaction. Carbon monoxide is produced as the oxygen is striped from the iron ore. Limestone, carbonate of soda, potash and lime might be used as a flux or **slag** forming agent to absorb impurities into a slag that can be separated from the liquid molten metal.

Ore

Through chemical decomposition, oxidization and other natural processes happening over geologic time, few metals are found physically in metallic form. Most of the earth's retractable metals are dispersed as small flakes or **inclusions** within an ore of some type. Since gold, copper, silver, and metals of the 'platinum group' are not very reactive chemically, early man was occasionally able to find bits of these "native metals" just lying upon the ground. Mankind likely first encountered metallic iron however, in the form of a deposited meteorite. With the low grade copper ores accessible today, soap bubbles and pine oils are frequently used as **reagents** to detach the metal from its crushed ore slurry. The cyanide process (cyanidation) can be used to extract (smelt) gold, copper, zinc or silver from their low-grade ores. Mercury dissolves gold and can form **amalgams** with several other metals as well. Easily separated from its crushed ore the gold can further be separated from the amalgam (in small samples) by squeezing it through a rag of chamois leather or by baking it in a potato. The mercury itself is usually recovered later by the distillation of its gas.

Some example melting points:

Tin	449° F	232° C	Sn, #50
Lead	621° F	327° C	Pb, #82
Antimony	1,167° F	630.6° C	Sb, #51
Magnesium	1,202° F	650° C	Mg, #12
Aluminum	1,220° F	660° C	Al, #13
lava		800°C	
Silver	1,763° F	962° C	Ag, #47
Gold	1,947° F	1,064° C	Au, #79
Copper	1,984° F	1,084° C	Cu, #29
lava		1100° C	basalt
Silicon	2,577° F	1,414° C	Si, #14
Nickel	2,651° F	1,455° C	Ni, #28
glass	2,700° F	1,500° C	soda lime
Iron	2,800° F	1,538° C	Fe, #26
Titanium	3,034° F	1,668° C	Ti, #32
Platinum	3,215° F	1,768° C	Pt, #78
kaolin	3,275° F	1,800° C	porcelain
Vanadium	3,470° F	1,919° C	V, #23
glass	4,200° F	2,300° C	silicon
Molybdenum	4,753° F	2,623° C	Mo, #42
Tungsten	6,192° F	3,422° C	W, # 74
Carbon	-	-	C, #6

Allotropes (an element's differing forms of atom arrangement) of carbon have the highest thermal conductivities of all known materials and they don't melt. Carbon undergoes **sublimation** at about 9,980 °F (5,530 °C) which is to say that the element transitions from a solid to a gas without passing through a liquid phase.

Carbon is also the fourth most common element in the universe by mass, forms more recognizable compounds than any other element and is the chemical basis or building block for all known life.

The **thermal conductivity** of a material is a measure of its ability to conduct heat. The opposite, or reciprocal of thermal conductivity is called **thermal resistivity**. Both are measured by an SI unit known as "watt per meter-kelvin" or (W/(m·K)). The thermal conductivity of most liquids and solids will vary with temperature. In general; as temp rises so does thermal conductivity. The fact that diamond has the highest hardness and the highest thermal conductivity of any material, has something to do with its inter-molecular spacing. Who would have imagined that a diamond would transmit heat, 67 times faster than stainless steel ?

Thermal conductivity (approximate @ 25° C)
W/m °C (watts per metre Celsius)

Diamond	1000	Nickel	91
Silver	420	Iron	68
Copper	400	Steel (low carbon)	48
Gold	312	Steel (stainless)	15
Aluminum	240	concrete	0.8
Tungsten	180	water	0.58
Sodium	135	wood (oak)	0.17
Zinc	112	air	0.03

Thermal conductivity and electrical conductivity are two very different things. A blacksmith would probably not be concerned with the electrical conductivity of a metal, but it will be mentioned here anyway. Silver, gold and copper are at, or near the top conductors in both categories.

Electrical Conductivity is the reciprocal (inverse) of electrical resistivity. "Free electrons" in the outer shell of a metal's atom, allow the conduction of electrical current. Too many valence electrons in an outer shell, will absorb energy and cause resistance. Metals that have a single valence electrons are the best conductors because these are free to move and cause a strong repelling reaction in other electrons. Silver, gold, and copper are the best conductors, and they each have a single valence electron.

- Electrical conductivity is affected by temperature as well. Increasing the temperature decreases the conductivity; conversely, reducing temp of conductor reduces the resistance). That's why rooms for supercomputers are refrigerated, for example. Or for the same reason that "heat sinks" or little fans are attached to cool the CPUs of personal computers.

Standard electrical resistivity (the opposition to the flow of electricity) is usually officially calibrated across the opposite faces of a one-meter cube of material and described as an **ohm meter** ($\Omega \cdot m$). [$\rho(\Omega \cdot m)$ @ 20°C]. Electrical conductivity is usually measured by **siemens per meter** ($S \cdot m^{-1}$). [$\sigma(S/m)$ @ 20°C].

Electrical Conductivity of some Materials

Silver	63 x 10 ^6	Zinc	16.6 x 10 ^6
Copper	59.6 x 10 ^6	Nickel	14.3 x 10 ^6
Gold	45.2 x 10 ^ 6	Lithium	10.8 x 10 ^6
Aluminum	37.7 x 10 ^6	Iron	9.93 x 10^6
Magnesium	22.6 x 10 ^6	Lead	4.87 x 10 ^6
Mercury	1.044 x 10 ^6	Water (fresh)	0.0005 - 0.05

Some alloying elements

Referencing the melting point table shown a few pages back; antimony, vanadium, molybdenum and tungsten are rare and seldom used as is, but instead are used in small amounts to make **alloys**.

Antimony is a metalloid, not a full flung metal. Like gravel compliments the integrity of concrete, antimony when combined with tin acts to significantly harden lead. Hardened lead alloy can be found in the cast bullets used for reloading ammunition, automotive battery terminals or in linotype (the special, continuously reused lead alloy once used only for typesetting). The biggest use for antimony today is in the production of lead acid type automotive batteries and to harden the lead wheel weights used when mounting and balancing new automobile tires.

Vanadium, molybdenum and tungsten serve mainly as steel alloys or as catalysts. **Vanadium** is useful in tool steels like drill bits, where it facilitates higher possible working temperatures without sacrificing temper (hardness). **Molybdenum** improves steel by restricting its expansion and softening at higher temperatures and was commonly used in artillery pieces and tank armor. Molybdenum also improves the corrosion resistance and weld-ability of steel. **Tungsten** is a rare element but having a very high melting point, found use in light bulb and x-ray tube filaments. Used in cutting tools and abrasives, tungsten –carbide tipped implements are almost three times harder or stiffer than plain steel. **Chromium** is yet another metallic element which is found alloyed within steel – to make "stainless steel".

- **Stainless steel** is markedly resistant to corrosion and oxidization under heat. Chromium at proportions between between 10 and 30 %, might be the dominant element to influence this alloy; but on occasion copper, nitrogen, sulfur, phosphorus, selenium, nickel, molybdenum, titanium, aluminum, niobium and of course carbon might be added also. There are more than 100 grades of stainless steel. These are generally divided into groups: austenitic, ferritic, duplex and precipitation-hardening. Further explanations are a little esoteric to dive into here.

Like with iron, the aluminum and copper items we see or use are seldom pure, but are alloys; hardened or strengthened by the addition of a different element. Bronze and brass are significant alloys of copper for example. The early Greeks made many weapons, implements and impressive hollow - yet free standing sculptures with '**arsenic bronze**'. While pure copper is so ductile as to be useless for certain applications, adding arsenic (a notoriously toxic metalloid) created a significantly harder and mechanically superior metal.

Tin (a rare and very soft metal) gradually replaced arsenic as the preferred alloying material for casting bronze. Aside from tin and arsenic bronzes, some other popular copper alloys in the past were leaded tin bronze, manganese bronze, silicone bronze, phosphor bronze, aluminum bronze and beryllium copper.

The distinction between bronze and brass is not always clear. But in general brass is more yellowish or golden because of the addition of zinc. **Zinc** (Zn, #30) has an usual history and wasn't even recognized as being a metal until fairly recently (1740's in western civilization). Under heat zinc can boil, turn into a vapor and escape as a gas before copper even begins to melt. The geology of zinc deposits is complex. Zinc and lead are often found together in the same deposit. Sphalerite (zinc sulfide (ZnS) or "zinc blende") can coexist with galena (lead sulfide) and cadmium sulfide.

Duralumin (or "duraluminium") was a very influential alloy of aluminum. Although an obsolete name now, Duralumin was developed (1909) before WW1, in Düren (of Westphalia, Germany). The yellowish* light alloy of aluminum, manganese and copper (about 4%); proved to be far harder and stronger than plain aluminum. German dirigibles and airplanes (like the all metal, Junkers J1) used this alloy successfully during 'The Great War ', but the British and the Americans ignored it, until afterwards. In 1943 the Americans improved upon the Al-Mg-Cu mixture by adding Zinc. Nowadays almost all aircraft use improved derivatives of Al-Zn-Mg-Cu alloy.

* Not to be confused with the yellow-green iridescence of **zinc-chromate**. A tinted undercoat found on the skins and metal parts of many WW2 aircraft. "Zinc yellow" successfully passivates (protects from corrosion) metals like tin alloy, galvanized steel, cadmium, magnesium and aluminum.

Forge, Furnace & Foundry

"There is a difference between smelting and melting". There is also a difference between "forge", "furnace" and "foundry" although sometimes these terms are used interchangeably. Each might perform very different functions when heating metals.

A **forge** is a furnace or hearth where metals are heated prior to hammering them into shape. The act of forging is heating and hammering metal into a shape.

A "**bloomery**" was the earliest form of furnace capable of smelting iron from ore. That does not mean that iron melting temperature is achieved; only that the metal becomes red hot or orange hot, and then malleable enough to be formed with hammer blows.

"**Blast furnaces**" are merely modern, refinements of ancient bloomeries. Improved blast furnaces though, reach heat that melts metals. The heat of fusion is a phase change from solid to liquid.

A **foundry** is a factory where metal castings are made from hot molten metal.

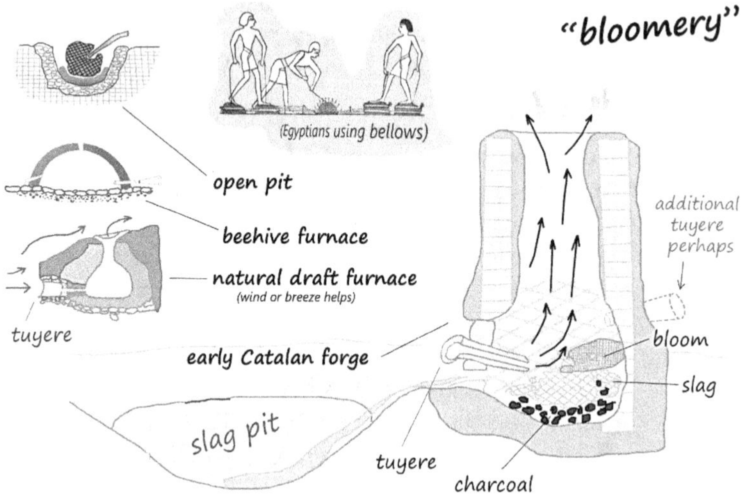

A bloomery can come in several shapes. They were not large at first, only a few feet or a meter high sometimes. Made or dirt and clay, the bloomery furnace was usually destroyed to get the sponge (or bloom) out. With the beehive type bloomery, alternating layers of fuel (wood charcoal) and ore were built up; to a point where the dome shaped stack was packed in tightly with soil. Then holes were cut in places to allow combustion air to enter and the exhaust gasses to leave.

A **tuyere** or cone shaped mouthpiece (of fired clay usually) would usually be poked down into the middle of the fire and then air from the bellows would be forced through that.

Bloomeries used forced air – from **bellows** (with the exception of the natural draft type – who's opening was faced into the wind). The ancient bloomery did not get hot enough to melt iron, but it got the iron and slag mixture, red hot; hot enough to extract and pound the "**bloom**" or "**sponge**" with a sledge hammer to drive impurities (slag) out of the sponge. In practice, the sponge would be brought out, pounded upon and shaped for a while, and then reheated once or twice – to repeat the process.

Cruder and more impure than 'pig iron' that comes from a blast furnace, a hot sponge from a bloomery was hammered, reheated, pounded, twisted and pulled similar to the way a baker would kneed bread dough. **Wrought iron** is the iron product produced by all that labor. Under a magnifying glass or microscope, its stringy grain looks very unique and distinguishable.

Authentic wrought iron is very rare today – its main source being from antique implements. But some people do play with the ancient process in their own back yards – and film youTube videos of the event. Instead of wrought iron which is hard to come by, a modern blacksmith today would normally work with low-carbon or mild steels because they are malleable and ductile, where as high carbon steel is not.

*[A **low carbon** steel today contains about 0.05–0.15% carbon. **Mild** steel is about 0.15 –0.3% carbon. Larger proportions of quickly make the metal harder and more brittle. **High carbon** steel might contain between 0.6–2.0% carbon].*

During the smelting (using either bloomery or blast furnace) carbon monoxide is created and actually needed for reduction of the metal; (pulling out or reducing the oxygen from the ore). Early European bloomeries were relatively small, primarily due to the mechanical limits of human-powered bellows and human powered sledge hammers. **Water power** changed that.

> *"The use of waterwheels, spreading around the turn of the first millennium and used to power more **massive bellows**, allowed the bloomery to become larger and hotter, with associated **trip hammers** allowing the consolidation forging of the larger blooms created".*

- Above and below are two historic, and therefore publicly owned (public domain) images; both are of water wheel powered bellows. The stock photography agency that eagerly watermarked these images neglected to credit the original artist and publication. The images have been slightly enhanced here, for clarification. At top the rotational mechanical force is being converted into a 'reciprocating inline motion' by means of a 'slider crank' linkage. A physical weight acts to close the bellows.

- Above, two alternating bellows should supply a more continuous stream if air. A suspended 'spring pole" lifts the bellows open. Pegs or protuberances on the rotating shaft press the bellows down and the air out.

Below is a medieval water powered stamp mill to crush ore into smaller bits, which facilitates and quickens metal smelting.

- (This image and a few others nearby, came from a minerals and mining book named, **De re metallica; (**meaning 'On the Nature of Metals' /Minerals). Written and published in Latin, in 1556. It was first translated from Latin by the 31st (US) president and his wife. He a mining engineer and she a geologist.

* This 17th century etching depicts both a small trip-hammer and bellows for a small forge, being powered by a waterwheel.

Bloomeries evolved into blast furnaces. Having changed little since the Middle Ages, modern blast furnaces grew bigger, still work by gravity, but now are continuously fed from the top by conveyor belts. One blast furnace might be set up for iron, but elsewhere another for copper or another for lead and zinc.

Just a big chemical reaction chamber, a blast furnace for iron works slowly, but continuously, year after year without being shut off. It might take an individual atom of iron a week or two to work its way down to the bottom of the furnace; where ultimately molten metal flows in one direction and slag another. An oxide like **hematite** (iron oxide ore) coke or coal (fuel) and limestone (flux) are continuously fed in at the top. The temperature in the foyer (base of furnace) hovers near 1800 °C.

The name "blast furnace" (or Fr: "*haut-fourneau*") reflects the fact that air (heated air in modern times) is forced into the bottom. In the following illustration, two different cutaways of furnaces are shown. One shows a bellows & tuyere but the other does not. By design, the **external pressures** caused by the downward weight of continuously fed ore and fuel, actually help to decrease the melting point of the metal. *"Steel will melt more easily and more rapidly when heated at higher pressures"*. For that reason the basic design of blast furnaces has changed little, in the last 3 or 4 centuries.

Increasing the percentage of carbon also helps to reduce the melting point of the metal, making life a little easier.

A "**draw kiln**" for the cooking of, or calcination of limestone to make cement, looks and works very similar to an early blast furnace. Perhaps at times the same structures have been used for both purposes. Less limestone would have been used in a blast furnace, because there the limestone is used only as a catalyst. But in a draw kiln, limestone becomes the intended product (cement clinker).

DE LA CONSTRUCTION DES HAUTS-FOURNEAUX

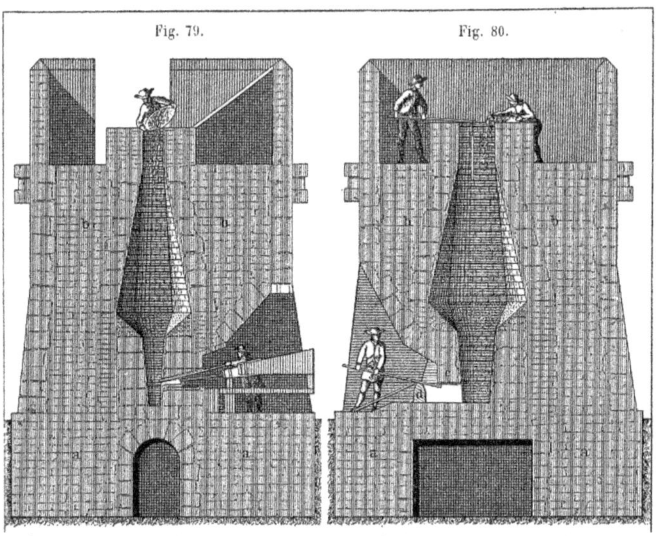

17th century blast furnace. From Manuel de la métallurgie du Fer, Tome 1, by Adolf Ledeburg, published Librairie polytechnique Baudry et Cie, 1895.

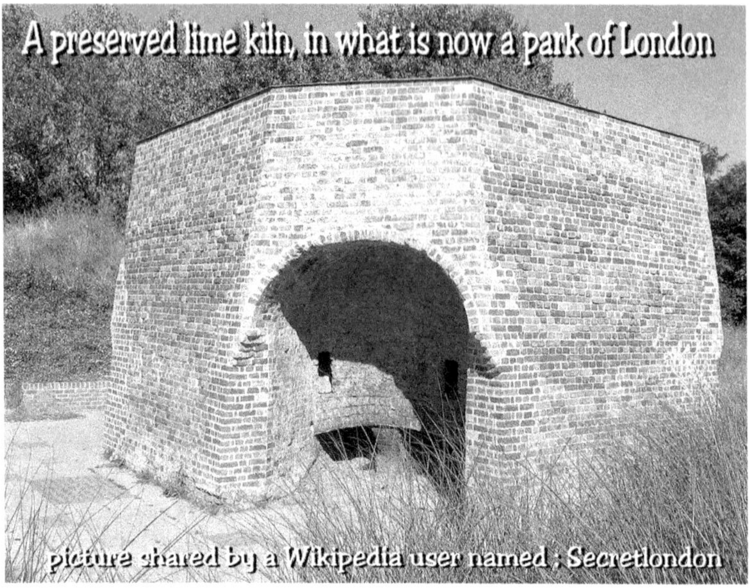

Above is an image of a **draw kiln** (or lime kiln); used for making lime for **cement** and mortar. Alternating layers of crushed limestone and fuel (wood or wood charcoal) were stacked in from the top. The inside is funnel shaped, like an inverted cone. Not much different, is the blast furnace below.

Disposition d'un haut-fourneau au commencement du XVIIIe siècle.

A detail below of the wider engraving (previous image), shows either fuel or ore being dragged by mules (upper left), into a shed that is at the same elevation as the top of the blast furnace.

Also shown is a **water wheel** and headrace that spins a shaft that powers two bellows, continuously.

Spanish steel has been famed since before Hannibal chastised Rome, during the 2nd Punic War (218 - 201 BC). A millennium later and Spanish Toledo steel would be celebrated as 'state of the art'; some of the finest steel in the world. When first exposed to it in the 16th century, even Japanese sword makers were amazed with the quality of Toledo steel. Their own steel was more brittle and less resilient by comparison. Although some of the alloys used in Toledo steel account for much of its quality, advancements in forge technology, coupled with an improved source of forced air contributed to production as well.

https://loc.getarchive.net/media/us-steel-duquesne-works-blast-furnace-plant-along-monongahela-river-duquesne-24

The above image is typical of modern blast furnace. This one in Pennsylvania was built in 1896, and was finally turned off in 1984.

In some locations, a device called a "**hydraulic trompe**" would begin to replace the waterwheel driven bellows or wooden piston compressors, to supply the forced air. Eastern Spain (**Catalonia** - French/Spanish Pyrenees) was one such area, where hydraulic trompes would be married to a style of hearth known since the 8th century, as a **Catalan forge**.

The trompe's origin and date of invention was lost somewhere back in the Middle Ages. But where flowing water permitted, tromps (from the 1550's) were even being used in several Italian churches, to provide **pipe organs** with compressed air.

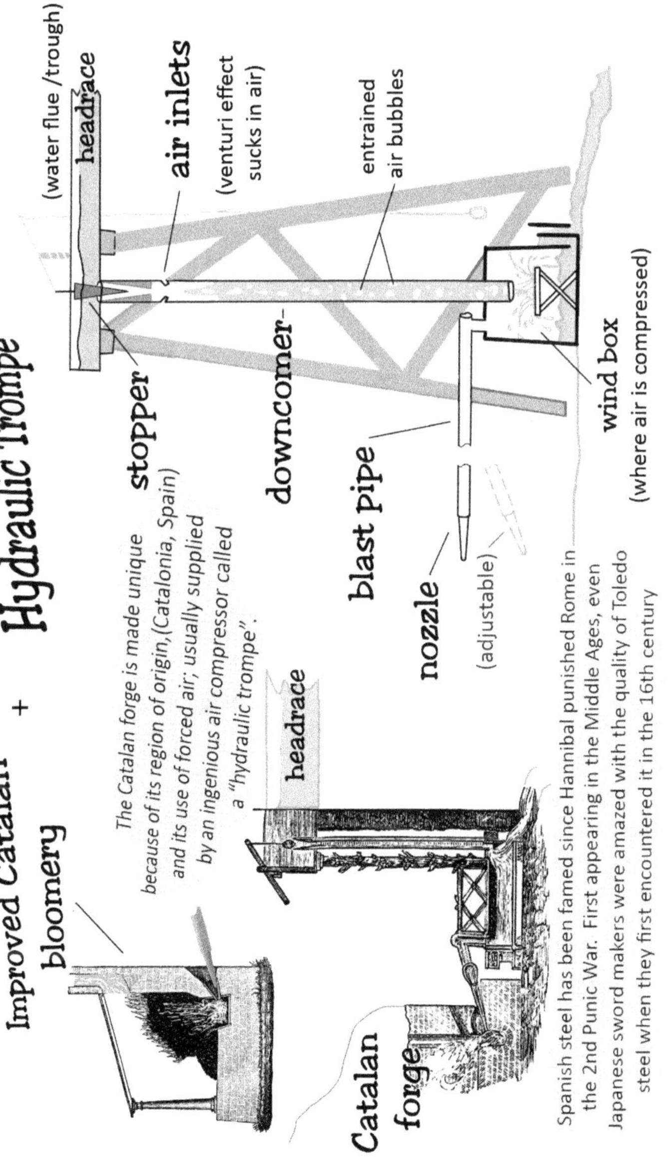

The function of a trompe is brilliantly simple. It has no moving parts. It compresses air with the help of falling water. Fresh air is drawn to or entrained into the dropping water, as a **venturi effect** is created, that sucks more air into the pipe. The cascade of liquid and gas within the pipe, falls and impacts with the bottom of a box or trap. Both are confined and push against each other. The pressure trap (wind box) only has two exits; a high one for air and a low one for water. Restricting the escape of water by making it rise, causes air-pressure to rise accordingly. Piping is needed to conduct the compressed air to the furnace, or organ.

Finery Forge

Before the 17th century, England mined and smelted some domestic iron ore, but the bulk (over 80%) was higher quality ore imported from Sweden. At the time Sweden was a powerful Baltic empire that controlled much of Europe's iron output, in the late 1600's.

From 1608 English settlers in **Jamestown** (N. America) began shipping ore, then later pig iron, back to England. Still, Sweden was much closer to ship from than America. A war against Sweden by an alliance of five neighboring nations broke out (The Great Northern War / 1700-1721); which was to ruin Sweden's iron export capacity.

Where before only a little pig iron was being imported from Maryland and Virginia, by the 1720's England turned in desperation to the American colonies. It's Navy was England's biggest consumer of iron. Overnight almost, English investors began spending large amounts of capital to construct blast furnaces near high grade ore deposits in the America colonies. What the English wanted and got was quality imported pig iron. They also wanted to sell some of this metal back to the colonies - in the form of manufactured goods (wagon wheels, household items, firearms, blacksmith anvils, nails, etc)[4].

The British did not wish to allow American industry to grow to a point where it could eventually compete with English goods. Parliament passed the *Iron Act of 1750* which provided for duty-free importation of colonial pig iron; but which restricted colonist from erecting refining forges to produce finished products. Obviously this

4 https://revolutionarywarjournal.com/iron-forge-in-colonial-america/

another reason for resentment in the colonies.

By the time of the American Revolution the colonies had a well developed iron industry; their combined output making them about the 5th largest producer in the world at the time.

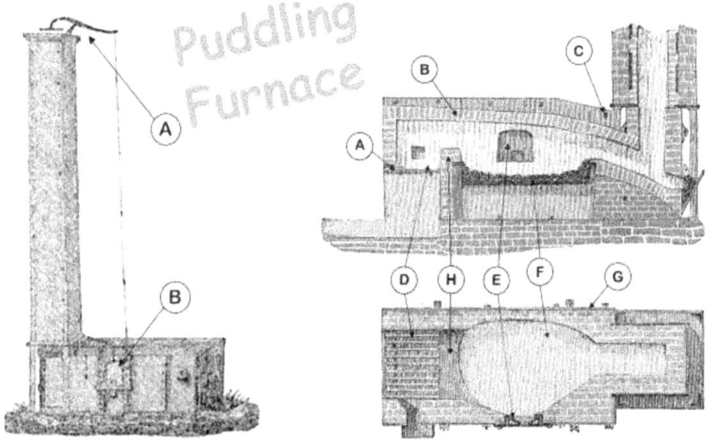

"The Manufacture of Iron, in All Its Various Branches", F. Overman, 1854

The **finery forge** (where pig iron was refined into a better metal) was succeeded by the **puddling furnace.** Developed in England during the 1780's, the puddling furnace allowed molten pig iron to be stirred around and reduced more effectively, while doing so without adding more carbon. The fuel and fire were offset from the hearth; only the hot gasses passed over the top of, or contacted the metal in the puddle. "Bar iron" (malleable wrought iron created without charcoal) was the initial product. But steel from puddling furnaces was to come later. These furnaces weren't very large.

High in the Andes of Bolivia and Peru, the Incas were smelting silver, copper and gold long before the Europeans ever arrived there. They constructed earthen "**huayra** or **huayachina**" furnaces on the windward sides of high hilltops – where the wind drove the conflagration. Little is known about these furnaces because in the last 5 centuries they have deteriorated and have eroded away.[5]

5 https://www.sciencedirect.com/science/article/abs/pii/S0305440312004141

Since most were build above 4,000 m (13,000 ft) or well above timberline – it seems evident that any decent fuel had to be transported uphill to the furnaces.

When the Spanish arrived in the 16th century they were very eager to exploit and export the region's valuable mineral resources. From about 1530 the Spanish employed European furnaces technology. **Silver** collected from the region outweighed the **gold** retrieved, by about 100 fold.

Much of the gold seized or taxed by the Spanish was actually a lustrous combination of copper and gold; that was called "**tumbaga**".

The Spanish Peso, also known as "the **piece of eight**" was the world's first international currency and remained so for about 3 centuries. It was just about the only coin used in colonial America, because the English would not permit colonials to mint their own money.

The crude hand-struck coin to the left is a "**cob** coin" minted by one of the several vice-royalties in New Spain (Central and South America). "*The freshly mined silver was first smelted into bars. Pieces were then sliced off and trimmed until the right weight was reached. The resulting disk, or planchet, was placed in an anvil mounted die. The coiner then struck the upper part of the die with a hammer, inscribing the crudely shaped coin with a year date".*

A silver **peso** was worth 8 reales (smaller more uniform coins in the old country / reale = plural for royals). Much smaller gold coins (escudo s) were minted in about five different denominations. The two-escudo (or $4 coin) was originally nicknamed the "**doubloon**" (from double) by Caribbean pirates.

This image (from the Library of Congress), has been altered in the low right hand corner; to show some piglets. Taken somewhere between 1890 and 1901, at the Iroquois smelter in Chicago – it shows a work crew ready to receive and control a hot stream of molten iron. The process is similar to irrigating a garden. The furnace is uphill – in the background. The liquid iron will rush down the main shute or canal and begin to fill up the bottom row of sand molds first. Using shovels, hoes and rakes, the bottom crew pushes or pulls sand to control the fluid, and then progressively works its way uphill. Another crew at the furnace monitors the flow rate. The cooled ingots were called pigs because their arrangement was reminiscent of suckling piglets. Pig iron is a crude, intermediate product, that will be reprocessed or refined into a better metal, at a later date. The individual ingots are sized to a practical weight for handling.

Pig iron is brittle. It is crude and generally too full of impurities like silicon (Si) or silica (SiO_2) and excess carbon, to be useful. Carbon is deliberately added when smelting, to lower the melting point. Later that carbon needs to be removed.

> *"The melting point of steel is 1300-1540°C (2370-2800°F), depending primarily on the chemical composition of the steel, which can affect the strength of the bonds between atoms".*

Generally, the higher the carbon content in the steel, the lower its melting temperature. The stronger the bond strength between atoms, the higher the melting temperature.

> *"The melting temperature of steel is also affected by external pressure. As the* ***external pressure increases****, the melting temp of steel decreases. That is, steel will melt more easily and more rapidly when heated at higher pressures."*

Setting outside as a museum display (at Kelham Island Museum, Sheffield) is this century old, open mouthed Bessemer furnace.

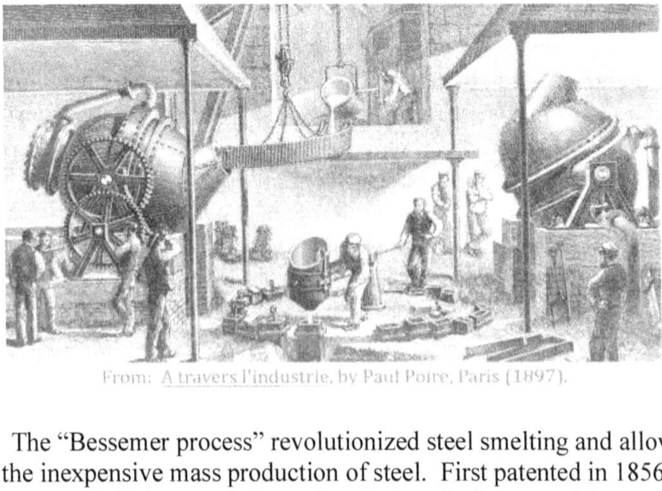

From: A travers l'industrie, by Paul Poire, Paris (1897).

The "Bessemer process" revolutionized steel smelting and allowed for the inexpensive mass production of steel. First patented in 1856, the process (which essentially squirts air into the bottom of the cauldron) is claimed by some to have single-handedly permitted the industrial revolution [6].

Steel became a dominant construction material solely because of this invention. The drastic reduction in steel's price prompted customers like Americans, to buy British steel train rails. At a time of energetic westward expansion when such rails were in very high demand. *"Steel rails lasted ten-time longer than iron rails and could support longer trains with heavier loads".* Also huge buildings like skyscrapers suddenly became practical because of the availability of strong and affordable steel beams. Alas, the Bessemer Process became obsolete by the 1960s and was replaced by newer technologies - like Electric air furnaces, among others.

THE "DAW" PATENT COMPRESSOR.

FIG. 1.

THE "DAW" CLASS E SINGLE STRAIGHT LINE COMPRESSOR.

A 19th century industrial air compressor. Big steel mills, manufacturers and large mining operations required machines like this. Cities like Paris ran dozens of steam air compressors this big or bigger; constantly, to provide compressed air to shops and industries within reach. **Compressed air** was a utility; just like electrical power, water and septic is today. Miles of pneumatic tubes ran underneath certain cities. But fire was needed to create the steam, which created the pressure. *<See chapter on steam>*.

6 https://dozr.com/blog/bessemer-process

Carbon

There are several categorizes of steel; as in carbon steels, alloy steels, stainless steels and tool steels. Carbon within the fuel accompanying iron ore in a blast furnace, plays a crucial role in removing oxygen from the ore. Ferric oxide (rust) for example is elemental iron mixed with oxygen). During initial smelting, oxygen is removed as carbon monoxide or carbon dioxide. Molten metal flows out as crude iron (pig iron). Again, pig iron is brittle and not very useful because of impurities, like silica. Pig iron has a high carbon content, ranging usually somewhere between 3.8 and 4.7%. Further smelting will be needed to purify the metal (later on and in a different furnace).

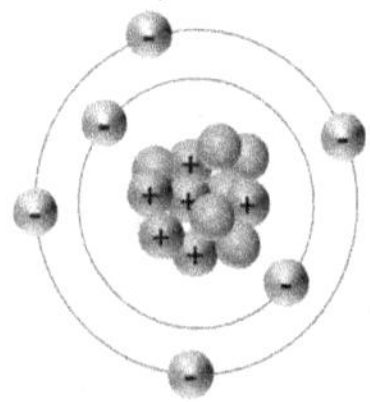

All forms of life on this planet are based on the carbon molecule ("**C**" - element #6). Because its valence shell possess 4 electrons, carbon is uniquely able to form up to four stable covalent bonds. This characteristic enables carbon to form more compounds than all the other elements combined.

There is nearly an unlimited number of carbon compounds on the planet.

Despite what climate alarmists might think, carbon is not plentiful; it makes up only about 0.025 % of the earth's crust and about 0.04% of its atmosphere (when combined with heavier oxygen as CO_2). And since it mixes or attaches itself so well to so many other things – carbon is rarely found in a pure state. Carbon burns, but what is often observed during combustion is the other impurities mixed with carbon that are burning first. Burning charcoal or coal for example still normally leaves some incompletely burned carbon behind in the ashes. Technically, carbon is usually not flammable. Much depends upon the type and physical state of the carbon, as to how burnable it will be. If it does combust, it decomposes and combines with oxygen to form CO or CO_2.

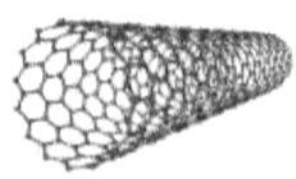

Some relatively pure allotropes of carbon are crystalline diamond, graphite and non crystalline amorphous carbon (like soot and coal) and fullerenes (like Bucky balls and carbon nanotubes). Carbon has eight known natural allotropes altogether.

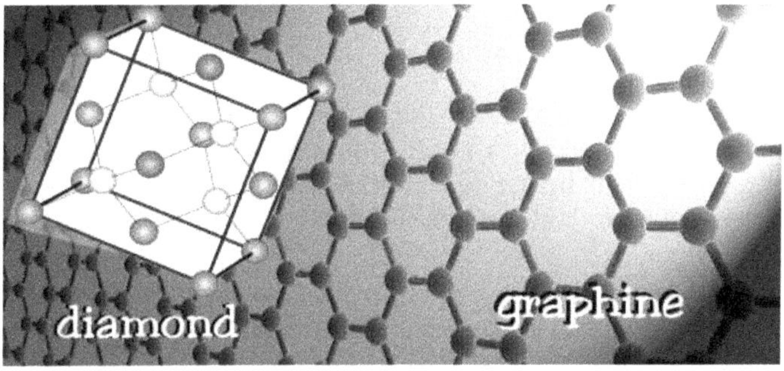

The mineral **graphite** has a crystalline structure that's composed of many individual layers of '**graphine**'. A single layer of graphine has its atoms arranged in a hexagonal lattice. The mineral called **diamond** on the other hand, has a "cubic crystal structure". A diamond crystal or "diamond cubic" is reminiscent of a box within a box; a nine sided polyhedron within another nine sided polyhedron (18 carbon atoms total). While a diamond is the hardest natural substance, it may not be the toughest. A diamond can cut through a steel hammer, yet a steel hammer can shatter a diamond into dust.

"**Crystallography**" is the science of determining the organization and bonding of atoms within crystalline lattices. Beyond regular light diffraction, nowadays x-ray light diffraction is also used to study crystalline arrangements or other large complex molecules like proteins and DNA.

Through magnifying lenses, highly skilled diamond cutters examine and measure raw diamonds for a good while before they undertake actions to shape and polish those stones into faceted gemstones. The slightest mistake can ruin a gemstone. The multiple cleavage planes running through the rock, begin at the atomic (diamond cubic) level.

Carbon nanotubes are the topics of great speculation and the objects of research in the material sciences. They can have high or even adjustable electrical conductivity, very high thermal conductivity and very high tensile strength. Carbon nanotubes either already are, or soon will be important components in optics, semiconductor electronics and composite materials.

Charcoal, Coal & Coke

Charcoal and coke both are the products of **pyrolysis** (the burning of wood or coal respectively in a low oxygen or inert environment). The water and some lighter, low molecular weight volatiles are driven off. Charcoal burns hotter and cleaner than the original wood. And makes less smoke.

Coal is a rock, a fossil fuel of decayed plant and animal matter, formed by geologic time (slow) and high pressure. In many instances coal makes a far better combustion fuel than wood, because it is a denser source of energy.

Coke is made from coal. But as with charcoal from wood, the energy density of coal can be improved by the thermal decomposition of pyrolysis. What is left behind has less mass, but is a more compact or dense form of potential energy. Coke is used extensively in the steel mills because it burns hotter and is more efficient than coal. However coke will not burn by itself alone. LOX (**Liquid Oxygen**) must be forced into a modern furnace, to maintain the combustion of coke.

An unaided charcoal fire can reach 900°C, but a powerful forced air supply can raise its temperature to nearly 2000°C.

Coal-tar creosote is that oily black-brown sludge that's painted on or soaked into telephone & power poles and rail road ties (sleepers). Coal-tar creosote is superior wood preservative and pesticide; a byproduct from the destructive distillation of coal, when making coke. Other creosotes made by the pyrolysis of plant material might be used in some antiseptics.

^ **Blast furnace slag** is a waste product from making pig iron and it is full of alumina, lime and silica. Since slag possesses some of the very components of Portland cement itself, it and fly ash can be used to replace up to 60% of the Portland needed to make a concrete.

^ **Fly ash** is the molten remains of powdered coal that was burned in a power plant to generate electricity. Like blast furnace slag, fly ash is a pozzolan rich in silica, alumina and calcium.

Cast iron has a high (2.14 to 4.0 % by weight) carbon content compared to steels. Cast iron is an intermediate product between pig iron and regular steel, and in addition to excessive carbon it also has other impurities like manganese and silica. Traces of impurities such as sulfur and phosphorus might exist. Cast iron (also called gray iron) has a lower melting point (between 2102°F - 2372°F or 1150°C – 1300°C) than traditional steel. The lower melting point combined with its high fluidity when molten make it an amenable metal to use for sand casting.

> *"Cast irons have become an engineering material with a wide range of applications and are used in pipes, machines and automotive industry parts, such as cylinder heads, cylinder blocks and gearbox cases. It is resistant to damage by oxidation".*
>
> *"The microstructure of cast irons can be controlled to provide products that have excellent ductility, good machinability, excellent vibration damping, superb wear resistance, and good thermal conductivity".*

There's "ductile" cast iron, "malleable" cast iron, "gray" cast iron and "white" cast iron. Apparently the Chinese produced some cast iron as far back as the 6th century BC; but the Europeans didn't produce it until the 14th century. Cast iron made it to England by about 1500.

Before the Industrial Revolution (1750- ish onward) the Dutch were casting state of the art cookware from copper, **brass or bronze**. Their pots and kettles were sold all over Europe, but were expensive. Along comes an English entrepreneur in the 1750s, who observes how things were being done in the foundries of the Netherlands. He goes back to England and spends some time learning how to cast his own cookware (especially what are now called "Dutch Ovens") from cast iron. *[see: Abraham "Darby's furnace" in the Roman builders chapter – with A-frames].*

The principle of a lidded container to bake in, dates back to Roman times (or before) where the utensil was made of ceramic and called a **"clibanus"**. It too was made to sit in or near a fire.

So after pig iron and cast iron (in terms of refinement), then come regular, common steels, that are subdivided into general types. The common steels are differentiated by their carbon content.

Low carbon steel (or mild steel) has a general carbon content between 0.04% and 0.30%. **Medium carbon steel** has a carbon range of 0.31% to 0.60%. They may contain manganese ranging from 0.6% to 1.65%. **High carbon steel** (or tool steel) generally has a carbon range between 0.61% and 1.50%. (Remember, pig iron typically has a much higher carbon content of about 3.5% to 4.5%).

There are hundreds of "steel grades", "alloy steels" and "stainless steels" to consider. Alloy steel is common carbon steel with added elements like: vanadium, aluminum, manganese, silicon, and nickel, chromium, copper, tungsten and titanium. These change the melting point of the metal and produce specific desirable properties in the final alloy. Professional metallurgy is high science.

Charcoal is still used

Yes, charcoal was the fuel of choice for metal smelters two or three millennia ago. The Greeks especially, created some amazing bronze sculptures and weapons using this fuel. But even today charcoal is desirable for smelting because of its versatility. Basically charcoal is just a great reduction agent. Charcoal contains a high level of 'fixed carbon' – a type of carbon that doesn't vaporize easily at high temperatures. Charcoal conducts electricity, making it useful in the electrolysis process where electric energy facilitates metal ion reduction. Charcoal is seen as a more sustainable / eco-friendly reducing agent than some other fuel alternatives.

The Riace Bronzes
National Museum of Magna Grecia

Two larger that life sized statues are known as the Riace Bronzes. They were discovered in 1972 by a snorkeler, offshore of the ancient city of Reggio Calabria (a Greek outpost 25 centuries ago when they were cast).

The Greeks were masters of bronze casting. Although the Greeks probably cast as many bronze sculptures as they chiseled from stone, the stone statues remain today whereas most of the bronze statues do not. Every time a new war came along, bronze statues were hacked to pieces to provide the valuable metal needed to forge new weapons and body armor.

These two hollow statues were produced by a process known as "**investment casting**" or the "**lost wax process**". The procedure involves creating a wax pattern, encasing it in a ceramic mold, melting out the wax, and then pouring molten metal into the mold cavity.

Bronze and brass also have peculiar properties and interesting histories. For example, the first bronzes were arsenic bronzes. Arsenic bronze is much harder than the original – excessively ductile copper. But **tin** gradually supplanted arsenic as the preferred alloy in bronze. While arsenic vapors are unhealthy and can attack the eyes, lungs and skin, another reason for preferring tin (which was very scarce and had to be imported) was that it probably afforded more control over the forging and casting processes. Tin made an alloy that would fill a casting faithfully without leaving voids.

Corinthian helmet / circa 500 BC

The iconic Greek Corinthian bronze helmet is an enigma to modern intellectuals because they can't quite determine how it was constructed. Our understandings of the construction of other helmet designs of the period are less controversial. Neither the process of casting nor forging by hammer stroke alone adequately explain how the classical Corinthian helmet was built. The best explanation seems to be that it was a product of both.

Bronze and brass, are historically noteworthy in that these alloys have always been heavier, more valuable and easier to work with than iron alloys. Roman officers preferred bronze swords to the new, cheaper iron weapons that rusted. For centuries the best cannons were forged from bronze because it was more durable, more appropriate. Early iron cannons exploded. To this day brass is still the best metal for pistol and rifle cartridges. Unlike other materials, brass (crucially) expands and contracts almost instantly; sealing off the chamber and preventing gasses from escaping backward in the breech. Brass is softer, creates less friction and jams less in the breach of a weapon.

Before the Industrial Revolution, iron smelting used **charcoal** as both the heat source and the reducing agent. But on an island like England, wood became a limited resource that was used for everything: housing, heat, substantial ship building, leather tanning, etc. By the 18th century the expansion of iron production was hindered by the lack of charcoal. England was dependent upon rival nations like Sweden and Russia to provide their iron. But by the 19th century, Britain became a world leader in metallurgy.

> *"Use of coal in smelting started somewhat before the Industrial Revolution, based on innovations by Sir Clement Clerke and others from 1678, using coal* ***reverberatory furnaces*** *and* ***cupolas****.* (Once synonymous, the terms now differ: reverberatory furnaces heat materials indirectly through reflected flames, but cupolas use a vertical design to melt metal with direct contact from burning coke and air). (The term "reverberation" is meant to

imply a conjugation or reflecting of heat).

"In 1875, Britain accounted for 47% of world production of pig iron and almost 40% of steel. Forty percent of British output was exported to the U.S., which was rapidly building its rail and industrial infrastructure. The growth of pig iron output was dramatic. Britain went from 1.3 million tons in 1840 to 6.7 million in 1870 and 10.4 in 1913".[7]

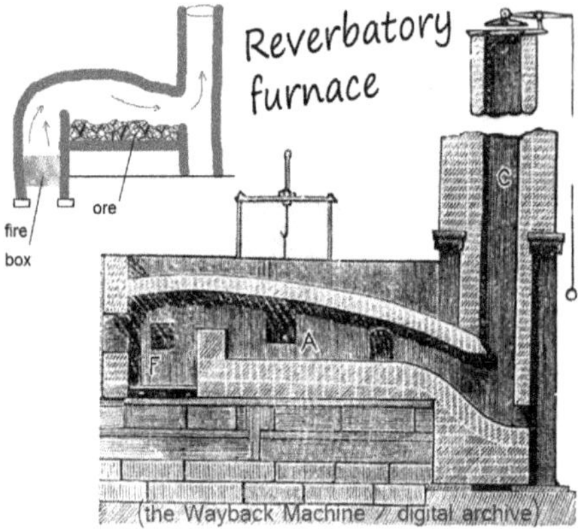

For a given amount of heat, coal took much less labor to mine than cutting wood and then converting that into charcoal. In England coal was more abundant than wood. The use of coal for smelting was largely responsible for starting the Industrial Revolution.

From the 1670's, coal reverberatory furnaces and puddling furnaces operated by the flames playing over the ore and charcoal or coke mixture. This reduced oxides in the metal and had the advantage of preventing impurities such as sulfur ash in the coal, from migrating into the metal. A puddling furnace was a small, specialized type of reverberation furnace that was used to create wrought iron or steel from the pig iron which came from a blast furnace.

A fact of life is that most coal deposits are contaminated with impurities. Coal which contains sulfur is especially undesirable. In modern times "**coke**" is the preferred fuel because it is more efficient, and also because it has had most of those impurities removed, during the pyrolysis from coal.

more furnaces

7 https://courses.lumenlearning.com/suny-hccc-worldhistory2/chapter/changes-to-iron-production/

RECTANGULAR ASSAY FURNACE.

*Caption above reads: "Rectangular **Assay Furnace**"*

A pair of tongs, and a jewelers crucible are inexpensive and handy at times. A propane torch can supply the heat. For larger projects, gas or electric furnaces can be used.

Anything less than a quality crucible might be destroyed before your metal sample ever melts. There are two general types of crucibles in the professional small metal works world also; ceramic and graphite. Both types have virtues and faults. Ceramic crucibles are much more affordable and tougher in terms of chemical resistance. But they are very delicate when hot.

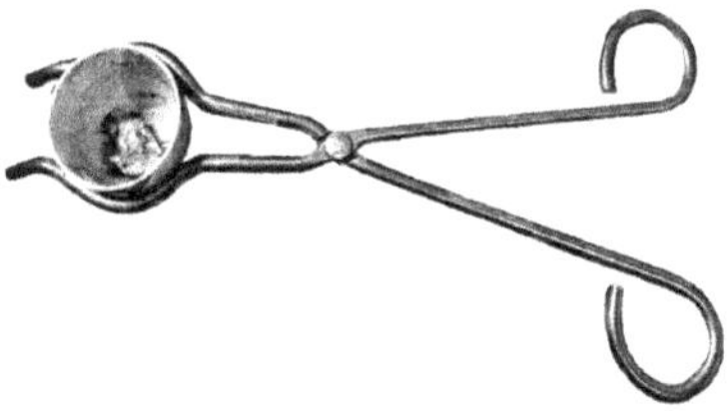

Ceramic crucibles are preferred for tasks like sintering, or for jewelry making where they have more durability because acid or base compounds might become involved [8]. But these ceramic utensils are fragile if heated or cooled too slowly; they break easily unless handled cautiously.

8https://www.betterceramic.com/graphite-vs-ceramic-crucibles/

Graphite crucibles really excel in thermal conductivity and heat resistance. So they can be heated and cooled back down fast. The material composition of such a crucible might be – 50% fire clay, 30% graphite, 12% silicone carbide and 8% borosilicate glass. Borosilicates have very low coefficients of thermal expansion (more resistant to thermal shock), but the reason for adding the compound, is to reduce the oxidization of the crucible. Oxidization is a weakness with graphite crucibles; more prevalent when propane furnaces are used, rather than with electrical furnaces probably. (More air can get sucked into a propane furnace).

To the side is a small picture of some little, portable, disposable, but generally safe to handle hydrocarbon gas canisters. Blue is the common color for portable **propane** bottles. Yellow is the color adopted for MAPP gas. The squat canister is green. It is intended for camp cook stoves, but it holds the same propane as is in the blue bottle canister.

MAPP (Methylacetylene + Propadiene + Propane) fuel gas is no longer produced in the U.S.. The canisters found on the market seem to be exported from predominately from France. MAPP gas was initially created (by Union Carbide) to safely imitate acetylene gas, when placed in a more portable fuel format. MAPP gas can also be used with pressurized oxygen, where when combined the two are safer than acetylene torches for underwater cutting. MAPP gas burns hotter than propane, reaching temperatures up to 3,730°F (in air), while propane typically reaches around 3,600°F. *[With pressurized oxygen MAPP flame temperature can reach 5,720 °F - which is still less than acetylene @ 6,020 °F]*. Propane gas is perfectly adequate for typical brazing and soldering. MAPP gas canisters heat a little hotter but are considerably more expensive than propane canisters. Bottled propane also has a higher energy density than MAPP gas, making it a more efficient fuel option.

Above is a typical commercially sold LP powered crucible furnace. This kit came from China. It has one air/fuel input duct or hole. Another version might facilitate two input holes, to admit more fuel and flames - that will swirl around and therefore heat the crucible faster. Therein lies another problem. The combustion oxygen (drawn into venturi holes just before the LP hose enters the furnace) is quickly deleterious or destructive of graphite crucibles. Ceramic crucibles might resist the erosion longer but would be slower to heat and would cause the consumption of more fuel.

The quick erosion caused by oxidization in a furnace at these high temperatures is stunning. If a homemade steel crucible were used in a LP furnace, its surface will quickly deteriorate. Large flakes of oxidized iron will fluff off and crumble. Typical ceramics do worse. A ceramic coffey cup (if it ever reaches 1000 °C without failing first) will crumble and shatter at the slightest vibration or pressure.

This image begins with a mold which will soon be filled with refractory. The outside is a plastic flower pot, in the center is a 3lb coffee can. An **armature** of wire separates the two. The can sits on a spacer. A long bolt connects the can to flower pot. The smaller cans will form a hole, where afterward combustion air can be forced in by bellows or electric hair dryer. The lid in the center above is taller than necessary.

For a hobbyist wishing to melt metals that come in a raw condition (like beer and soda cans painted with inks and polycarbonate, polyester or acrylic resins) a gas furnace might be more desirable. The impurities initially need to be burned off.

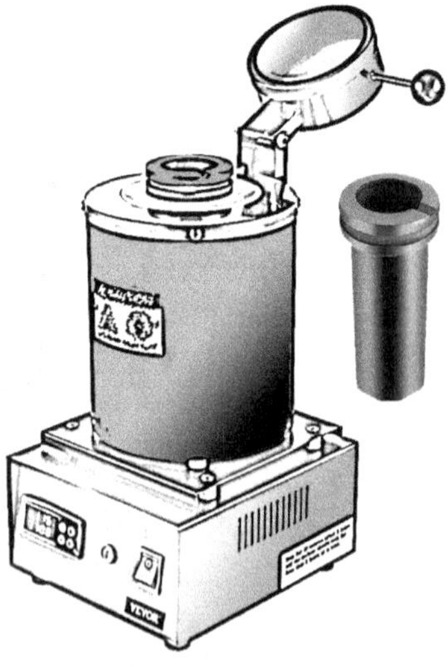

But for more refined metals like coins or jewelry, an **electric powered crucible furnace** would be more appropriate. At least the electrical powered version will introduce less oxygen and should therefore be less destructive to a graphite crucible. Small electrical crucible furnaces are more expensive than LP gas versions of the same size. They work slower, taking about 20 minutes to heat up. They also also need to be cooled down for a period after use. But they are gentler on the crucibles, which should be considered. Crucibles for these furnaces are shaped to make handling easy when they are dangerously hot. The grove at the top enables a firm grip with a pair of tongs. One kg, 2kg and 3kg crucibles will fit the same e-furnace.

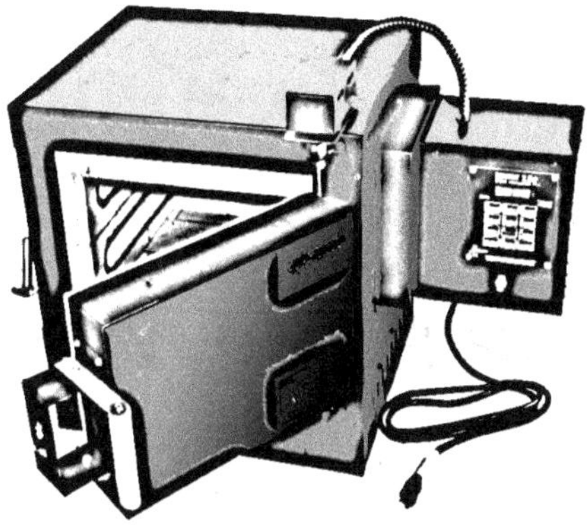

Higher grade, more expensive electric **annealing** or **heat treatment furnaces** can be had. These might come equipped with swinging or guillotine doors, thick fire bricks, thermocouples and better digital controls. These probably come sporting 240 volt power and might reach 2400 °F (1315 °C). [After that would come industrial grade machinery using three-phase power @ 380v or 400v perhaps]. The goal of an annealing furnace is to create a metal implement that is hard enough to withstand hard use but tough enough not to crack. Heat treating changes the physical or chemical properties of a metal; through a process of heating, cooling and reheating. Complex heat treating cycles may vary depending upon an alloy's mechanical properties.

To heat treat something like a knife or sword for example, the metal is elevated to its **"critical temperature"** (not melting temperature which causes a phase change). Here significant changes in its properties occur; crystalline structures change. The temperature is maintained until a new microstructure is consistent throughout the metal. Then the hot metal is **quenched** (quickly cooled), otherwise it would return to its original form. At this point the metal is stressed, hard and brittle. So finally the blade must be "**tempered**", which means being heated again to a temperature below critical temperature.

Blacksmiths with old coal forges and hand powered bellows did this kind of hardening work all the time. Undoubtedly though an electric annealing furnace should afford a modern metal worker with more control over temperature and more free time to use on other task.

A **foundry** is a factory where metal castings are made from hot molten metal. In the next image are **cupola furnaces**, as would be used perhaps to cast a bell, an ornament or an artisanal sculpture. Such furnaces could generally be used to fill molds with molten metals (like bronze, brass, aluminum or cast iron). Traditional metal casting techniques include lost-wax casting, plaster mold casting, die casting, and sand casting (and others).

- *"The first known cast object is a copper frog that dates back to 3200 BCE, found in present-day Iraq".*

Fig. 62.—Foundry Cupola with Drop Bottom

Cupola furnaces / public domain images

The surface of the sun and the earth's inner core are both about 5,778 K (**5,505 °C**). Which is not that hot in the great scheme of things. The core of the sun is much hotter, assumed to be about 15 million °C. The fission bomb "Little Boy" dropped on Hiroshima generated a heat of about 299,726 °C at its core. There are no instruments like thermometers or thermocouples to physically measure even these relatively low temperatures but instead must rely upon idealized thermodynamic theory to extrapolate these numbers.

Traditional mercury-in-glass thermometers typically register to about 212°F (100°C). But these old fashioned thermometers if built a little differently, might read up to **392°F (200°C)**. **Thermocouples** (since 1821) use two dissimilar metals to generate a voltage proportional to the temperature. But these have limits too. Of the five common types, Type-K alone, is capable of reading up to 2,502°F (1,372°C). Thermocouples are susceptible to wear and tear, they melt and degrade.

Pyrometers are newer remote sensing thermometers that from a distance, detect the thermal radiation emitted by an object and then extrapolate or estimate its temperature. Pyrometers are not precise, but they are generally adequate for many applications.

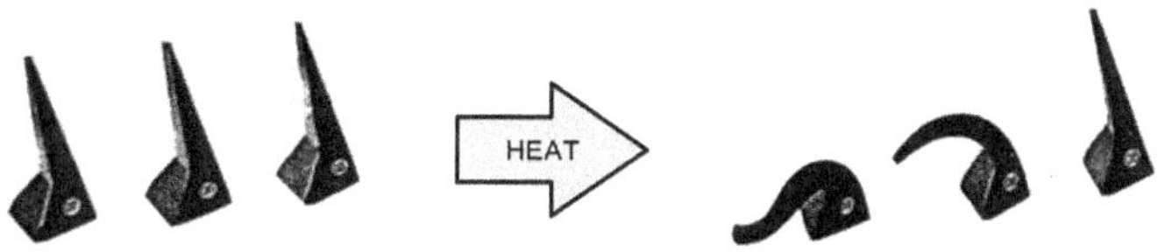

Kiln cones or **pyrometric cones** are made from carefully formulated ceramic compositions, that bend reliably at specific temperature ranges. They are still popular and widely used in firing ceramics, despite the emergence of digital pyrometers.

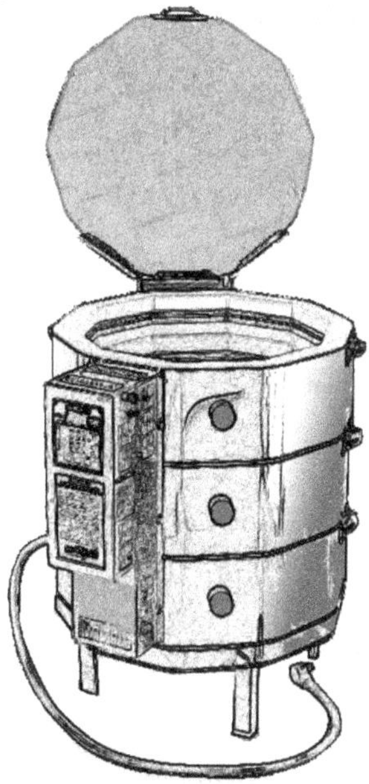

Ceramics must be fired slowly and carefully. If the job is rushed the products will most likely be ruined. Working with traditional ceramics means initially shaping them from moist clay and then allowing them to to dry before the first gentle "bisque" firing. If the ceramic survives this first treatment, then it can be "glazed" or not and subjected to a hotter and final firing.

When the water soluble clay is being shaped, it might also contain impurities and trapped air. Drying the clay never removes all the water, because some is chemically bonded. Ceramics are fired slowly to allow air, steam and other gasses from organic materials to escape gradually, which helps prevent cracking or exploding during the firing process.

When fired, ceramics are usually stacked tightly together but not touching – to allow for gas circulation. So one fault can destroy several nearby pieces simultaneously.

The chemical and physical properties of dried clay are changed as it is baked into a ceramic; the process can not be reversed. Kiln firing temperatures generally range between 1,000°C to 1,400°C, depending on the clay.

Once fired at higher temperatures, some ceramics become porous and absorbent while others turn impermeable to water. It depends upon ingredients in the original clay. A clay like kaolin contains mica, quartz, and feldspar. When cooked these minerals become "**vitrified**"- basically turned into amorphous glass.

Porcelain is a type of high density, fully vitrified ceramic made from kaolinite (kaolin) that is durable and water resistant. Sometimes translucent and yet very strong, porcelain was simply called "China" after Marco Polo introduced the pottery to Europe.

Fired earthenware like a terracotta brick is inwardly coarse and porous. But it can be made impervious to water by applying glaze. A **glaze** is made of ground up minerals mixed with water, applied to a ceramic's surface, that vitrifies when heated enough. Popular glazes have been based upon wood ashes (contains potash and lime), potash (potassium-sodium) feldspar, soda (sodium-calcium) feldspar, tin, lead and salt.

Beehive kiln, Porth Wen, Wales
CC BY-SA 2.0 / attrib: David Dixon

Bottlenecked kiln, Oxfordshire, England
CC BY-SA 2.0 / Stefan Czapski

The two 18th century kilns shown above would have been appropriate for firing pottery, bricks, glass or floor tile. These two examples are cleaned up but similar kiln sites around the world are dilapidated and surrounded by years of cracked debris strewn about. Given their locations, these two kilns probably used coal.

The next image shows a Japanese "**climbing kiln**". It is essentially a modification of a Chinese "**dragon kiln**" (which can be very large). These type kilns are built on a hillside, so that the heat will travel up-slope. While the fire is fed at the bottom, the temperature in the top chamber can get very hot. Firings can take weeks in this type of kiln. The rawest / least cooked ceramics will be in the lowest chambers, while the most matured ceramics will be toward the top. In this image each chamber has a door, so that when the kiln it cool, the finished products can be removed, and the train of ceramics advanced up the ladder, so to speak.

Nomadic N. American Indian tribes did not make pottery but most of the **civilized tribes** (Pueblo Indians) did. For example the Hopis in Arizona have been making pottery since about 700 CE. The Hopis are famed for their pottery yet they've never used thermometers, timers or kilns. They just dig a big hole in the ground and very carefully fill it with clay ceramics stacked with fuel. The preferred fuel is mainly sheep and goat dung; for its slow, even, and consistent heat. But occasionally a little juniper (cedar) or coal might be added to control temperature. The potter constantly manages the airflow to achieve the desired balance between oxidation and reduction. If organic material (like dung or wood) comes into direct contact with the hot pottery in a reducing atmosphere, the carbon from the fuel is absorbed into the clay, turning it black. Temperatures in these pits reach from about 1200°F to 1600°F . Not that hot, but it gets the job done.

Metal Market Spot Prices

Jewelry and precious metals have an intrinsic value that paper money, plastic credit cards and cryptocurrency like bitcoin will never have. But still, many factors dictate what jewelry and precious metals are worth. Their values can vary widely. Up one minute, back down the next. Any currency is only worth what someone else is willing to trade for it. In times of economic calm and quite the value of collectibles like jewelry and precious metals may stagnate or change very little. But with even the prediction of oncoming hard economic times (foreshadowed by currency devaluation, depression, impending warfare) the worth of those valuables can skyrocket.

Gold and silver values fluctuate from day to day in the marketplace because that is how the money lenders and the powers that be would have it. These and other precious metals are treated like stock commodities. It is normal for something like the value of an ounce of silver to slip 6% in value, overnight. The next day some broker holding silver futures, might cash in to gather a profit if he had guessed right. A week later and the price of silver probably returns to its previous value. For every winner there is also a looser in this game, that many people love to play.

A "**spot price**" is the current value at any given time, for a commodity on the market. A price that changes minute to minute, day to day. For the day February 1, 2026 these momentaneous spot values were valid:

Gold = $4,903.85 /oz
Silver = $79.64 /oz
Platinum = $2,154.90 / oz
Copper = $0.38 /oz ~ $6.00 / lb
Nickel = $0.50 / oz ~ $8.00 / lb

(Silver and gold can easily be melted in practically any crucible furnace. Separating the two metals is not that easy though. The alloy formed would best be separated with nitric acid (which would dissolve the silver and not bother the gold)).

Platinum has a very high melting point. It might take a professional refiner to separate gold from platinum. Several years ago platinum was worth twice as much as gold. Now the situation is reversed.

Quote: *"The decline in platinum prices is influenced by a combination of factors, including reduced industrial demand, the shift towards electric vehicles, increased recycling and the use of alternative materials, a strong US dollar, geopolitical uncertainties, and variations in mining production".* [9]

Since 1967 south Africa has minted one of the world's most recognized gold bullion coins; the Krugerrand. Actually Krugerrands come in four denominations: 1 oz, ½ oz, ¼ oz and 1/10 oz. But those ounces are Troy ounces, not what Americans call an ounce.

1 Troy ounce = 1.09714286 Avoirdupois ounces
or: 31.1 grams vs 28.35 grams

A 1oz. gold **Krugerrand** is a bit thicker and a bit wider than an American quarter. Krugerrands are 22 carrot; meaning 91.67% gold and 8.33% copper (or silver). Pure gold (24 carrot) is soft and wears easily. The alloy hardens the yellow metal, allowing it to stand up to daily use.

The value of a Krugerrand fairly well follows the spot price of gold. But some coins depending on date of coinage, may have more numismatic value. An official gold bullion coin of the United States, the American Gold Eagle and the Canadian Gold Maple Leaf coins also follow the same make up (22 carrot), weights (in Troy ounces) and denominations (4) as the Krugerrand.

** Counterfeit Chinese gold ingots and coins are something to be wary of. They have been / or were once making gold coins and bars that have the correct weight and dimensions, using tungsten plated with gold. Tungsten (worth about $10 per pound) has almost the exact same density as gold. The counterfeits fail the "ping test" though, if one knows what to listen for.* [10]

Once upon a time the American dime, quarter, half dollar and dollar were made of silver and were truly worth their "melt value". Nowadays the coins in circulation are "clad coins"; meaning faked, with a layer of copper sandwiched between two layers of **cupronickle** (copper-nickel alloy). Fairly worthless because of the inflation engendered by politicians, it's costing more to mint these imitative coins than either their face value or melt value is worth. The same thing with European coinage, except that their alloys differ slightly.

9 https://goldbuyingfacts.com/why-platinum-prices-dropped/
10 http://plancksconstant.org/blog1/2015/09/lessons_in_buying_precious_metals_ping_test.html

("German silver" is cupronickel mixed with 2% zinc / "Nordic gold" is an alloy of 89% copper, 5% aluminum, 5% zinc, and 1% tin – that is used in several Euro coins).

The Swiss franc and American nickel (5 cent piece) are actually solid homogenous cupronickel. Since 1866 the standard composition of the American nickel has been 75% copper and 25% nickel; making it the most valuable regular U.S. coin by metal content. A quarter by comparison contains less nickel (8.33% nickel, 91.67% copper). The production cost of a nickel coin exceeds its face value, with estimates around 8.52 cents per coin due to rising metal prices. But don't go melting nickels in your crucible furnace. Some congressmen made sure to pass legislation to make that illegal and worthy of a $10,000.00 fine.

> It is illegal to melt U.S. pennies and nickels due to their metal content being worth more than their face value. However, melting other coins, like silver dimes and quarters, is generally allowed if not done for profit.[11]

There is little worry of **pennies** being hoarded or destroyed for their metal. Pennies haven't been made of real copper for 40 years. Before September 1982 pennies contained 95% copper. After 1982 they have all been made of 97.5% zinc, with a fake, electroplated copper surface. If one could find real 95% copper pennies still in circulation (they are rare) it would take about 145 (@ 3.11 grams each) of them to make a pound. Good luck. *(After 230 years) the U.S. Treasury Department (2026) announced that it has stopped producing pennies. It was costing 3.69 cents to mint a fake copper penny).* Not to worry; there are still **114 billion pennies** in circulation.

Old silver coins are still considered legal tender, but considering their collectible worth, it would be irrational to spend them for their face value. *"Individuals are generally allowed to melt other coins, such as silver dimes, quarters, and half-dollars, as long as they are not doing so for profit. For example, if you're melting coins for personal use, such as creating art or jewelry, and not for the purpose of selling the metal, it is typically permissible".*

Odds & Ends

Most modern days farriers might be equipped with LP fueled furnaces because of the convenience. Sometimes horseshoes need to be bent into shape. Ranchers might require portable heat at roundup time when branding or general maintenance on cattle is preformed.

11 https://www.govmint.com/learn/post/is-it-legal-to-melt-us-coins

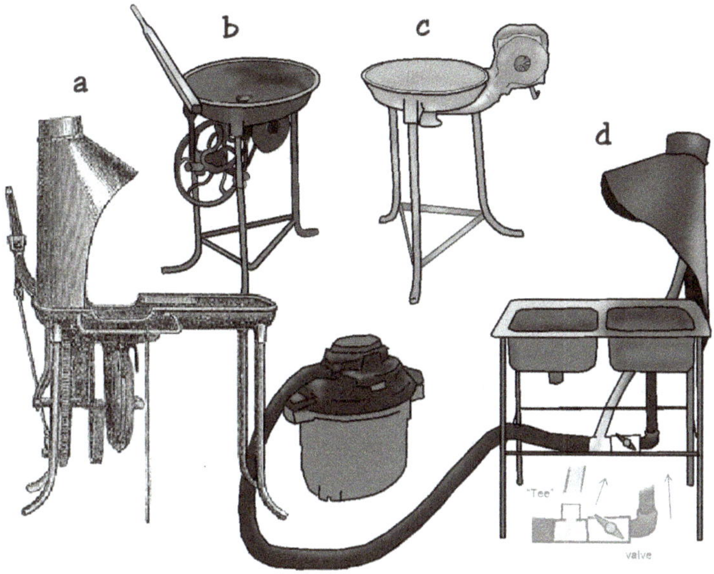

Blacksmithing 101

Above are some small, hand cranked blacksmith forges. Portable concentrated heat sources of the previous century. Example (a) is a hooded forge that was advertised sometime after the 1870's. (The 1st Sears and Roebuck catalog wasn't published until 1894). As with (a), example (b) also has a lever cranked, turbine blower that forces air up into the fire from the bottom. Example (c) circa 1920"s, uses a hand powered - geared turbine fan. Example (d) was an idea that incorporated a kitchen sink. First published in a 1941 Popular Mechanics magazine, a metal sink is mounted onto a metal stand. One side of the double sink is to be lined with refractory cement, the other can be filled with water, for quenching. This illustration attempts to show that the air is being blown in by force from a "shop vac". A PVC valve adjust the amount of draft that goes into the fire pit; some air is always bypassed to go up through the hood, pulling smoke upward. The refractory mix for the sink lining can be made of Portland cement, sand, fireclay and Perelite (a puffy, almost weightless, amorphous volcanic glass / often added to potting soil). Equal portions (or 25% less on the cement and Perelite.

By heating metal until it is soft, the blacksmith can easily bend and shape it, cut it, weld it and punch better holes than he can cut with a drill. Before the advent of modern welding technology, an indispensable metalworking tool was the anvil.

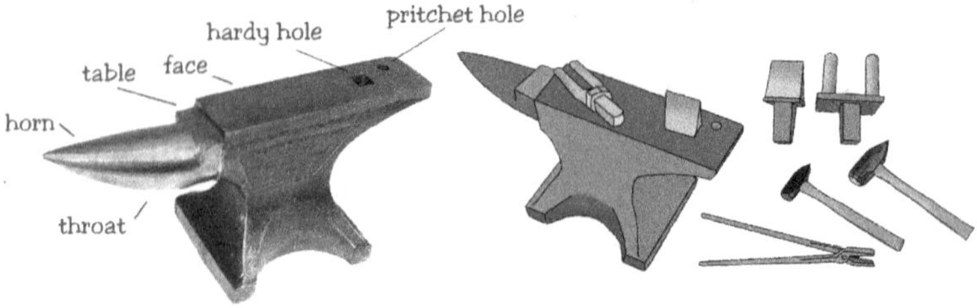

Shown above are a couple of "London Pattern anvils". A shape that has been popular and recognizable for at least two centuries, although some of its features date back to medieval times. In addition to a good heavy anvil (can cost $4 to $8 / per lb.) a blacksmith might have a vise, a pair of tongs, an assortment of cross-peen hammers and a few hardys. The features on a London pattern anvil aren't placed there by accident.

- A **hardy** is an accessory which fits in the hardy hole; it has a square base so it won't rotate by accident.
- The **face** is hardened steel where most of the forging and shaping take place.
- The **table** (or chipping block) should be of softer metal than the face and is where hot metal should be cut.
- The pritchet hole is where punch-outs are completed. (Holes in hot metal are initiated by punching them upon the face from both sides with a round punch, before being moved over the pritchet hole for completion).

Hot chisels are for cutting hot metal and are made of mild steel. Cold chisels are made of hard steel and are used for cutting cold metal only.

Fuel for the forge is typically hardwood, charcoal, coal or coke. A high quality <u>soft coal free of sulfur</u> is considered the best choice.

- There is much for a blacksmith to know – insofar as how to temper, how to anneal, how to harden. A skilled blacksmith can distinguish between grades of steel by observing the sparks thrown off when grinding it. And develops the ability to judge temperature by observing the color and glow of the underlying heated metal as well as the color of the oxide or scale formed on its surface. Hammering a piece of steel upon an anvil at a little above the critical temperature has the effect of reducing the grain size and refining the steel. The hammering strokes preformed by a blacksmith are not thoughtless or random but are instead precise and calculated. Maintaining a small and tidy fire in the forge and knowing how to apply heat to metal uniformly without causing oxidization are important skills too.

Stoves & Heaters

Above are a few wood fired stoves and kitchen ranges. The two small stoves in the center can heat water in a pot or fry food in a pan, but they have no ovens. The two larger stoves have ovens, with baffles and control levers to force the exhaust from the firebox to circulate under, around and then back over the oven compartments. Both kitchen ranges have splash-backs and warming ovens; one looks to have a water warming reservoir. These examples are simple; but other models were sometimes much more ornate and had many more features.

Controlling effectively the fire, with all the compartments, dampeners and draft levers, took skill. Needless to say that at one time these were the best central heating units that a home could have. During cold months; during warm months however, cooking on one of these ranges might not have been so pleasant.

In the colonial days of 17th century America, few people had stoves; they usually cooked in open-hearth stone or brick fireplaces. They cooked in or over the fire in copper, brass or (after 1705) cast iron utensils.

Benjamin Franklin invented a smarter type of stove in 1742. It was not very popular, probably because it was complicated to install. Cool air from underneath the floor was drawn into the baffle, where it was warmed and entered the room. The fire's exhaust was compelled to furl around that same baffle by also going below floor level before siphoning around the back and up the chimney.

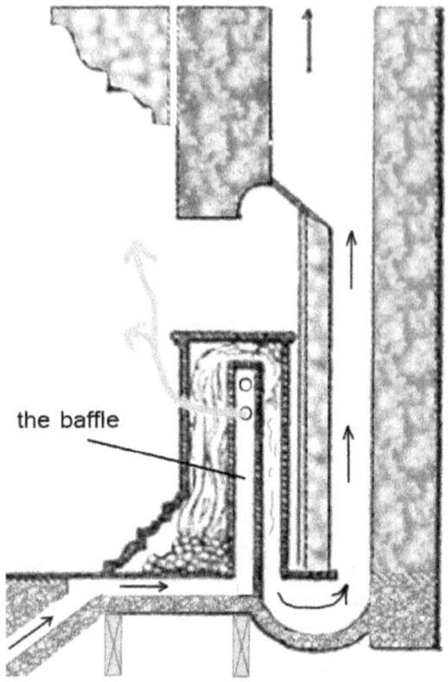

Above is a restored Roman ruin, in a villa near Caen, Normandy, France. Most of the floor has been peeled away to reveal a "**hypocaust**" (an under floor crawl space where hot air circulated). Many public buildings, homes of wealthy Romans and especially Roman public baths were heated this way. A fire was lit at one end and the smoke and exhaust was drawn through to exit somewhere on the other side. It was the sole occupation of some slaves or servants to maintain the fires.

public domain / Published 1803 / John Augustus Atkinson

A traditional masonry fireplace might be recessed into a wall and might loose up to 90% of the heat up the chimney. But a traditional "Russian stove" by comparison is usually a free standing unit within a room and made of massive amounts of masonry. Notice teenagers(?) sleeping on it at top right.

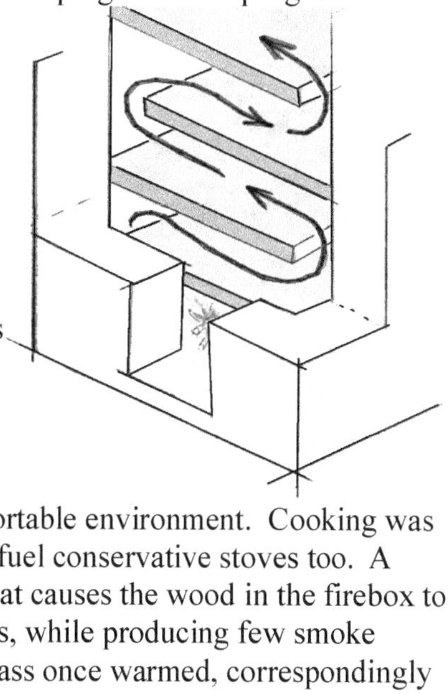

On the inside of such a "masonry heater" the hot exhaust gas is forced to furl around several (vertical or horizontal) masonry baffles before it can escape. By the time the gas does escape, most of its thermal energy has been absorbed by the masonry mass. Only a small fire in the morning and another in the evening were necessary to maintain a comfortable environment. Cooking was preformed on or within these fuel conservative stoves too. A siphon or suction is created that causes the wood in the firebox to burn at very high temperatures, while producing few smoke emissions. A large thermal mass once warmed, correspondingly takes a long time to release its heat. From the 15th century in Eastern Europe, the Baltic and Central Asia, these masonry stoves/ heaters were very popular and common.

The "**rocket stove**" shares similarities of principle with the Franklin stove, Roman hypocaust, Russian ovens, Russian "samovars" and "Kelly Kettles".

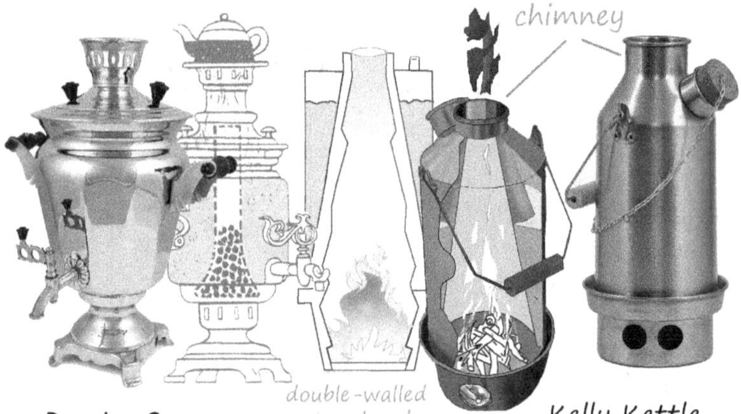

Russian Samovar Kelly Kettle

Credit for the rocket stove's invention apparently belonged to permaculture guru Dr. Larry Winiarski, in 1982. The rocket stove is an improvement over traditional wood fire cooking methods; in the congested cities of 3rd world countries where excess smoke pollution can be prevalent. A rocket stove roars to life and burns fast like a little jet engine. It requires only twigs or small materials for fuel, makes very little smoke and will boil water or heat a meal in mere minutes. The design is simple and easy to replicate with many different materials.

Like the Russian stove / mass heater, the **samovar** is also an old and widespread idea. It is an urn that heats water for tea service. Hot coals of charcoal are dropped down the chimney and water is held in a jacket surrounding the chimney flue. Many different styles, some very ornate have been made over the centuries.

The similar but more rustic **Kelly Kettle** was thought up by a couple of New Zealand infantry men during WWI. The first camp kettles were made in Ireland, followed by production in England and a knockoff version in New Zealand. The best ones are still made of stainless steel, because of its resistance to damage by fire. A kit of stack-able camping utensils might accompany the kettle. Now those products labeled as Kelly Kettles seem to be coming from China.

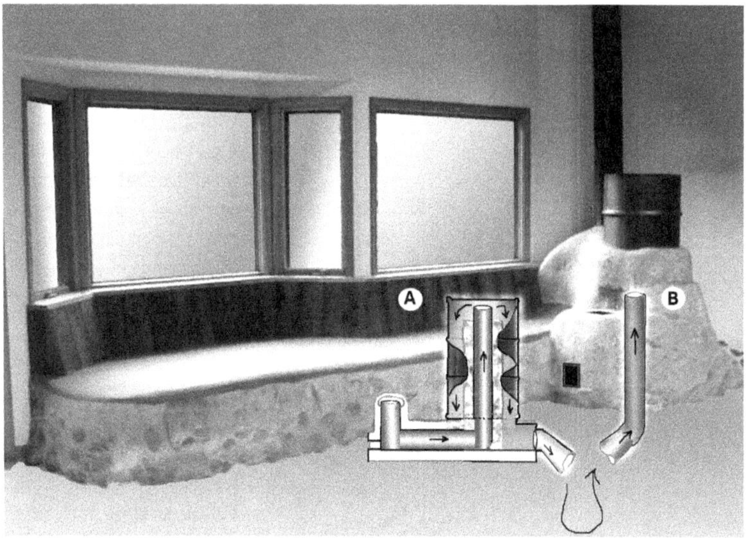

This image is one of a "**rocket stove mass heater**" because it incorporates a simple variation of the rocket stove and it is a mass heater because the whole length of the masonry "banco" (bench), has a loop of stove pipe running through it. Notice the two points of interest . **A)** Inset: a 55 gallon steel drum is placed over the end of an upright stove pipe – but the gap between the stove pipe and the end of the drum is crucial to achieve good draft (the gap should be no more than an inch or two). **B)** A tall exhaust vent or chimney will pull much better than a short one. To burn hot and efficiently the rocket stove part needs good gas circulation. These mass heaters use far less wood than regular heating stoves.

Fifty Five gallon steel drums are not very attractive and the only reason that builders leave them exposed and sticking out, is in hopes of feeling heat sooner once a fire is built. It would still heat admirably if the drum was completely covered with tile or masonry - like a more typical mass heater.

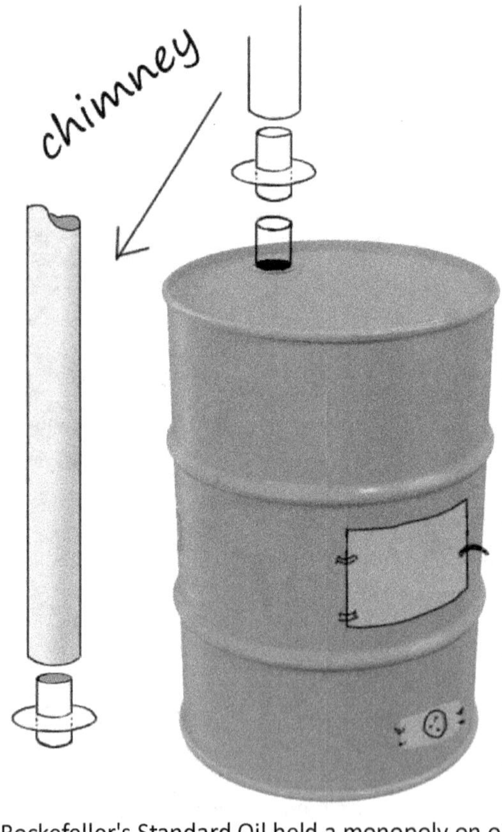

Used drums can make useful barbecue cookers, smokers or simply cheap heaters for workshops or garages.

The **55 gallon steel** drum became standardized somewhere during WW2. Optimized for transportation of liquids, 4 barrels fit a standard forklift pallet and when bundled these pallets could be stacked high in a ship's hold, without crushing.

The barrel of oil quoted in the stock market is smaller [**1bbl = 42 gal (U.S.)**]. Notice the two b's in the "bbl"notation. That hearkens back to a time between 1870 – 1911, when John D. Rockefeller's Standard Oil held a monopoly on oil and gas. At the time barrels were made of wood. Standard Oil had their shipping barrels made 2 gallons oversize (to allow for leakage or evaporation) and then had them painted blue, to underscore their authenticity.

Ch 2 Waterworks

A mechanism like shown above, has been used for 5,000 thousand years and is still found in some parts of the world. It might be called a "shadoof" in an Egyptian well, or part of a "dhekli irrigation" system when used for crops or gardens in India.

^ Water is heavy. One **cubic foot** of water (7.48 gal) weighs **62.4 pounds** (28.3 kg). One **gallon** (3.78 L) weighs **8.34 lbs** or 3.78 kg. (Because 1 cc of water weighs 1 gram).

* The word **waterworks** normally means the whole complicated system of supplied water to a town or city. Here on a lesser scale

though, the word is intended to include all those little subjects indicative of potable water, on a more individual level.

The human body is mostly water. Almost 70% of a person's weight, is reducible to water. Without frequent liquid replenishment, an average person can last only about 72 hours before they expire from dehydration.

A sudden loss of about 2% in body weight (water mostly) indicates that **dehydration** is setting in. Along with the sensation of thirst, the body begins to defend itself, and to conserve all water. The kidneys begin to decrease the volume of urine produced, the sweat glands emit less sweat, the body's temperature rises, the blood becomes thicker and blood pressure goes up.

Long distance endurance runners represent some of the most capable of humans, in terms of dealing with dehydration. They have honed stamina to a fine point; they have conditioned their bodies to survive more extreme degrees of exertion and water deprivation than normal. Records show that some runners can tolerate the loss of as much as 8% of their original weight, without suffering dire consequences. A limit for fluid loss, that most people might not survive. That 8% represents perhaps 5 ½ pints or 3 liters from the athlete's body, lost as sweat as it evaporates and cools down the body.

> * Without this temperature regulation (the evaporation of sweat) we would all expire. Hyperthermia (the opposite of hypothermia) is medically defined as a core body temperature reaching 37.5–38.3 °C. A potentially fatal heatstroke begins to occur when a person's internal temperature surpasses 40°C (104 °F).

You can go for a week or two without eating any food. Fat people could live even longer, because they have more fuel stored away that can be metabolized. But a source of fluid to drink is far more crucial to your day to day survival than food. A fact perhaps lost to most of the public. Should their domestic water pressure suddenly vanish, the total drinking reserve in many a person's abode might consist of no more than a bottle or two of beer in the refrigerator, a ½ filled cup of cold coffee and whats held in a toilet down the hall.

Presuming a situation develops where the electrical grid is attacked or damaged and not expeditiously repaired, then water delivery will disappear too. That could turn deadly. Because almost all **potable** (drinkable) water today, whether it's from a private well or municipal utility, relies upon the constant motive transfer pressure, produced by electrical pumps.

By exploiting gravity, **elevated water towers** ensure consistent water delivery in the short term, when the municipal electrical pumps themselves might be turned off at any given moment. Depending upon the number of faucets turned on, a tower does not take long to empty. Without electric power to the city pumps, the whole neighborhood will soon be out of water.

Other people might be reasonably near sources of **surface water** like rivers, lakes and streams. Rural outliers living lengthy distances from such water resources however, might be compelled to pull water up from the ground. Homeowners fortunate enough to have a well, might benefit from having a photovoltaic powered submersible pump as backup. A hedge against long term grid power outages. Otherwise they might be able to pull the pipe stack out of a well, and then dip water from the cased well.

The WINDLASS

Until the last century or so, all wells were dug by pick and shovel. A good well was lined with masonry and usually had a windlass with its bucket installed at the top. Nowadays wells are predominantly drilled with machinery and are smaller in overall bore diameter. There are millions of wells today. But without the help of electrical power, most people have no clue as how to retrieve water from a normal well. It can be done.

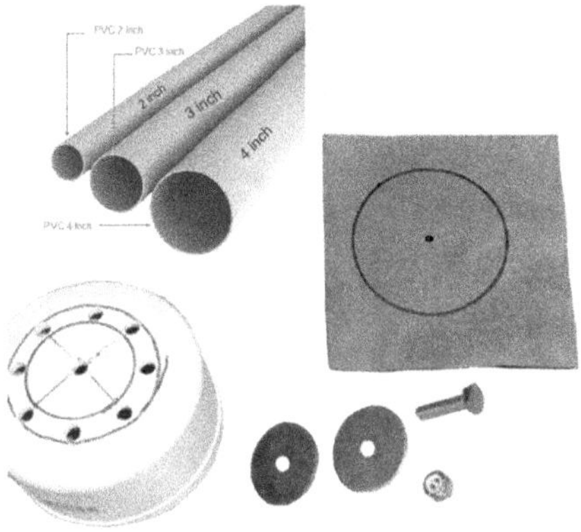

Well or 'Bail' bucket

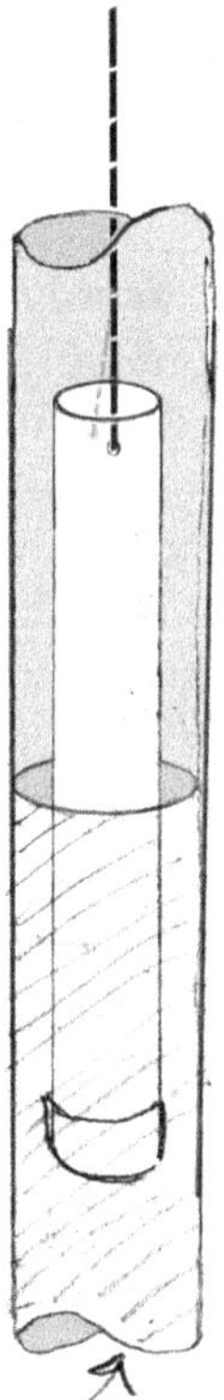

A modern day improvisation of an older day tool. It is simply dropped into a well. It fills with water and then is pulled out by rope. Where some versions might use ball or check valves, this one uses a flap-valve made from a piece of any thin, rubber like material. No electricity required.

The volume of a cylinder = (Π r^2 h).
Therefore a 36" piece of 3" PVC pipe should hold a little over a gallon.
(3.1459)[(1.5)(1.5)](36) =
254.46 in ^3 , or 140.74 fl oz, or 4.16 L

(1 gallon = 128 fl oz = 3.78 L)

An antique; made from tin (galvanized tinplate); and more valuable as a rusted keepsake than as a tool that still gets used. Some Amish communities may still fabricate them.

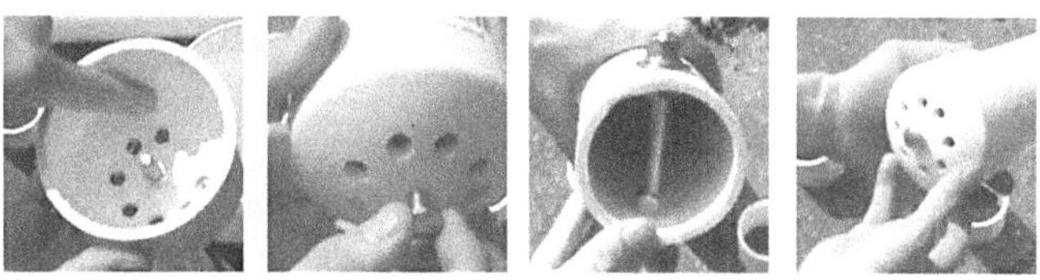

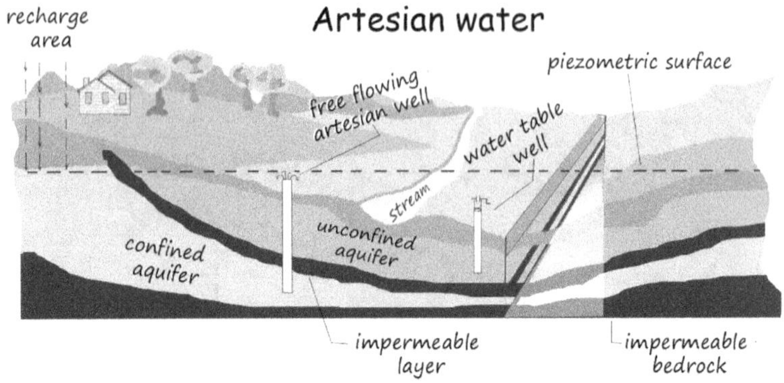

A normal "**artesian well**" indicates that the well taps into a confined aquifer; which is separated from the shallower water table aquifer. A '**free flowing**' artesian well can burble up out of the ground if or when the imaginary '**piezometric surface**' of a confined aquifer, somehow exist above the ground level where the well taps into it.

When water is trapped in a **confined aquifer** (an underground layer of water-bearing sediment, below an impermeable sediment layer, then its water can not can escape vertically; its only predictable path of movement would be in the horizontal plane. Depending upon the tectonics of an area, a lower aquifer (there night be several) might be saturated to an altitude that's higher than the wellhead. Rain water feeds or recharges this confined aquifer, somewhere out beyond the interception of new **impermeable** sediments beneath foot. A decent difference between the piezometric surface **elevation** and the wellhead elevation can allow an up-welling of water from the deep. ~Sometimes water from a new well can squirt up out of the ground in a fountainhead that can be several feet high. Usually these are plugged or capped, perhaps with a valve installed. A shallow surface **water table** is subjected to normal atmospheric pressure and as a result, hand cranked 'suction pumps' will work; but these are only effective down to a depth of about 25' or so.

Exploratory "**wildcat wells**" are drilled in remote or unfamiliar regions; in search for oil or mineral resources or for geological datum. A normal wildcat might be drilled to a depth anywhere between a 2,000 to 15,000 feet or more down. A geologist studying the evidence usually sees several differing sediment layers; sometimes several impermeable layers (or none), sometimes several confined water aquifers (or none) and occasionally salt water, instead of fresh. Artesian water from wildcats wells is seldom exploited. Usually the isolated well is simply plugged off, as per regulation.

Wells

Wells come in all shapes and sizes. Historically wells were dug by hand. After Edwin Drake sparked the first oil boom (in 1859 Pennsylvania) the technology for punching holes in the ground soon improved, significantly. But even today digging a water well by hand is ordinary in some countries. In the U.S. drilling contractors or their machinery would not become affordable or practical to hire until the 1940s-50s. Even then in remote areas, ranchers and farmers intending to erect windmills to provide water for livestock or crops, might have dug their wells with pick & shovel.

Much depends upon the height of the **water table**. Even in what seems to be a bone dry desert or prairie, water can sometimes be found relatively close to the surface. If the land is flat and nearby the type of dry stream or creek bed that tends to flood dangerously after a rainstorm, then the odds are good that the water table is high.

In the past some people placed importance upon having a "diviner" or "water witcher" search out the likeliest spot to dig a well. The scholarly consensus though, is that such ideas are akin to voodoo, and that an aquifer is contiguous and runs in almost every direction. That in the general area, you'd probably hit water at the same depth, no mater where you dug.

Deepest Wells

The digging of the deepest drinking water well ever dug by human hands, began in 1852 and took 4 years of continuous labor. The Woodingdean Water Well (in Brighton and Hove, England) at an amazing 392 m or **1,285' deep**, is slightly deeper than the Empire State building is tall. The well was intended to supply a newly constructed poorhouse and an industrial school with cheaper water than could be had by pumping it overland from an neighboring waterworks. The well was originally intended to reach down no more than 400' (which was sea level for that site, only some 2.75 km from the coastline).

The project didn't hit water where they expected to but with the help of practically free labor from the Dyke Road Workhouse, they changed directions a little and kept digging. Digging, brick laying and winching materials up or down continued round the clock, 24 hours a day, under dim candlelight. The diggers at bottom usually stripped down naked because of the stifling heat; air had to be piped in so that they could breathe. A one way trip to bottom or top took a work-crew about 45 minutes to complete, if they hurried. When water was reached (1862) it was a surprise.

At the bottom one day, a bricklayer noticed that the mound of dirt on which he was standing, began to travel upward like a piston in a cylinder. The enormous and very expensive project was only used for four years after completion. Then it was capped and abandoned. By that time water piped from the waterworks of another town became more practical and affordable than trying to extract the resource from this deep well.

A claim to the **biggest hand dug** water well (not the deepest) is made by a tourist attraction in Greensburg, Kansas, called "The Big Well", begun in 1887. Alternating crews of 10 to 15 farmers, cowboys and youngsters worked from sun-up till sun-down to earn about 50 cents a day. The well is 109 feet deep but it is an unusual **32 feet in diameter**, even after being "cased" or lined with a thick layer of rock masonry. The well water served a train line and as a municipal water supply for Greensburg until 1932. Since 1939 though it has preformed as a popular tourist attraction.

In other parts of the world where the water table allows, the ancient "**sinking method**" of digging a well might be used. Here, a ring of brick & mortar (preferably) casing is built on top of the ground; perhaps 20' in diameter. Layers of masonry are added until the walls of the ring get fairly high. Then men jump inside and begin removing dirt. As they dig under the edges of the masonry casing, it begins to sink under it's own weight. Then more rings of masonry are added, the casing gets heavier and the digging resumes – until water is reached. Many modern wells still use the same principle. But now stacks of pre-cast rings of concrete are preferred. Sometimes wells like this are so big that excavating machines can be lowered into the whole – to preforming the digging.

Modern Wells

There are several reasons aside from water, for drilling a well. Oil and gas, rock mineral exploration, geologic surveys and geothermal steam energy are also rational motives for drilling deep into the ground. Drilling deep wells demands using big, industrial, expensive machinery. There may be multiple, random water aquifers discovered in the deeper boreholes. Also, the deeper a big commercial well is drilled, the more numerous and more expensive any problems will suddenly become. Hard rock layers, shifting movement or collapse of sediment layers in the borehole, excessive heats, poisonous and corrosive gas pockets are the types of problems that occur with growing regularity, the deeper one drills.

A famous natural gas well in Oklahoma, known as the The **Bertha Rogers Borehole**, was the deepest well in the world for a while, reaching 31,441 feet (5.9 miles; 9.6 km) on April 13, 1974. But the rig hit molten sulfur, which flat out halted operations. The liquid sulfur froze (solidified) around the drill pipe and quickly caused the rest of the (almost 6 mile long) drill stem to be twisted off. Plugging the hole in essence; making it nearly impossible or unfeasible to drill further in that particulate hole. Not a complete loss however because the **exploratory well** was then plugged at 13,200' (2.5 miles down) and natural gas from the discovered zone above that point was collected for the next 23 years.

The **average depth** of exploration oil wells in the U.S. is about 7,800 feet. The deepest oil well so far, was drilled offshore, to a depth of 35,050 ft. in 2009. The deepest well of all is the **Kola Superdeep Borehole**, @ 40,230 feet, 7.6 miles or 12.2 km down. The boring on this Soviet scientific research project began in 1970 and lasted until 1992. The hole wasn't sealed until 2005.

The Soviets did make some unforeseen discovery s. In 1989

the heat in the hole reached and then surpassed an unexpected 356°F (180°C), they found liquid water at unexpected depths, they found microscopic fossils from single-celled marine organisms dating back 2 billion years and then, using seismic waves, they did not find a clear transition from granite to basalt (a crust to mantle borderline known as a 'Conrad discontinuity') as they had expected to.

The **deepest water bore** (a well initiated for the intended purpose of retrieving drinking water) is in Montana, with a depth of 7,320 ft. Usually water wells are not drilled nearly so deeply. But in special circumstances where fresh water is scarce, governments, municipalities or cooperative farmer organizations might join together to pay the high expense of drilling a deep well.

Five volcanoes make up the Big Island of Hawaii. Mauna Kea is the tallest (6.3 miles from base to summit) but the slightly lower Mauna Loa has a bigger base, forms ½ of the Big Island alone and is still active. All the Hawaiian Islands are made from almost entirely igneous lava (or basalt). Yet an estimated 90% of the state's drinking water comes from ground water and wells. Mostly that water comes from deep municipal wells that were drilled through stubborn "blue rock" (grayish basalt : mohs 6). Yes, about 10% of the population does drink "**catchment water**" (rainwater that runs off a roof and into a cistern). The point is however, that even on an island made of raw rock, underground aquifers can be found. It might be very expensive to tap into them though.

Likewise the dept to water of an aquifer in Arizona's Verde River Basin or underneath the Coconino Plateau (a hardscrabble region between the Grand Canyon and Flagstaff) can be a deep 3,200 ft. below land surface. Cities in arid regions are sometimes compelled to dig deep wells.

Private water wells

A well depth of 50' or less is considered a **shallow well.** For potable (drinkable) water a well's depth should extend to 100' or beyond. "*Wells with a depth of less than 100 feet lack enough soil thickness to adsorb all of the contaminants seeping into the ground*"; (like bacteria, motor oil, pesticides and fertilizers).

In the US an estimated 42 million people, living in about **12 million homes** - drink and bathe in well water. The average well depth for private homes is somewhere between 100 to 800 feet.

To have a contemporary domestic water well completed in 2024, will cost about $15 to **$25 per foot** just to drill (but perhaps $65/ft if rock is encountered). An additional $5 to $10 / ft. cost should be expected to finish the well with casing. In 2023 the average cost to install a new water well was between $3,750 and $15,300 (in the U.S.). If combined with a new septic system, the average cost for both might range anywhere between $6,000 and $20,000. Afterward the expense of electrical service, the pump, a pressure tank and piping to the house would be added to that.

Even if someone's property has a high water table it is generally still advisable to drill down beyond 100 ' (for sanitation) before setting the casing pipe. **Casing** is a large diameter, thin walled (preferably metal / electrically ground-able) pipe which once it is set - keeps the bore-hole's sediment from sloughing off and collapsing. A quality casing job includes cementing; forcing cement down between the metal pipe and the muddy edges of the borehole. Several feet of small penetrations will be made at the base of this casing, for water to seep in. The water will be sucked through a smaller diameter 'well pipe' that fits inside the casing. The well pipe inside the casing, need only reach down past the static water level a little. But fresh water can only seep in from deeper, at the bottom of the casing. A **'foot valve'** (at the bottom of the well pipe) should never set on the bottom of the well, where it could draw sediment.

Take for example some imaginary well on a random property, which has a high 20' water table. The well should be drilled and cased to about 120'. But the well pipe and foot valve on its end should probably stick down 40'. Because the (20') depth of the water table might vary with the seasons or drop during a drought. Hydrostatic pressure constantly pushes water up the casting, to the level of the water-table. For deeper wells, the seasonal fluctuations in water table are estimated accordingly and submersible pump is installed in a safety zone where it won't run dry nor be close enough to the bottom of the well to suck up sand or sediment.

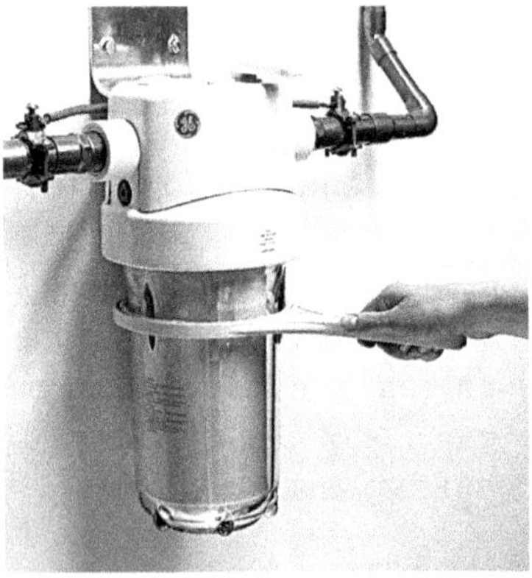

The deeper you drill a well, the more minerals will be present, and more minerals = **hard water**. Commercial water filters and water softeners can be tied in to the plumbing before water reaches the faucet. A **well water filter** is made up of a series of filters. The first stage of such a filter usually removes particulate larger than 1 micron (like dirt or sand). The final stage of the filter might reach down to less than 0.04 microns, for trapping bacteria. *[Most bacterial size range from 0.2 to 2.0 µm in diameter and 2 to 8 µm in length.]*

Filtration

There are several ways to clean, filter or sterilize water. Using ultraviolet light, ozonation, filtration, distillation and chemicals are all common means to disinfect water. "Flocculation" occurs when a liquid is allowed to rest and then particles settle or fall out of suspension. For wine and beer and some other liquid mixtures, "chill filtration" is flocculation assisted or accelerated a bit by reducing its environmental temperature. Racking or decanting is the act of liquid transfer between containers; intended to leave sediments behind. Boiling water for a suitable period is an age old way to insure drinking water safety.

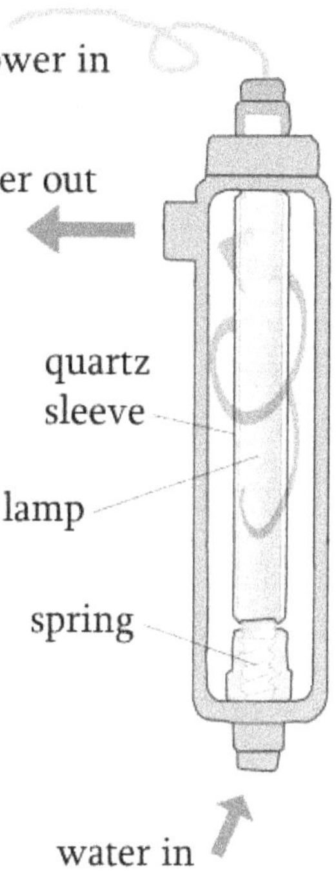

Ultraviolet radiation (invisible UV light) does not necessarily kill protozoans, viruses or other 'bugs', but it can 'zap' or scramble up their DNA. It makes then unable to reproduce correctly and therefore quashes their pathogenic threat. Water running to a sink or through an entire household can be sterilized by installing an inline UV lamp. The electric lamp is housed inside a metal tube, then water flows through the tube and around the lamp. The lamp itself is housed in a quartz sleeve, which is sealed with 'O-rings'. A tiny bit of mercury is vaporized to make the UV-C light.

Ozone (O_3) can also be used to sterilize water. When found in the upper atmosphere, ozone filters out much of the sun's own ultraviolet light and so protects earth organisms from this potentially damaging radiation. Ozone for water purification can be manufactured by using one of two different methods. This ozone can then oxidize and purify water by being forced to bubble up through that water for a period. Oxidizing decontaminates by weakening and rupturing the cell membranes of bacteria, viruses, fungi and parasites.

Ozone is unstable and needs to be manufactured at the point of application, by an ozone generator. One type of ozone generator uses UV light to bombard pure oxygen (O_2) into single oxygen (O) atoms. These then have an affinity to join with oxygen gas (O_2) again, to create O_3. The second common method of making ozone is by using a 'Coronal Discharge' (CD-type) ozone generator. Such a mechanism creates ozone, while emulating lightning - by generating high voltage.

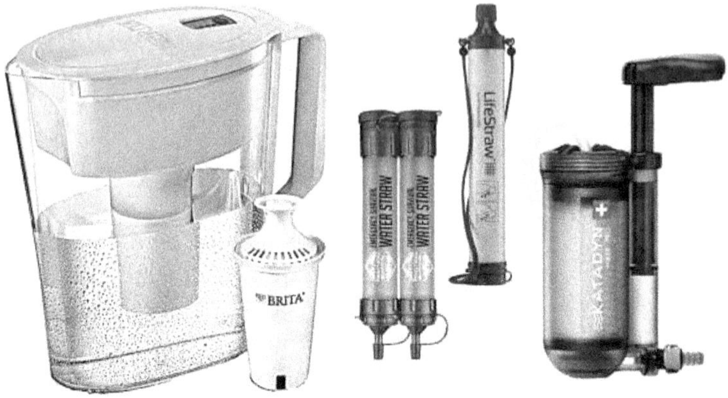

There are quite a few portable water filtration systems on the market. Some may use just activated charcoal (carbon) to remove chlorine, some metals, pesticides and most cysts (protozoans and single-celled organisms).

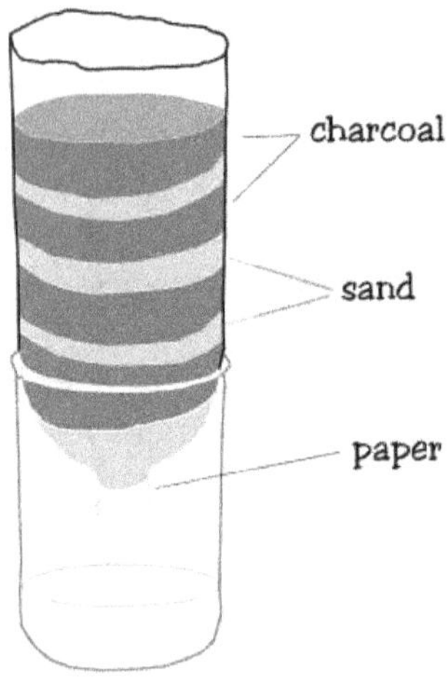

Some filters work by gravity, some by suction and others by pump pressure. Most might use a combination of activated carbon along with some new type of 'hollow fiber membrane' in their filter. Some of the drinking straws for hikers boast filter membranes that can stop objects down to 0. 2 microns in size. (Giardia, E. Coli and Cryptosporidium require at least a 1.0 micron filter). But the lifespan of these filtration systems is limited by the amount of water cleaned, before the filter becomes clogged. A good water straw is limited to about 1,000 liters (264 gallons) before it becomes too obstructed to suck or push more water through.

"Water, water, every where, Nor any drop to drink".
(Samuel Taylor Coleridge, 1798)

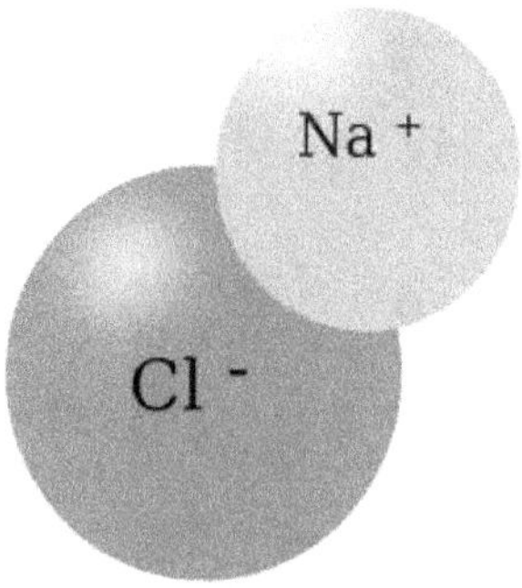

Removing salt from seawater is difficult. Usually, only reverse osmosis or distillation can accomplish the chore. Diffusion is the net transport of matter from a region of high concentration to a region of lower concentration. When nature attempts to equalize the concentrations of matter on both sides of a semipermeable membrane, this is called **osmosis**.

Reverse osmosis uses pressure to overcome osmotic pressure. Forcing salt water through a membrane is an example.

Seawater ranges between 27,000 to 35,000 ppm total dissolved solids. A high pressure (about 800 psi) and a reverse osmosis **membrane** that has a pore size of about .001 microns, is needed to pass the water but block the salt in this type of system. The process is doable but inefficient.

Reverse osmosis systems work well for both industrial and household scale applications. Reverse osmosis filters out nearly all contaminants and minerals too (which may or may not be a good thing) and also lowers the pH. But these systems are *expensive and wasteful* (they intake a high volume of water and waste most of it – to make a little pure water). These filtering systems are complex and costly to install and maintain.

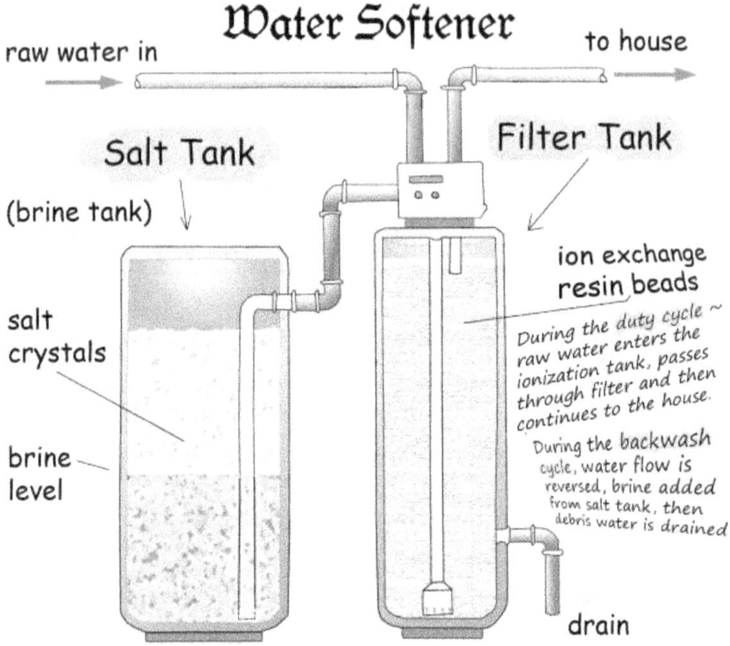

Water softeners are not water filters in the typical sense. A filter is intended to remove things like bacteria, chemicals, and sediments. A water softener however is intended to to remove minerals like calcium and magnesium, which tend to make water hard. A comprehensive water treatment plan might call for both a water filter and a water softener, in tandem.

"**Hard water**" is an expression that might indicate that water smells bad, taste strange or that is generally unfit for the task of laundry – because clothes become stiff or scratchy. Also, water heaters have a tendency to precipitate these minerals out of suspension, which are then deposited and buildup inside pipes or that grow sometimes as actual rocks inside the heater tanks.

Softeners work on the principle of **ion exchange**. Typically small, specialized porous microbeads of acrylic or polystyrene will fill the filter's tank. Raw mineralized water enters the tank and then the beads capture or latch onto the Mg or Ca charged ions in the water. The water now free of minerals, continues onward to the faucet. But as time passes and more water flows past the beads, they become saturated with these minerals. At this point the filer needs to be backwashed (flow reversed), like the sand filter of a swimming pool would be. But the minerals cling tenaciously to the filtering media.

The larger tank (usually, of a two tank softener design) does nothing except hold salt water. During a backwash cycle, salt water brine is introduced during or in-between rinse cycles. Slowly the filter becomes "recharged" as new sodium ions displace the unwanted mineral ions that were attached to the plastic beads before. All the wastewater is ejected to a drain.

Programming the control box of a water softener, is critical to optimum performance. The composition of the local water, the length of the regeneration cycle and the frequency of those cycles must be considered. A typical water softener might be programmed to flush and recharge once a week. Depending upon conditions this backwash/recharge cycle may take 20 minutes or so, and perhaps be preformed late at night.

The process is also wasteful, discharging perhaps 40-60 gallons per session, of salty, mineralized wastewater. That dirty water must go somewhere; it's probably not a good idea to send it into a flower bed.

Softened water typically contains about 12.5 milligrams (mg) of sodium per 8-ounce glass. Although this might be considered a modest proportion of sodium by the FDA, it may nonetheless be a concern for someone with high blood pressure.

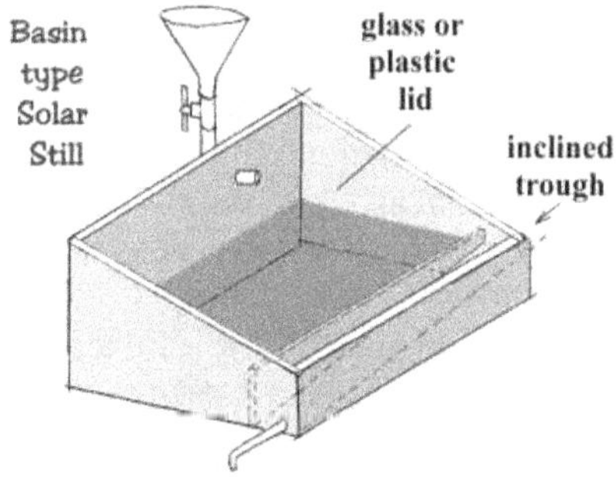

Solar Still

The concept of extracting fresh water from salty or brackish water may not be new, but the successful practice of doing so is. Relatively.

In centuries gone by, many seafarers have gone thirsty, or may have died from lack of drinking water. It is VERY dangerous, potentially deadly to drink seawater. If someone is guzzling seawater because of thirst, they are only compounding the problem; they are depleting the fluids from inside their body's own cells. It is back to osmosis again. The salinity of seawater is almost four times that of our bodily fluids. The kidneys will dehydrate the body more as they attempt to remove excess sodium - by making urine.

Evidently the first solar stills ever made were constructed in the Chilean nitrate mining community of Las Salinas in 1872. A Swedish engineer named Charles Wilson apparently constructed a series of stills that were eventually capable of purifying 23,000 liters of clean water from brackish water, per day.

Some years later the "**Saltpeter War**" erupted between Bolivia, Peru and Chile in 1879. The struggle was for ownership and control over the extremely important and lucrative sodium nitrate deposits residing underneath Las Salinas and other locations in the same harsh and dry Atacama Desert. Sodium nitrate and sodium nitrite were valuable historically for their utility in meat preservation. But in the 19th century, world powers also needed lots of **gunpowder.** Such nitrate deposits are rare.

The European political powers in particular were acutely dependent upon steady shipments of South American saltpeter. By 1909 however, a German chemist named Fritz Haber discovered a way to make ammonia from nitrogen and hydrogen. Reacting ammonia with nitric acid (also produced with the Haber process) makes the fertilizer and explosive prerequisite known as - ammonia nitrate.

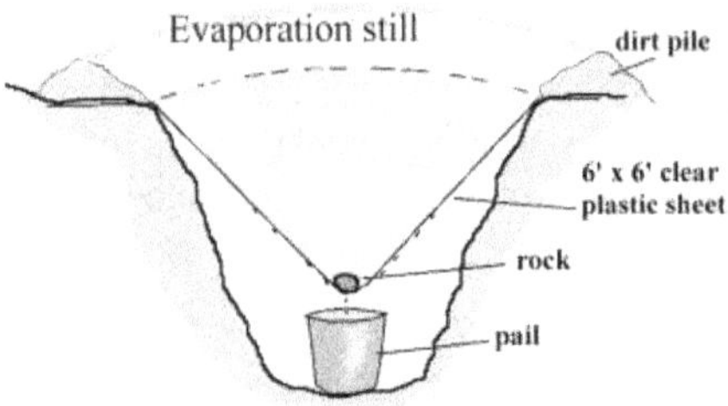

Back in the 1960's a pair of PhD's working in the soil hydrology laboratory for the USDA invented a solar evaporation still that could suck useful drinking water out of the ground. Even in the arid desert around Tucson, AZ. where they were located, they realized that the soil entrapped useful moisture. Such a solar still is made by digging a pit in the ground, placing a collection pot in the bottom and covering the hole with a sheet of plastic. Additional moisture could even be gathered by placing green vegetation under such a tarp. The same concept is now used to purification purify saltwater in an emergency. Little inflatable water still kits, are sometimes placed in the lifeboats or escape capsules of seacraft.

Chemical Purification

During WWII and on through the Vietnam War era, troops were issued (chlorine based) **halazone tablets** to purify drinking water. They came as part of a C-rations kit after 1945. The idea is that you drop a couple into a canteen of questionable water. But one must wait 30 minutes before it's safe to drink. You can still buy halazone tablets today; and iodine crystal tablets and chlorine tablets too, for the same purpose.

Iodine kills most, but not all of pathogens and protozoa s that might be in the water (they can't successfully fight *E.coli.* for example). A little bottle of iodine pills is useful to have in a "bug out" (emergency survival) kit. Iodine tablets in an unopened bottle have an almost unlimited shelf life, whereas halazone tablets have a very short usable lifespan once the bottle has been opened; just a few days. Most hikers these days though, prefer to invest in a water straw filter; not chemicals.

Shock chlorination should always be preformed upon a new well, or if ground water has intruded into a well or if the piping or submersible pump has been contaminated by being pulled out of the hole for replacement or repair. This disinfection is needed to kill coliform bacteria, algae, viruses and fungi. The preferred chemicals used to preform this water treatment task are either sodium hyporchlorite or calcium hypochlorite. **Sodium hypochlorite** ($NaOCl$) is commonly encountered as liquid laundry bleach. Aside from making white clothes more bright, this corrosive is commonly used to chlorinate water. **Calcium hypochlorite** ($Ca(ClO)_2$) or bleaching powder, is usually encountered as a loose dry powder or in capsules used to sanitize swimming pools. As a solid it is more stable; and in solution provides more available chlorine than sodium hypochlorite.

These days a chemical called **chloramine** is increasingly used in the US instead of chlorine. Chloramine is formed by combining chlorine and ammonia; it breaks down into ammonium chloride. As a disinfectant chlorine smells or taste stronger and act faster than chloramine. But chloramine is milder tasting and milder acting, forms fewer disinfection products and is more persistent in terms of purification longevity.

The standard water treatment shock treatment requirement varies from state to state but usually ranges between 50 and 100 parts-per-million (ppm). The proper method is to calculate the volume of water in a well (or tank, cistern, jug – what have you) and then read a chart to estimate how much chemical to apply. [50ppm = apx 1 oz per 100 gal]

The diameter of the casing, depth of well and static water level are used to calculate the volume of water in a well. The volume of a cylinder is: $V = \pi r^2 h$. (for example a 300' column of water, 6" in diameter, estimating 1.5 gallons/ft, means 450 gallons of water in the well need to be disinfection. Therefore about 4.5 ounces of bleaching powder is needed to reach 50 ppm. Or twice that amount to meet 100 ppm state requirements.
(With liquid sodium hypochlorite, a ratio of 0.2–2 mg per liter of

water, is an adequate safeguard at point of use disinfection). [12]

In practice, it is safe to overestimate the quantity of chlorine, because it is usually to be flushed through the system (house waterlines and faucets) anyway, until it can no longer be smelled. Chlorine also dissipates after 12 to 24 hours.

Iron & other **water contaminants**

Iron is fairly common in the earth's crust and unfortunately iron stain from well water is fairly common as well. There are three kinds of iron found in well water: ferrous, ferric and bacterial. "Ferrous and ferric" refer to two different iron oxidization states. **Ferrous Iron** (Fe^{2+}) is often called "clear-water iron", while **Ferric Iron** (Fe^{3+}) is known as "red-water iron". The first form might be dissolved in the water but go unseen, but the second form is already oxidized and is visible in the water. But the really ugly and oftentimes smelly reddish iron stains seen in some plumbing, bathroom and kitchen appliances might be the result of **iron-oxidizing bacteria**.

These feed on dissolved iron, resulting in reddish-brown slime and deposits. *"Iron bacteria feed on dissolved iron and can proliferate in low-oxygen environments, such as shallow wells or stagnant water".* While small colonies of these bacterium might exist naturally in shallow aquifers, they commonly find their way into a water well because of humans. Activities like the process of drilling a well, installing or withdrawing well pipe or pumps on the ground can lead to contamination by such bacteria. Iron bacteria are stubborn and very difficult to eradicate apparently. Shock chlorination can be effective treatment, for a while. Installing a "greensand filter"is the other common choice of remedy.

Greensand filters are often used in an array of water treatment schemes to remove iron, manganese, hydrogen sulfide, arsenic, and radium. Sometimes these unwanted materials are found together, but a single filter is usually incapable of removing more than one undesirable mineral at a time. Greensand filters are made from a greenish phyllosilicate called glauconite. The glauconite might be doped with a specific coating, which then helps it act as a catalyst. With the introduction of an oxidizer, the contaminants can be filtered out effectively.

---

12 Lantagne, D. 2008 Sodium hypochlorite dosage for household and emergency water treatment. J. Am. Water Works Assoc.

Pumps

There are multitudinous pump designs for forcing liquids to move. Even today new pump designs sporting unique features are continually added to the list. Some older historic pump designs though, using no electrical or petroleum power, were ingenious in their own right and might still be encountered today. Motivating liquids to move has garnered much inventive attention over the centuries.

Archimedes' screw-pump

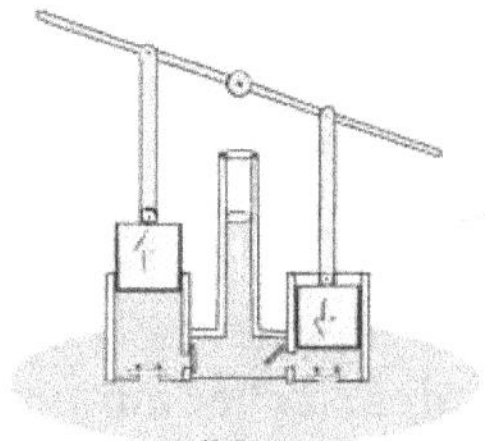

Ctesibius' piston pump

Chinese chain pump

In the 3rd Century BC, mathematician and engineer Archimedes of Syracuse invented the ***'Archimedes type screw pump'***. At the same time or soon afterward a contemporary of Archimedes, named **Ctesibius**, invented a primitive ***pressure piston pump***. Ctesibius who was also a Greek mathematician & inventor, was living in Alexandria when it was still part of Ptolemaic Egypt. Ctesibius is noted for his work and treatises on pneumatics and for his invention of a type of water powered pipe organ called a **hydraulis**.

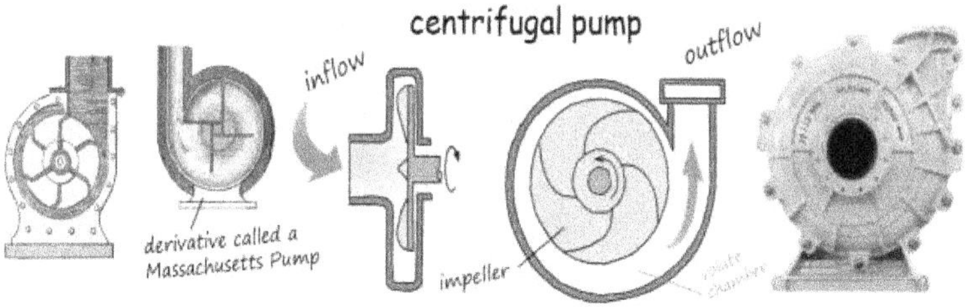

From modest beginnings some 2,200 years ago, the history of pump designs has taken many turns. **Chain link pumps** were first being used somewhere in-between 5th and 6th century China. The first drawing of a **centrifugal pump** was made in 1475.

Gear Pump

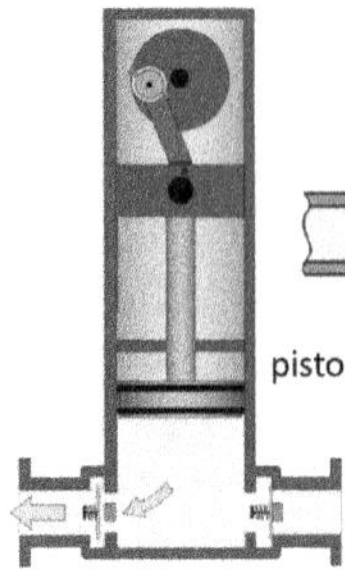

Reciprocating Piston Vacuum Pump

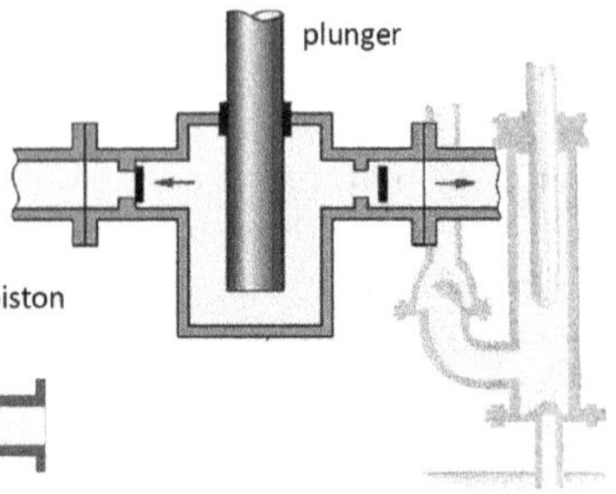

Plunger Pump

Gear pumps were invented near the end of the 16th century. The first mechanical air and **vacuum pumps** were invented around the 1650's. The **plunger** type pump, good for pond drainage and fire fighting was patented in 1675. These later became useful for high viscosity fluids.

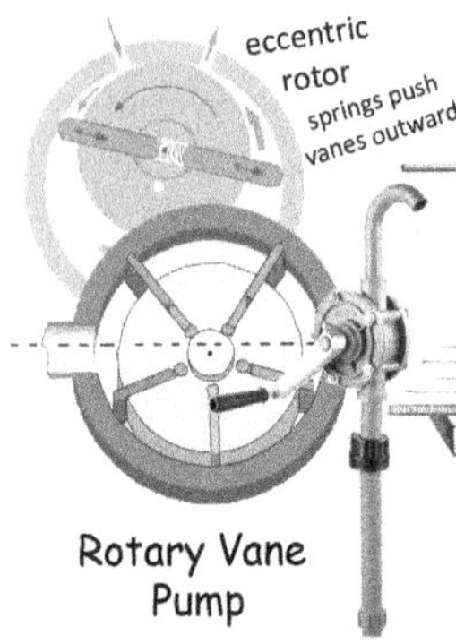

Rotary Vane Pump

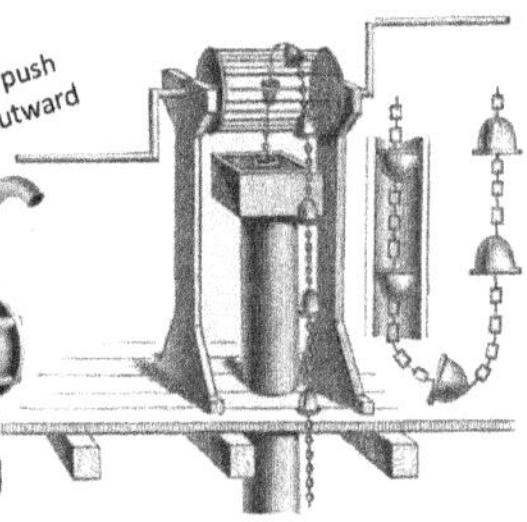

French Chain Pump

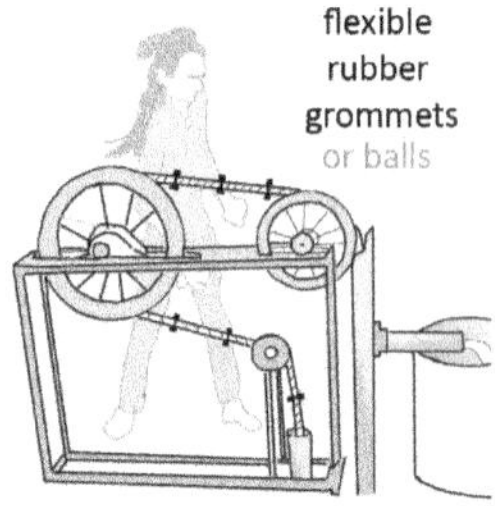

Rope Pump

Vane pumps, chain pumps (European) and **rope pumps** were described and talked about back in the 16th century.

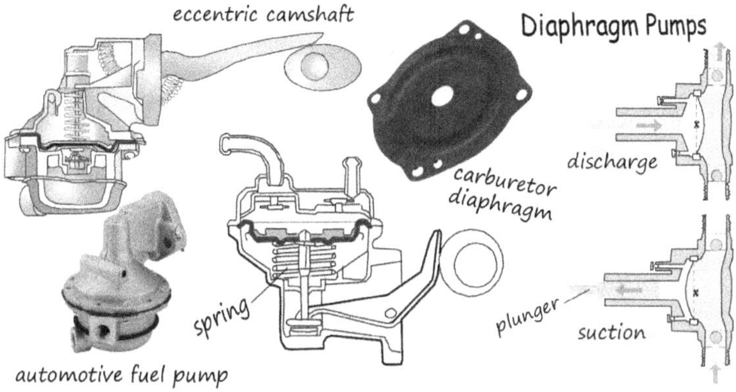

The **diaphragm pump** was invented nearly 3 centuries ago in 1732. The diagram part is usually formed from a flexible material like rubber. Almost every automobile made between 1900 and the 1990's used a mechanical fuel pump and/or a carburetor that employed a diaphragm. After the mid 1980s cars evolved to EFI (Electronic Fuel Injection) and this generally required a move to electrical fuel pumps, which were capable of supplying higher pressures than camshaft driven mechanical suction fuel pumps could. But even today those electrical fuel pumps still house a little diaphragm pump inside.

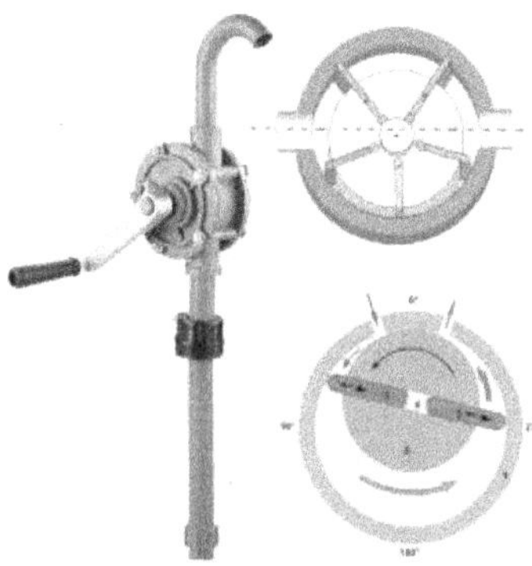

The rotary vane pump or **barrel pump** may look like centrifugal pump but it's much different. Here the vanes of the rotor are spring loaded and push outward, to the inside surface of the volute chamber. Also the rotors are eccentric (not centered) in the volute chamber. Normally these hand cranked pumps are seen refueling tractors or other heavy equipment from 55 gallon fuel drums at the back of some truck. Anyhow this type of pump was patented in Germany, in 1909. They develop a good vacuum and are self priming. Naturally one could be used to pump low viscosity fluids like water. They are however a little complex to maintain and unsuited for high pressures or for high viscosity fluids.

^ **Viscosity**: the friction in a fluid, produced between the adjacent molecules of the fluid. Ninety weight gear oil is more 'viscous' than kerosene for example.

The differences between a hand cranked "**pitcher pump**" and a **windmill pump** are slight. Both rely upon flap valves or check valves and both pumps have short strokes. The submersed windmill pump sits down in the well, pushing a column of water above it. The pitcher pump however, is merely creating an empty vacuum at the top of a pipe stem. Atmospheric pressure is causing water to rise and fill that vacuum.

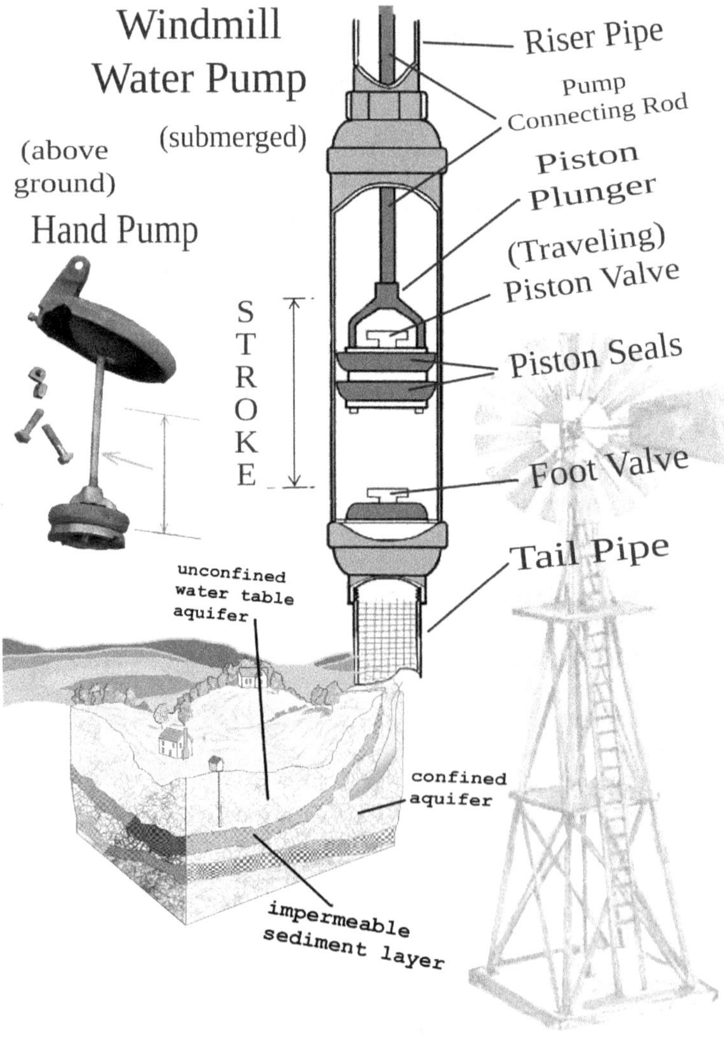

The Driven Sand Point Well

- Inexpensive, simple
- requires high water table
- modest yield, might require filtration

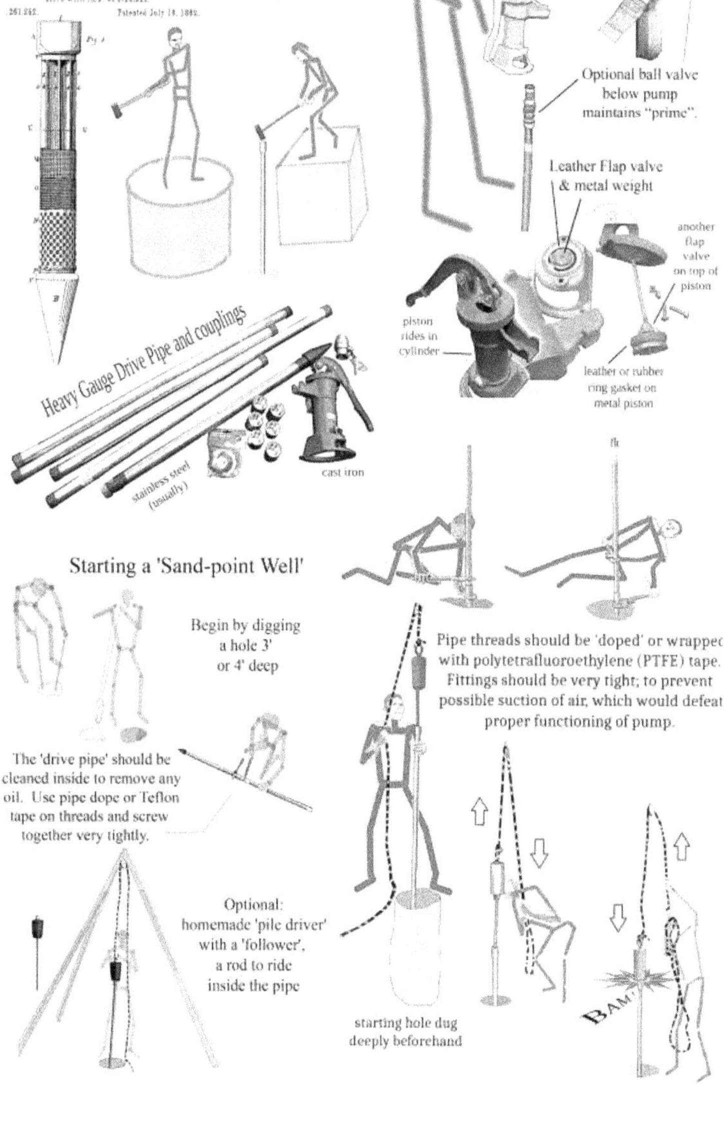

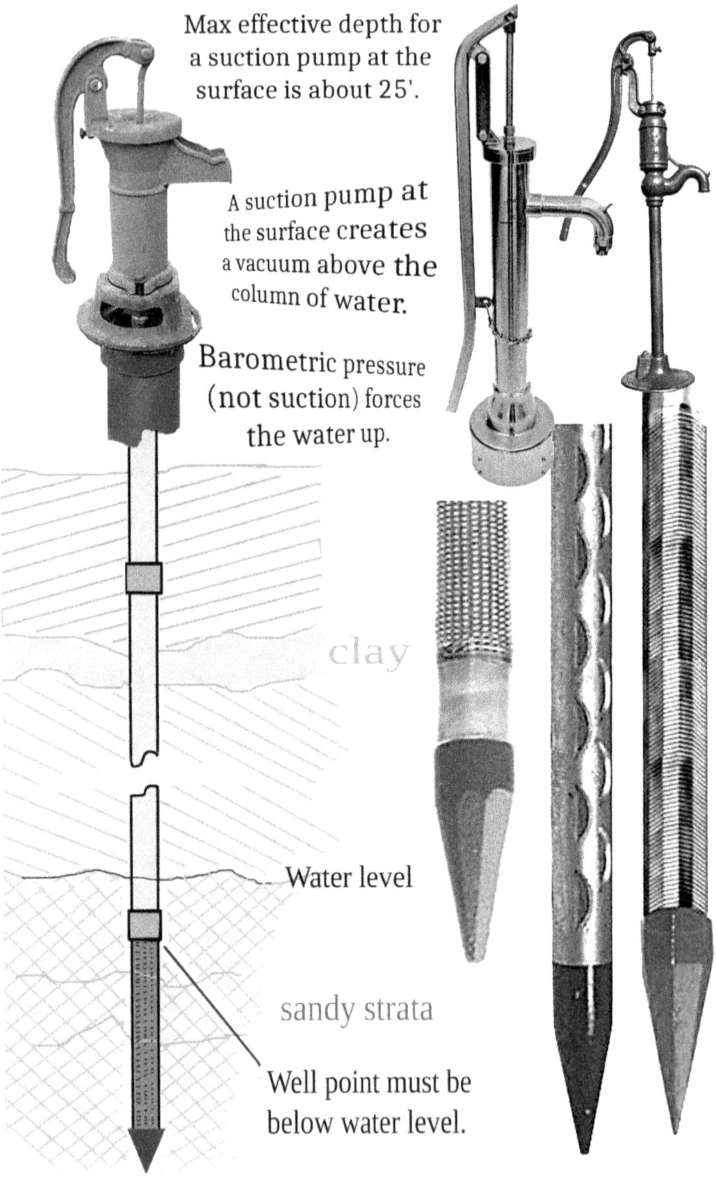

The "stand pump" (that **stands on the ground)** has a longer stroke and a bit better suction than the counter top, pitcher type pump. **Drive pipe** is heavy duty and comes in 1 ¼ ", 1 ½" and 2" diameters. The "Drive Well Points" are liberally perforated and screened, but still sturdy enough withstand being driven by a sledgehammer. Usually the pipe is galvanized but in a case where water pressure is used to dig, the pipe might be PVC.

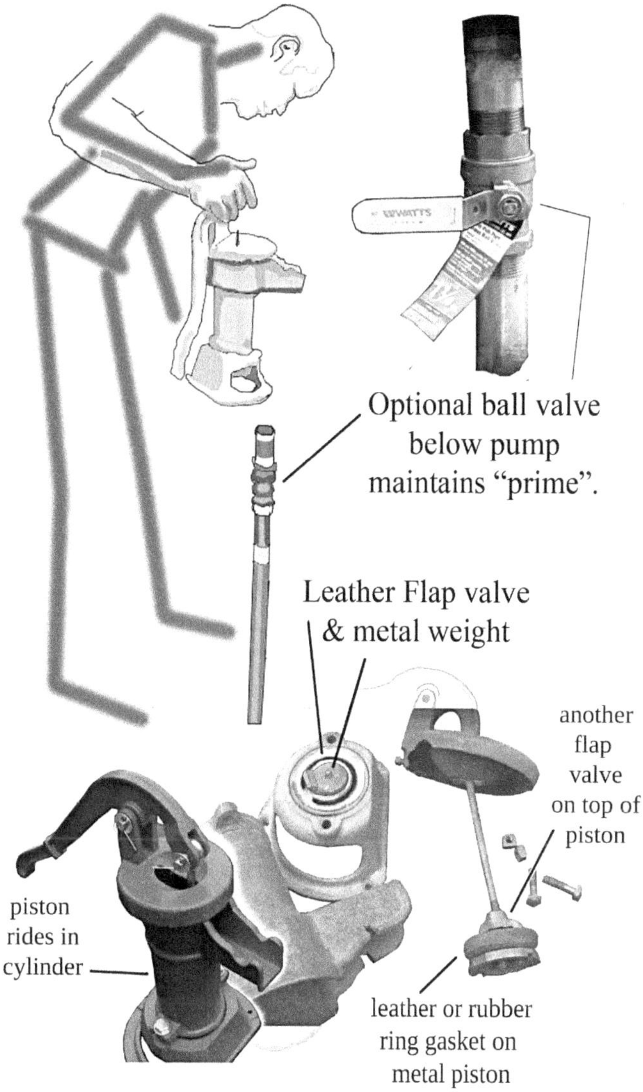

In the previous century, sometimes houses and their kitchens were built over a well, and the sink placed beneath a pitcher pump, both installed on the counter-top. Like anything else mechanical that requires maintenance, the "**leathers**" (**flap valves** and gaskets) in these pumps needed periodic inspection or possibly replacement.

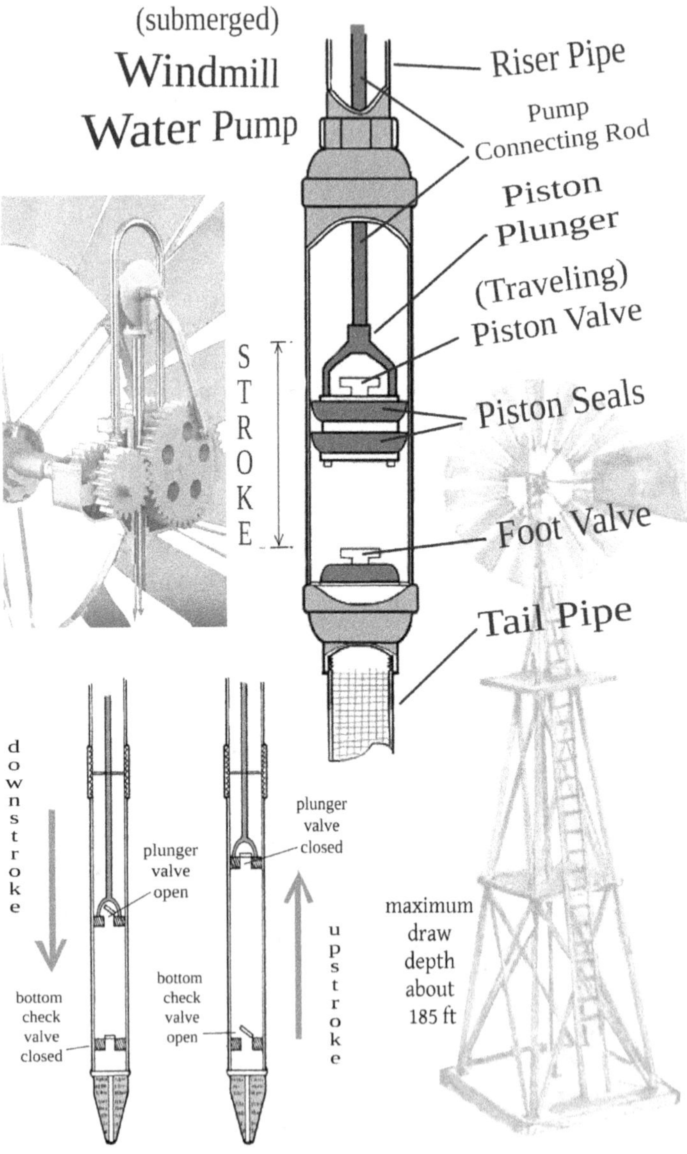

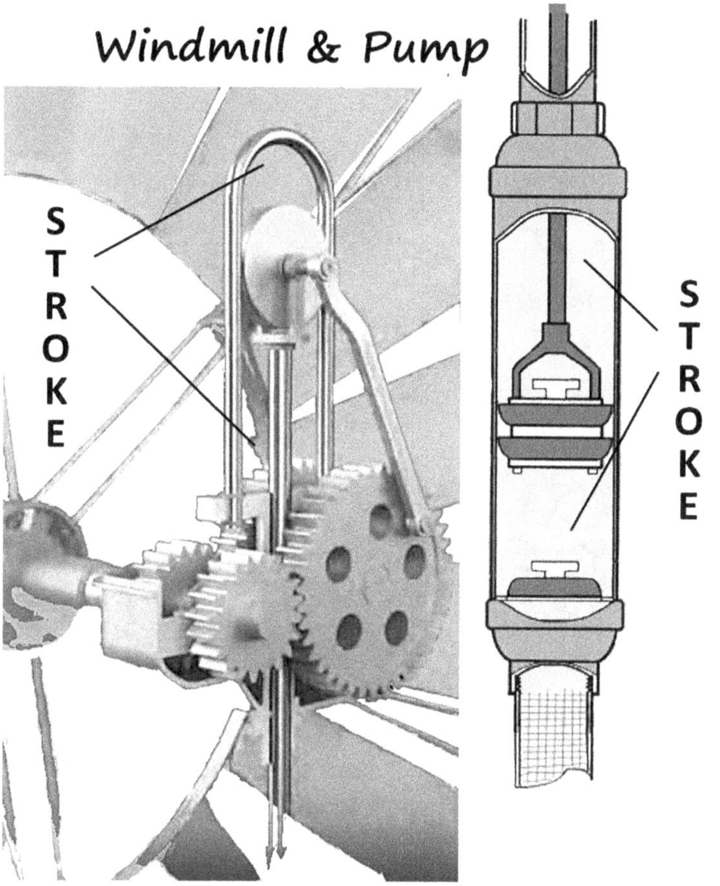

The gearbox of a windmill preforms more than one function. The weight counterbalances the propeller, the gears reduce the speed of rotation, while also transforming the rotational energy into a reciprocal mechanical motion. A long rod connects the gearbox to the pump below. The travel or stroke of the piston inside the pump is fairly short, perhaps 10 or 12 inches.

Check valves and plunger valves ensure that water has only one path to follow; up. There are many variables that can effect the outcome (pipe diameter, wind speed, wheel diameter) but, a modest windmill with a small 1 ¾” pipe and 6' wheel, might lift water 200'.[13]

---

13 https://aermotorwindmill.com/pages/windmill-pumping-capacities

Ctesibius's piston pump of the 3rd century AD, was still being used 16 or 17 centuries later.

It stands to reason that such pumps as these required a ready reserve of **nearby water,** before they could function. In most cases these pumps were tapped into fire hydrants or somehow into a cities' underground "main". But the luxury of having clean potable water delivered throughout a city did not happen overnight. Requirements for a cities water distribution system include at minimum a pumping station, a water tower, water mains, fire hydrants, and service lines. Taking a random city like New York City for an example, one might be surprised to learn that its first new water supply and distribution system was created **initially, for fighting fires**.

(book) Firefighting, Life of a Fireman — detail from a 1854, Currier & Ives print

(Notice, no horses. These NYC firefighters are pulling and pushing this water pump. It takes precious time to harness up a team of horses; then when they arrive those horses can't operate the pump. An 1854 illustration by Currier & Ives).

New York City was not the first or last city to grapple with installing a water system; but now it has one of the world's finest. It now has three aqueducts supplying water, beginning from sources as much as 125 miles away. But for its first 200 years New York residents depended upon only hand dug wells and a couple of small reservoirs (ponds) for all their water. That changed in 1831.

Quote:

"In April 1831, a new water supply and distribution system opened, for fighting fires. It included a well and cistern on 13th Street between Bowery (today 4th Avenue) and Third Avenue, which was then at the northern fringes of the city. The well was immense—16 feet across and 112 feet deep—blasted largely through rock, resulting in a quarry of over 175,000 gallons of water. A steam engine had the capacity to lift nearly half a

million gallons a day. An octagonal iron tank, 43 feet in diameter and 20 feet high was installed atop a 27-foot-high stone tower. Mains under Broadway and the Bowery delivered the water to hydrants on Pearl, William, Hudson, and a dozen other major streets, in six-, ten-, and twelve-inch pipes, delivering water to a height of 60 feet above the highest streets". [14]

Direct-Acting Steam Pumps

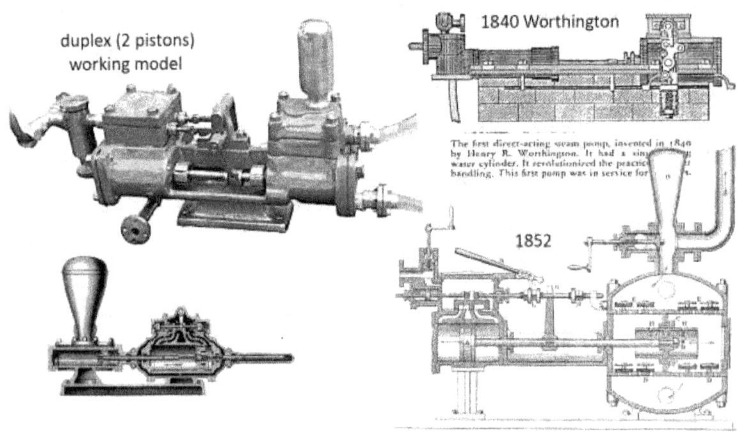

^ A mechanical engineer named Henry R. Worthington became a notability once he invented the first direct-acting **steam pump** in 1840. It was a "simplex" (single piston design). Aside from establishing a couple of pump manufacturing plants (in NJ and NY) he also patented the first "duplex" steam pump in 1859 and built the first "duplex waterworks engine". Original waterworks engines were huge and pumped water for a town or city. The smallish pumps pictured above have no boilers to make their own steam; that must be piped in.

The USS Monitor of Civil War fame, used two (of the 1852 variety, diagrammed in the lower right corner) Worthington direct-acting simplex steam pumps for pumping feed water, bilge draining, fire fighting, and as general service pumps.

14 Koeppel, Gerard T. (2000). Water for Gotham: A History. Princeton, NJ: Princeton University Press. pp. 135–37. ISBN 0691089760.

A massive waterworks pumping engine; a duplex (2 cylinders) built by Worthington (1870), for the city of Newark, N.J.

A NEW YORK WATER-TOWER IN ACTION.—DRAWN BY T. DE THULSTRUP.—(SEE PAGE [illegible].)

In the illustration above (by well known magazine illustrator Thure de Thulstrup (1848 - 1930)), a newfangled steam powered pump pushes water into an even newer invention called a "**water-tower**".

Steam powered water pumps were far superior to hand cranked pumps for fighting fires. But before any of this machinery can be used, the water must be available. As in an underground water system with taps (fire hydrants) installed throughout a city beforehand.

https://i.pinimg.com/originals/6d/95/1f/6d951fe03d8d29779c15ce50fb6a6d1a.jpg

Expensive, shiny brass and copper machinery like this was usually imported from Europe (especially from England).

Library of Congress Control Number 2012646404

Above: Detroit Fire Department's final ceremonial run using horses, on April 10, 1922

Skyscraper water

What are the mechanisms that enable very tall buildings to supply water to their highest floors ?
The **Empire State** building at 102 stories, was the world's tallest and most prestigious building, for 40 years. Then the first of the two World Trade Center twin towers (@ 110 floors) topped it out, in 1970. But then that record height lasted for only three years, before being surpassed itself by some other building. Today 53 other skyscrapers are taller than the notably famous 'Empire State'. The world's tallest building at present, is the "**Burj Khalifa**" (@ 163 stories) in Duabi, United Arab Emirates.
In a skyscraper, logistics and the laws of physics rule out any notion of having just one big water pump force water so high. Pressures, especially at the bottom would be unreasonably high. Piping and plumbing would need to be excessively thick, expensive and heavy just to survive the stress. (Threaded couplings are the weakest link in a vertical stack of drop pipe).

So what happens is that the buildings are divided into zones (like bottom, middle and top) where lower pumps send water up to a comfortable level and then a separate set of pumps begin anew to send water to heights above that. Within the Empire State building, pumps in the basement send water to reservoirs in the 20th floor. Then pumps on that level service levels between floor 21 to floor 45' where more cisterns or tanks reside and are filled. Then a new set of pumps take over from there; and so on. The Burj Khalifa has seven such levels; called **mechanical floors,** but each consisting of multiple levels themselves. Water tanks, water pumps piping, air conditioning and electrical substations occupy these 7 multi-storied mechanical floors.

Atmospheric pressure

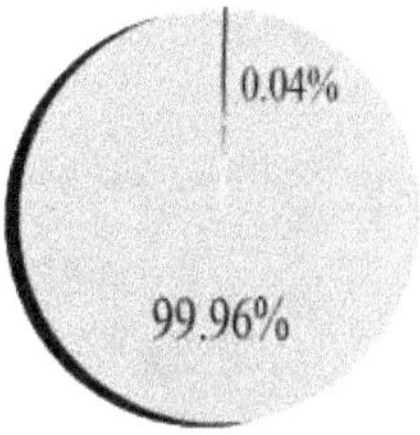

Although the earth's atmosphere may be invisible generally, it still has mass and is pulled towards the earth by gravity. Gasses get squeezed and compressed into smaller volume, whereas liquids are not compressible (when liquids do get pressed, their volume does not change). Just three gasses (**nitrogen** 78.08%, **oxygen** 20.95% and **argon** 0.93%) make up 99.96% of the atmosphere. The tiny remainder (four one hundredths of one percent) contains a mixture of 6 more gasses: hydrogen, helium, krypton, carbon dioxide, neon and xenon.

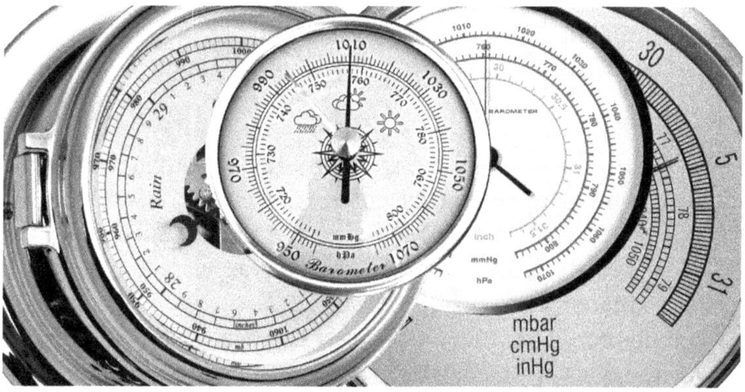

Aneroid type barometers. (Barometers were previously mentioned on pages 85 – 87 in Vol 1).

In 1643 an Italian scientist borrowing ideas from a preceding Italian scientist (Galileo Galilei), was credited with inventing the first **mercury barometer**. Galileo had already used water and air in an inverted glass vessel to study temperature. But 50 years after Galileo's experiments, Evangelista Torricelli filled a yard long glass tube with mercury, put his thumb over its top and then inverted the whole tube into an open dish filled with mercury.

So instead of running out of the tube, the mercury dropped a little from the closed end, but the rest of the tube remained filled. The new empty space above the mercury column varied in size though. From this beginning came the notions of air having weight, the concept of air pressure, probably the creation of the first sustained vacuum and the notion that a vacuum exists above our atmosphere.

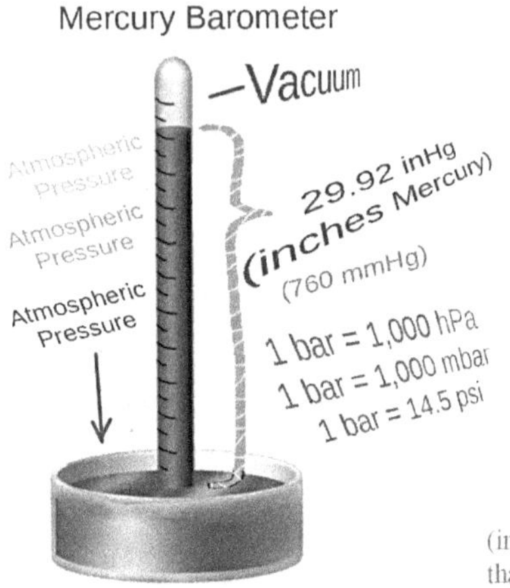

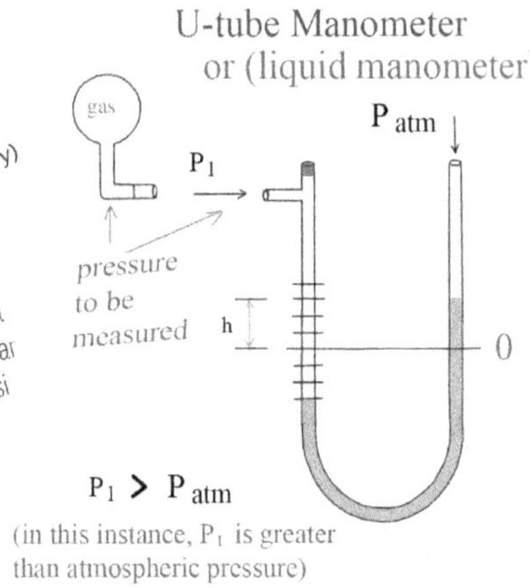

The use of mercury, which is far more dense than water, allowed for smaller, more transportable laboratory instruments to be constructed and used. Nonetheless the Renaissance men of the 17th century were still lugging fragile glassware and heavy bottles of mercury "over hill & dale" (*dale = valley*). So that they could record varying air pressures. These scientist soon discovered that air was less dense on a mountain top and more dense near sea level. They also demonstrated that air temperature decreases with height traveled upward into the atmosphere (about 0.07°C per 100 meters). An important correlation with weather was eventually noticed; that *"the height of the mercury in the tube changed in advance of variations in the weather. When the mercury level lowered, the weather turned poor. When the mercury rose, the weather became fair"*.

> Galileo also argued that **suction pumps** were able to draw water from a well because of the "force of vacuum". The pump creates a vacuum but atmospheric pressure moves the water.

Very similar to the first barometer was the first **manometer**, which is also a glass tube filled with liquid, used to measure pressures. The U-tube manometer is essentially the forerunner of the **analog dial pressure gauge**. With liquid in the tube and both ends opened to the atmosphere, then the levels on ether side of the U were equalized or 'zeroed'. Test were attached to one end, the other end remained open to the atmosphere, and movements observed.

Most compressed air equipment is designed to operate between 6 or 7 atmospheres (or bars ~around 90 - 100 psi) . These are the common tool type compressors for inflating tires or for driving staple-guns. They can only produce a maximum pressure not far above 7 or 8 bars. (7 x 14.5 = 101.5 psi).

Occasionally though, stronger compressors are needed; for filling SCUBA air tanks or welding gasses in cylindrical bottles. Here specialized air compressors are used to reach up to 200 – 300 bars. The strength and dependability of the containers are usually the limiting factor.

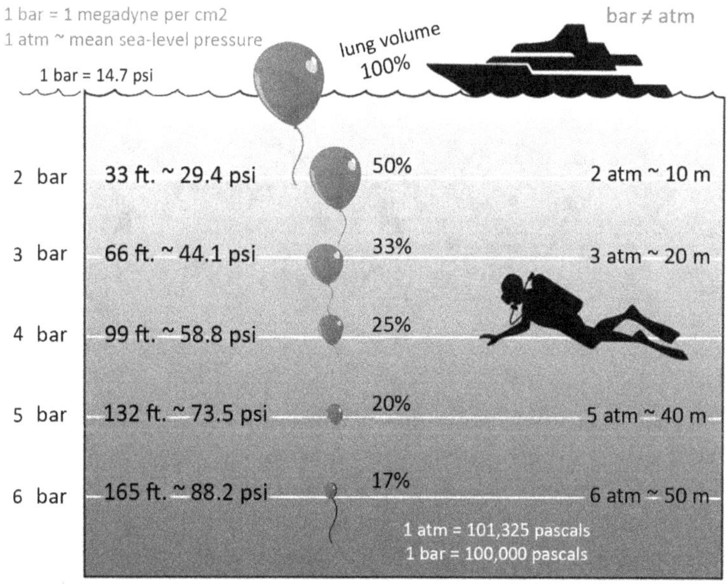

1 bar = 0.986923 atm (14.5 psi)

1 atm = 1.01325 bar (14.7 psi) or (760 mmHg)

The **BAR** is a metric unit of pressure and widely used. But the **ATM** is the SI preferred unit for – mean sea level atmospheric pressure on Earth.

^ While the purity of compressed oxygen in a welding gas bottle is above 99% pure and breathable, it is not carefully filtered for impurities as is medical oxygen or SCUBA air.

^ For recreational diving, Scuba divers never use pure oxygen because oxygen becomes toxic as pressure increases beyond a certain point. The deeper a professional diver goes though, he or she is sometimes compelled to use an esoteric mixture of other gasses. All divers strenuously wish to avoid "**the bends**" or **decompression sickness**. Which if breathing regular compressed air, can result from nitrogen bubbles forming in the blood or tissues - if one's body has been under pressure for a while and then ascends too quickly.

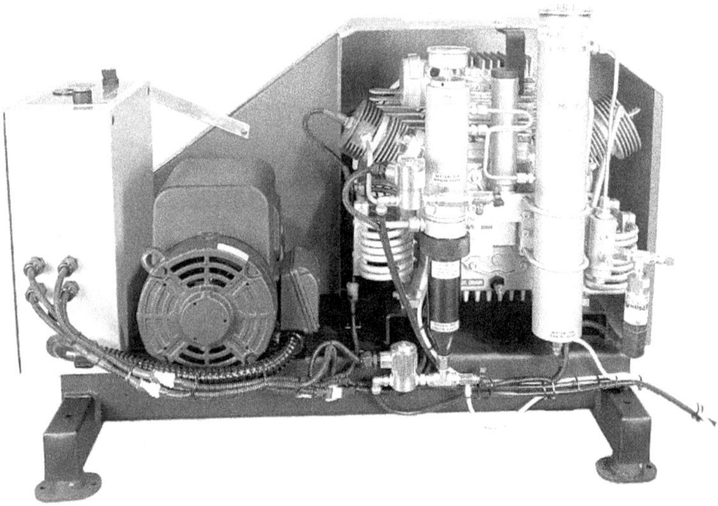

Above is a typical continuous duty, air cooled, three stage, three cylinder, high pressure **air compressor** with a Tri-chem ® breathing air purification system. Capable of filling an 80 ft^3 aluminum dive tank to **3,000 psi** in about 7.5 minutes. While this one is powered by 3-stage electrical power, other versions might be powered by single stage power, gasoline or diesel motors.

Intended to refill SCUBA tanks, the electrical models might be used in dive shops while the more portable petroleum models might be found on ships or private yachts. Specs for this one are: 10.8 scfm (18.4 m3/hr) charge rate, 9.0 F.A.D., and 5000 psi (345 bar) capable. (***scfm*** = *standard cubic feet per minute,* ***F.A.D.*** = *Free Air Delivery – which is an efficiency rating* *concerned with the delivery of input air).*

The Tri-chem ® breathing air purification system apparently incorporates a disposable cartridge of **molecular sieve** 13x, **activated carbon** and hopcalite. (**Hopcalite** ® is a trade name for a mixed catalyst of copper and manganese oxides; used to convert CO to safer CO_2 . Molecular sieve 13x is a crystalline aluminosilicate structure that features pores 13-angstrom in size; enabling, selective adsorption of molecules smaller than 13A).

Oxygen machines (a footnote)

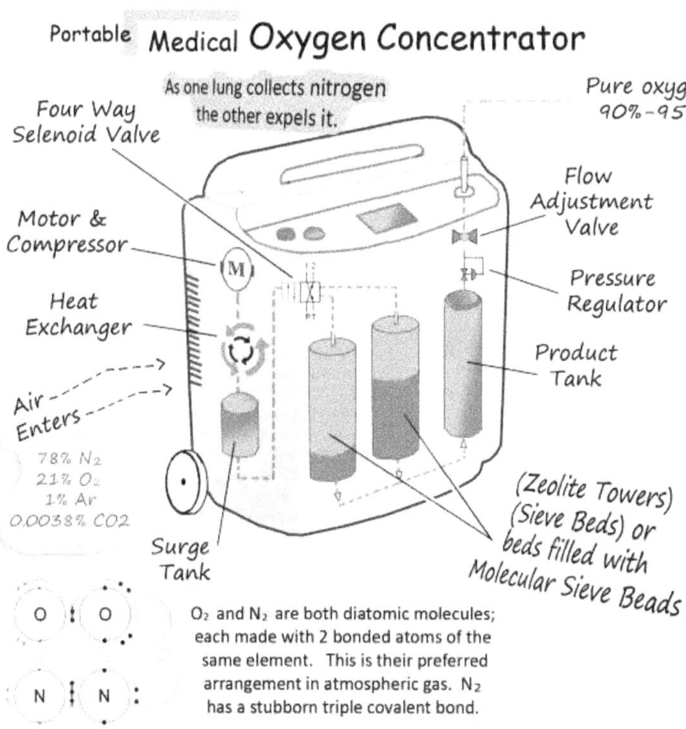

Two methods are generally used by machines to concentrate oxygen from the atmosphere; **membrane gas separation** and pressure swing adsorption. In the membrane scheme, using special impermeable polymers in their filters, nitrogen from pressurized air is persuaded to go one way and oxygen another. In the **pressure swing adsorption** (PSA) method, a gas like nitrogen might have an affinity for a porous crystalline aluminosilicate. So air under pressure is blown into a chamber where nitrogen is temporally absorbed, but the desired oxygen is passed on out. After a while the air stream is stopped, the chamber pressure is suddenly lowered and the nitrogen is detached and dumped out, before the cycle begins again.

→ back to the subject of water

Water Head

Library Of Congress photo, taken in Nevada County, California, 1866

Water head is the weight of water in a static column. Since fluids don't compress, the weight of water in a pipe is directly related to its pressure at the bottom (measured as **psi** or pounds per square inch). Yes, yes this is the 21st century and 'enlightened people' should use the SI system. But like the American engineers that first sent humans to the moon, American contractors like well drillers, generally don't often talk in terms of "hectopascals", just yet.

As a stream drops in elevation, its **head** is a measurement of that drop. Water weighs 62.427 lbs **per cubic foot**. There are 1,728 cubic inches in a cubic foot. A cube of water 12" high, 12" wide and 12"deep would have a psi of ((62.427 / 12) /12) or 0.433 lbs. per square inch.

Any column of water 1 ft. high, regardless of width, still has a water head of 1 ft. and a psi of 0.433 lbs/in^2. Water drop is simply multiplied by the constant 0.**43**3 to determine the potential **psi**. One newton is equal to approximately 0.22 pounds. Pressure (force /per unit area) in SI is measured as newtons per square meter (N/m2) or pascals (Pa).

In mountainous, western mining areas of the 19th century it was sometimes possible to witness electrical generating plants that used high pressure water to force turbines to spin. Nearby are two huge **Pelton wheels** attached to the same shaft that drives the generator. These turbines were built by General Electric and were employed from 1901, in the mining town of Victor, CO. *(Thanks to a user named Gomez, this image was once hosted by Wikipedia).*

Lester Allan Pelton patented the Pelton wheel type of **impact turbine** in 1880, but he got the idea from the placer mining gold fields of 1870's California. He made improvements to similar, existing Knight water wheels (which sometimes powered sawmills). Water in sluice-boxes runs all the time and waste energy. Some unknown, experimenting miner probably invented the first such turbine, by inverting a wagon underneath the end of a sluice-box, and then watching the wheel spin.

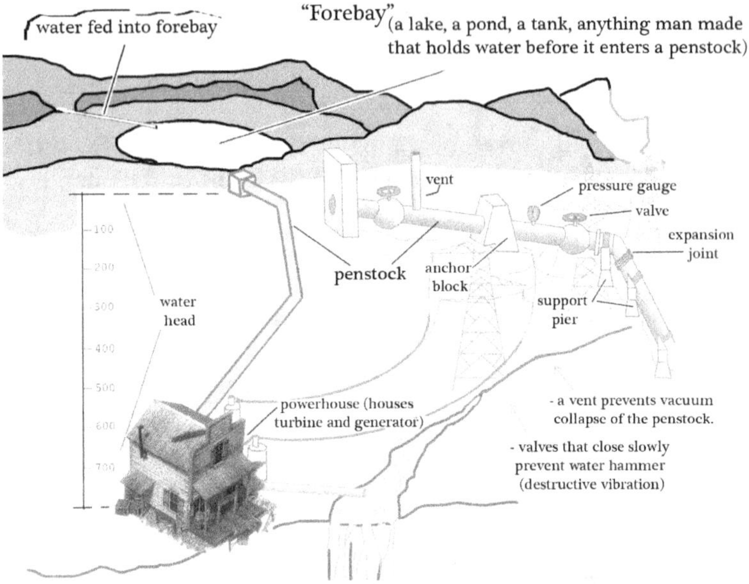

A **forebay** is an artificial holding pool of water downstream of a larger body of water. The larger body of water may be natural or human-made, but it is dammed and controlled somehow to feed the lower forebay tank or reservoir. Piping or water flues then conduct the water down to the work.

Hydraulic mining

A form of hydraulic mining has been practiced since Roman times, when they diverted whole streams to wash overburden away from ore. **Ground sluicing** was a labor saving, **placer mining** technique, where the heavier ore is ultimately left behind and exposed.

"The Monitor" by Henry Sandham / The Century Magazine January 1883

Two black arrows have been added to the previous public domain image. The upper arrow points to penstock piping, beginning its fall from a hill or mesa, far away. The second arrow points to a makeshift counterweight, which helps control the "**monitor**" (water cannon or nozzle). Notice the rocks piled atop the counterweight. Obviously it would take several men to move the heavy monitor and attached penstock to a new location, but small adjustments to the aim and focus of the monitor could be done by one man.

Nevada City, CA was the most noteworthy or notorious of locations where hydraulic mining was once practiced. Invented in 1853, by the 1870's water cannons were washing whole mountainsides away. To keep the forbay lakes and penstocks full was an engineering marvel in itself.

Above, the picture from California is dated 1866. The penstock is made of metal. Two (?) men in the center of the image, stare down to the edge of a canyon (where this water goes) some mile and a half away.

Quote #1: *"Engineers built a network of reservoirs, lakes, ditches and flumes extending as far as 40 miles to catch every precious drop of rain or Sierra snowmelt".*

Quote #2: *"A single monitor [water cannon] with an eight-inch nozzle like this could direct 16,000 gallons of water a minute," he says. "It could tear away 4,000 cubic yards of earth from the hillside every day." Into the sluices went the result, out came the gold, and the rest was dumped into the Yuba and sent downstream".*[15]

When the California Gold Rush began at the tail end of 1848, there were few people around and accommodations were rough. But in just a few short years, by the mid 1850's, big venture capitalist, were joining to form cooperative associations, and stock markets were speculating in both placer and hard rock mining. Mining technology advanced quickly. The engineering and the quality of construction back then, would be considered impressive by today's standards. Yet it was done a decade before the Civil War and isolated from all established industrial manufacturing. The rest of the nation, was on the other side of an unfriendly wilderness; perhaps 2,000 miles away by land or half a year away by sea route.

Some penstocks and open top flues were made of wood. Usually treated wood that was also milled with tongue and groove. The metal hoops in this image seem usually close, indicating high pressures. If a penstock holding water dropped 500' in elevation, then the water head pressure inside the pipe at the bottom would be 216.5 psi. That's more than twice the pressure that a **common water hose** is expected to contain. : *"A garden hose typically has a PSI (pound per square inch) rating*

15 https://factsanddetails.com › world › cat51 › sub325 › item2418.html

between 20 and 100 PSI, with 40-80 PSI being the most common". A tire hose is rated 200 – 250 psi.

Turbines & wheels

Water wheels are ancient, they come in many sizes and they can be fed or impacted by waterforce from multiple directions.

To install an '**undershot**' watermill requires the least amount of terraforming, but is also the least efficient angle of attack or approach to use. Hitting the wheel higher up, with a '**breastshot**' angle of motive water, is more efficient. More work is extracted from the water. More efficient still would be the '**overshot**' and '**backshot**' angle of approach, although these would also involve more breastworks and excavation labor to install.

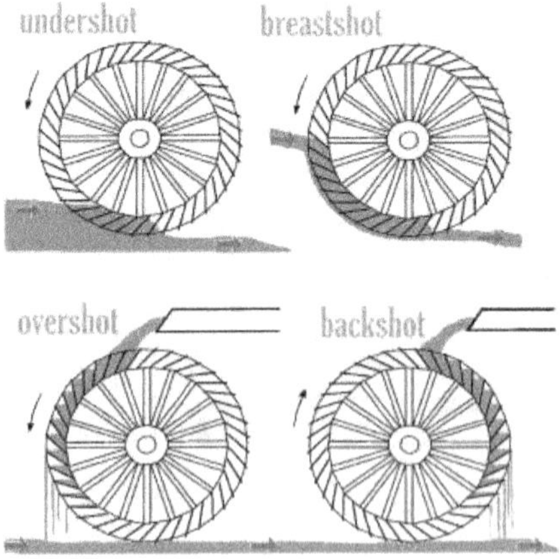

B
C
A
D
O
N
G
H
K
I
L
P
Q
M

A—Axles. B—Wheel which is turned by treading. C—Toothed wheel. D—Drum made of rundles. E—Drum to which are fixed iron clamps. F—Second wheel. G—Balls.

The two old etchings on the pages just before this, were taken from a noteworthy textbook named "De Re Metallica". This 12-volume work, was written by a metallurgist and mineralogist named Georgius Agricola and published in 1556; a year after his death. The extra time was needed to complete the numerous subcontracted illustrations. This was the first authoritative account of the European mining industry.

The famous book(s) were not translated from Latin to English for centuries. In fact it was finally translated at the beginning of the 20th century (1912) by the 31st U.S. President to be, and his wife; early graduates of the new Stanford Univ., both. Married in 1899, she (Lou Henry Hoover) was a geologist and he (Herbert Hoover) became a successful mining engineer.

* In the same 2 previous images, water seems to be the motive force powering the wheel. But in the second illustration, it is men walking inside a treadmill, that is supplying the motive force. The treadmill (not watermill) is powering a rope pump or chain pump, to extract water from a mine-shaft.

* Egyptians were lifting water with similar wheels in the 3rd century BC. It would take about 4 more centuries before the Greeks, Romans and Chinese would reverse the process, and use waterwheels to convert the power of flowing water into useful mechanical energy.

A water powered Japanese rice mill. Lithograph by M. Kellogg (1852-1854, during Commodore Perry's Expedition to Japan).

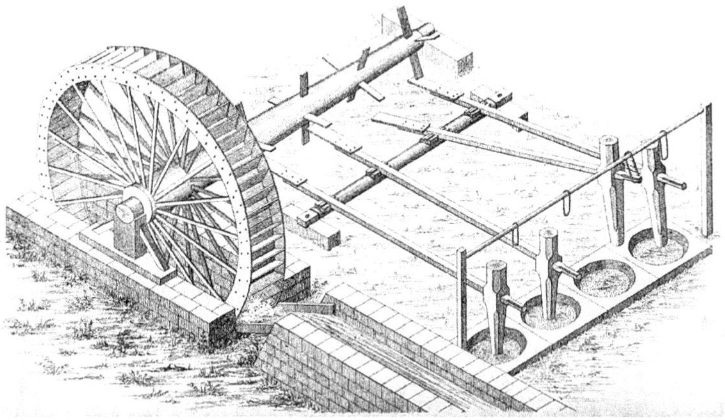

A water powered, rice mill in 1850's Japan. <above>

Water Turbines & harnessing rotation

The word "turbine" was coined from a Latin word for "whirling" or "vortex". The main difference between a water wheel and a water turbine is usually the swirl component of the water as it passes energy to a spinning rotor. Although the Romans might have been using a simple form of turbine in the 3rd century AD, the first proper industrial turbines began to appear only about 200 years ago. Turbines can be smaller diameter for the same power produced, spin faster and can handle greater heads (water pressure) than waterwheels.

A Frenchman named **Fourneyron** invented the first industrial turbine in 1827. The idea was brought to America and improved upon, in the form of the **Kilburn** turbine in 1842. By 1844 a conical draft tube addition resulted in the **Boyden** turbine. There were dozens of Boyden turbines in operation in northeast America by the time radical abolitionist John Brown raided **Harper's Ferry** in 1859. Located at the confluence of the Shenandoah and Potomac rivers, Harper's Ferry was a **national armory** and a beehive of activity where gunsmiths made small arms.

In 1859 at least 2 Kilburn and 5 Boyden turbines were driving the jack-shafts and belts needed to the power lathes, sawmills and other equipment necessary to keep 400 employees busy at the armory. Fourneyron's turbine and subsequent Kilburn and Boyden types were further followed themselves by increasingly efficient turbines including: the **Leffel** double turbine, John B. **McCormick**'s mixed-flow turbine, the New American and Special New American turbines.

All of these are known as ***outward flow reaction turbines*** (which are reminiscent of cinder, sand or fertilizer spreaders – but with water spraying out at the bottom).

Before electrical power came along, factories and machine shops powered their tools with belts & pulleys. The motive force was supplied by either **water turbine or steam engine**. [* **Steam** is dealt with in another chapter]. That force was distributed throughout a building, usually by a sophisticated system of mainshaft, jackshafts and countershafts.

These were axles fitted with sheave wheels, pulleys and leather belts. Usually the powershafts were suspended from the ceiling, but not always. Being made mostly of cast iron, the massive machinery often didn't need to bolted down to the floor either; it wasn't going anywhere.

.

Of course during the same time period (1860s - 1880s), before generation of DC or AC current was commonplace, state of the art farm machinery was belt driven too. The **leather belts** were simple, relatively safe, capable of being repaired or even fabricated on the farm. Up to the task of transferring the motive force, from transportable engine to transportable implement. Powered belt drive drums were commonplace on steam donkeys and steam tractors, even commonplace on newer gasoline tractors up into the 1950's. By then the clutched PTO (Power Take Off) shaft coupling would replace the belt drums.

> Between the 1880s and 1902 a vicious battle known as the ***"Harvester Wars"*** was waged between American farm equipment manufacturing rivals. The capacity to build harvesting machines, far exceeded demands. After much infighting between themselves, five large grain harvester companies merged together to form the International Harvester Co. Later (in 1912) the U.S. government filed an antitrust action against IHC, and the litigation dragged out for six years. In 1923, IHC's U.S. farm equipment sales were $150 million; triple those of second place Deere & Co. Now the situation is reversed and Deere is by far the dominant company. But now John Deere itself is facing **antitrust litigation**.

Originally a human that gathered grain in the field was called a "thresher". Later came the threshing machine, which is a piece of farm equipment that separates grain seed from the stalks and husks. It does this by beating the plant to make the seeds fall out. The words "harvester" and thresher are almost synonymous. Horse-drawn "combination harvester-threshers"were introduced in Michigan, in 1836. One century later and tractors were pulling "combines" in California.

Today the biggest self-propelled combines, are capable of cutting swaths 60 feet wide and have 800 horsepower engines. These harvesters might easily cost over $1 million each, have required satellite internet communications, air conditioning and other accouterments. The farmer might as well not be there; the stock market probably knows how much grain is collected before he does. He can't repair or work on "his" machine anyway. The machines are filled with wiring, electronic sensors and computerized chips. The company has soundly monopolized the repair of its products. Only an approved technician from the dealership can work on a machine. Deere is currently litigating multiple **"right to repair" class action lawsuits**. Some poor farmers are trying to buy Russian or Ukrainian hacker produced software, in an effort to diagnose their own farm machinery's proprietary problems.

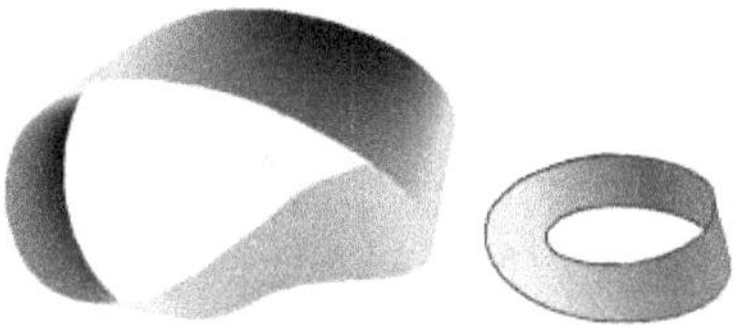

The perplexing **Mobius strip** is something of a paradox. The single edge of a Mobius strip is topologically equivalent to the circle and mathematically it is non-orientable. A physical Mobius strip can be constructed from a belt or strip of paper. One simply grabs the two ends and gives one end a half twist before taping or sewing the two together in a loop. The resulting surface then has only one side and one edge. Imagine a miniature gravity defying car driving around the surface of the strip. If the car began on the top side of the surface then its path after one revolution of the loop would place it on the bottom side of the surface. Consider a bug dragging a paintbrush while walking along the right edge of the strip and making two revolutions of the loop. We perceive two edges to the strip but realize there is only one. M.C. Escher incorporated the Mobius strip in some of his graphical art. In the real world recording tapes and typewriter ribbons have been spliced in the **continuous-loop** – Mobius strip fashion to double playing time or ink capacity. Large conveyor belts have also been wrapped the same way, to increase belt life by doubling the surface area.

Fig. 349. — *Batteuse Damey à manège direct placé sous la batteuse.*

A horse-powered thresher from a French dictionary (published in 1881).

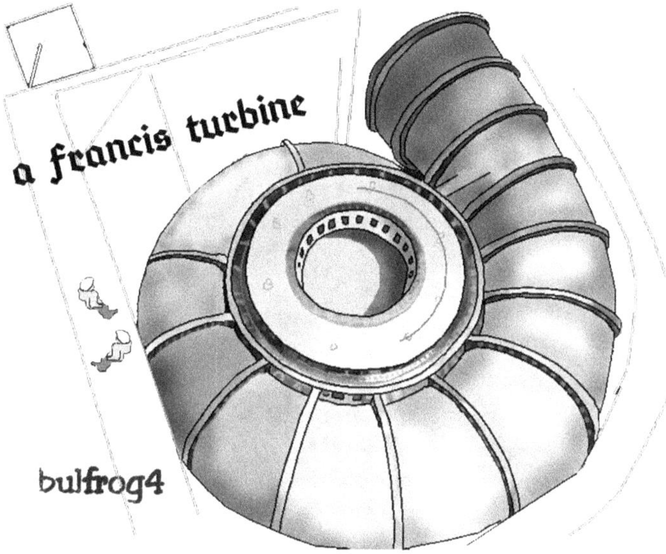

A different type of turbine called an inward flow (or **radial flow) reaction turbine** was developed by James b. Francis in 1849. In the snail shaped **Francis** turbine, water is sucked into a spiraling funnel that decreases in diameter. Used at the beginning of the 20th century mainly to drive jack-shafts and belts for machinery in big textile mills. Francis type turbines would soon however become the favored type of turbine for hydroelectric plants.

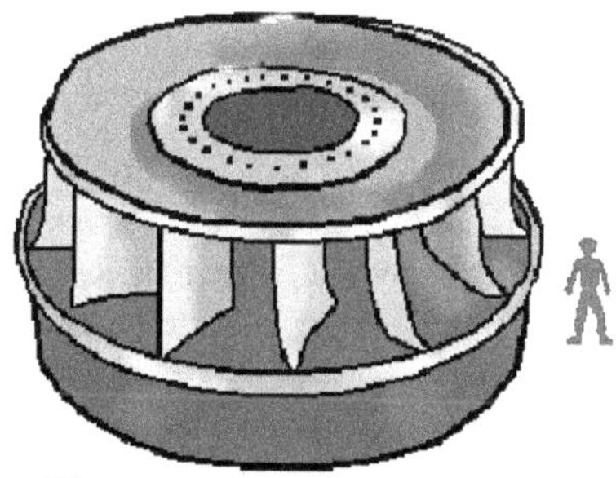

Francis runner

The next photo (of **Hoover Dam)** must have been taken sometime near 2003, because that's the year that the Federal Highway Administration began to construct the approaches to the new bridge that it had planned. Construction on the bridge (over the gorge of the Colorado river, 1,500' below the dam) would commence in 2005 and take 5 years to finish.

Of course, the top of the dam was part of a highway, used for the crossing the canyon for about 70 years before the bridge came along. Hover Dam ("Bolder Dam" for the years Roosevelt was in office) was one of those big civic projects taken up during the globally severe period known as the "The Great Depression" (1929-1939).

Other big public water projects during the period were the Parker Dam, Grand Coulee Dam and the Colorado River Aqueduct. All dreamed up by the Bureau of Reclamation, financed by Roosevelt's "New Deal" and constructed by contractors (like Six Companies, Inc) and sometimes helped by labor from the **CCC** (Civilian Conservation Corp). The CCC was designed to supply jobs for young unemployed, unmarried men ages 18–28 and to relieve their families back home with an allowance. Their work was mostly conservation and reclamation labor done on public lands.

The dam took 5 years to complete (1931-1936) which was a fast time, considering such a large project. Extreme summer heats (avg 119° F) killed many workers. Harsh times, harsh desert environment. Unions attempted to move in but were thwarted. People were standing in line to get work. The impetus to build the dam and others like it, was not to generate hydroelectricity to sell, but to use that power to pump water into irrigation canals so that farmers could grow food. The new reservoir behind this dam was named Lake Mead.

Hoover dam is home to 17 massive Francis turbines.

Turbine Installation at the Ganz Works, Budapest; before 1886.

Above is a Francis turbine intended to be installed in the vertical axis rather than the horizontal axis. Companies like Ganz Works made big things like ships, bridges, locomotives, turbines and electrical generators. Some of the earliest **electric locomotive railways** were established by 1881 (in Berlin, Germany) and by 1883 (in both Brighton, England and Vienna, Austria). Before 1900 some Swiss engineers had toyed with electrical locomotives and tracks, but by 1902 the little country had pretty much decided to electrify all of its rail lines. As of 2022 Switzerland had 680 turbine using hydroelectric plants that generated about 62% of its electricity. Coal and wood feed another 6% of its energy consumption. Canada has lots of hydro power too.

Although small Francis turbines are manufactured, its the large ones at hydroelectric power stations are impressive. Some produce more than a **million horsepower** each (1,341 hp = 1 Megawatt). The largest and most powerful Francis type turbines in the world are in the Grand Coulee Dam (Washington USA). The runners of the turbines there have diameters of 9.7 meters and are attached to generators producing as much as 820 MW each. (1 MW = 1,341 hp, 820 MW = 1,099,638.11 hp). These turbines like to be kept full of water at all times, but depending up on demand or water availability, might be producing just a few kilowatts at any given instant.

The world's **largest dam** (currently China's Three Gorges dam) is also capable of the world's largest electrical output, with 32 main generators producing an average 700Mw each, for a theoretical total of 22,500 MW optimal output.

But located between Brazil and Paraguay is the world's **second largest dam** (in terms of generating capacity). The **Itaipu dam** has 20 Francis turbines - powering 700 MW generators. In 2012 and 2013 Itaipu's annual electrical output surpassed that of Three Gorges, due to the amount of rainfall and available water.

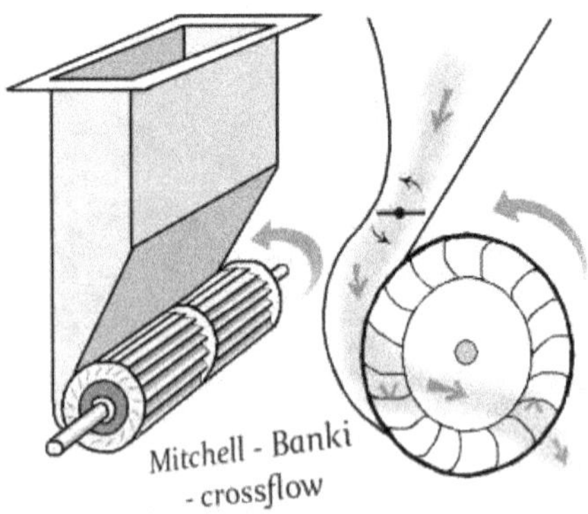

Yet another type of water turbine is tenuously referred to as a "**crossflow turbine**". In the early 1900's two individuals on opposite sides of the world independently contrived about the same turbine design. A Hungarian professor named Banki and an Australian engineer named Mitchell invented turbines that combine aspects of both a reaction (or constant-pressure) turbine and an impulse (or free jet) turbine.

The runner of a **Banki -Mitchell (or Ossberger) crossflow turbine** *(take your pick)* is cylindrical, and resembles a barrel fan that one might find in a forced air furnace or evaporative swamp cooler. The design uses a broad rectangular water jet that travels through the turbine only once but travels past each runner blade twice. The moving water has two velocity stages and very little back pressure.

Most suited to <u>locations with low head but high flow</u>, low-speed crossflow turbines like this have a flat efficiency curve (the annual output is fairly constant and not as much affected by fluctuating water supply as are some other designs). Large commercial crossflow turbines are manufactured that can handle 600 ft. of head and produce 2,500 hp.

Small homemade Banki – Mitchell units have been constructed that are capable of producing about 400 watts using a car alternator with 5.5 CFS (cubic feet/sec) of water from a stream with a head of only 33 inches. That would certainly be useful power for some off-grid cabin in the woods. These units can make considerable **noise**, so to keep vibrations minimized these turbines might be sheltered behind insulated walls and should be well balanced and spun at moderate revolutions per minute.

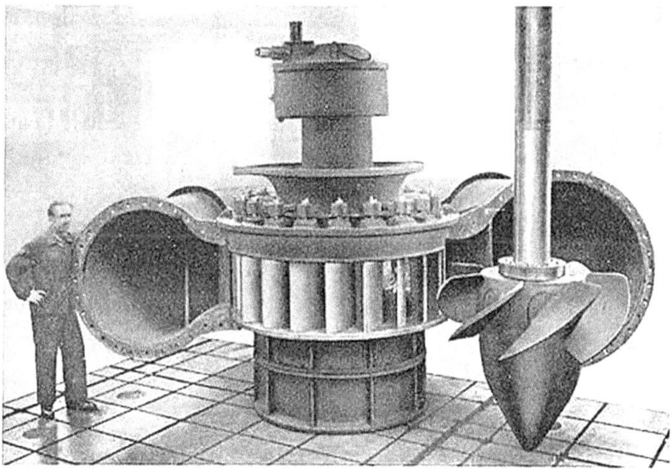

1 TURBINE KAPLAN DE 2750 CHEVAUX SOUS UNE CHUTE DE 34 MÈTRES

{2,750 horsepower Kaplan turbine under a 34 meter drop}

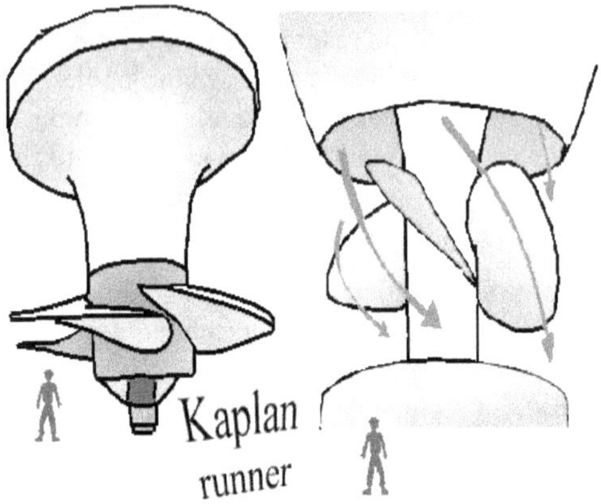

Another type of reaction turbine was developed by an Austrian in 1913 looks like a boat propeller. Some windmills are called **Kaplan turbines**. The blades or vanes on a Kaplan designed hydro turbine are adjustable, allowing the turbine to be efficient at different workloads or with varying water pressures.

Although complicated and expensive to manufacture, the Kaplan design is being encountered more frequently around the world, especially in projects with low-head, high flow watersheds. They can be found working in the vertical or the horizontal planes. Large Kaplan turbines have been working continuously for more than 60 years at the Bonneville dam. The Bonneville dam is on the Columbia River between Washington and Oregon, several hundred miles downstream from the Grand Coulee dam. Both dams were started at the same time during the depression and were initiated by Roosevelt's (FDR's) "New Deal". Small inexpensive Kaplan turbines (without adjustable vanes) can be made to work in streams with as little as 2 feet of head.

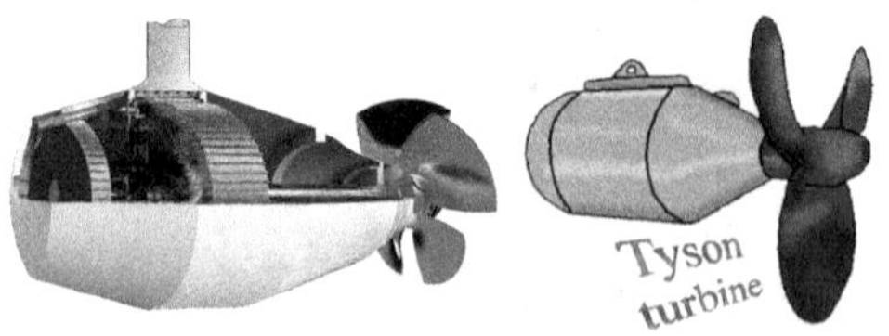

The so called "**Tyson turbine**" looks like it could qualify as a Kaplan turbine but this modern example of **micro hydroelectric technology** is small and encases its own generator in a waterproof housing. The whole unit is submerged into a stream or creek and is usually suspended from a small floating tethered raft. The stream can be shallow but obviously a high flow rate will encourage the best electrical generation.

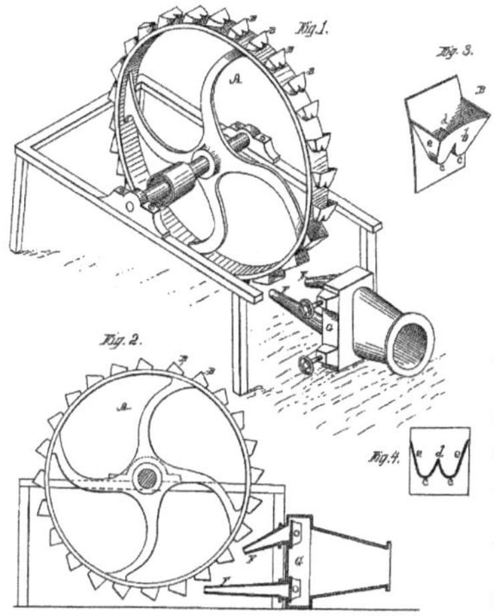

As stated earlier, the **Pelton turbine** was patented in 1880, but Pelton only borrowed the idea from an earlier wheel used in the hydraulic mining of the 1870s goldfields.

Pelton's real contribution was to improve efficiency by splitting the cup. By aiming the water jet at the center divider between parallel cups, the splash was diverted away at an angle, so that it did not interfere with the incoming jet.

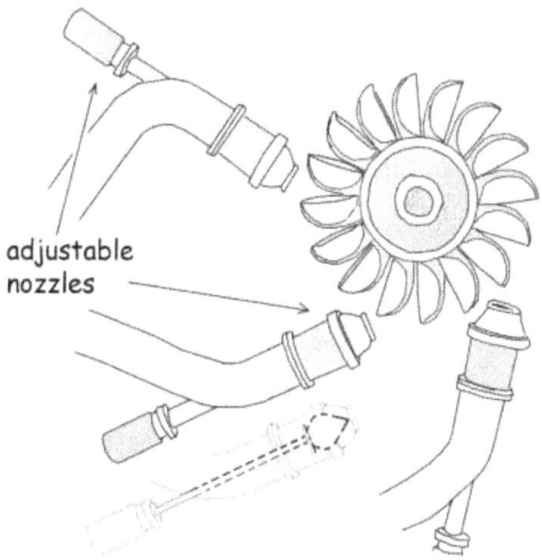

Today, Pelton turbine wheels come in all sizes, from 4 inch diameter "**micro wheels**" to large 10' to 12 ft. industrial size wheels. Sometimes water supply conditions allow for multiple jets to impact a turbine simultaneously thus creating more power. Being able to adjust the spray from a nozzle, allows a turbine to be fine tuned for optimum performance.

The big commercial Pelton wheels are obliged to be made of high quality metals, because they take a beating under continuous use. The corrosion or wear is mostly confined to the nose or tip end of the cups, where they are first impacted by the high speed water. The phenomenon is very similar to **cavitation** damage caused on a high speed boat propeller.

For the DIY (Do It Yourself) type of individual, another readily employable impact turbine in the micro hydroelectric line is the "**Turgo wheel**". Like the miniature Pelton wheels, these

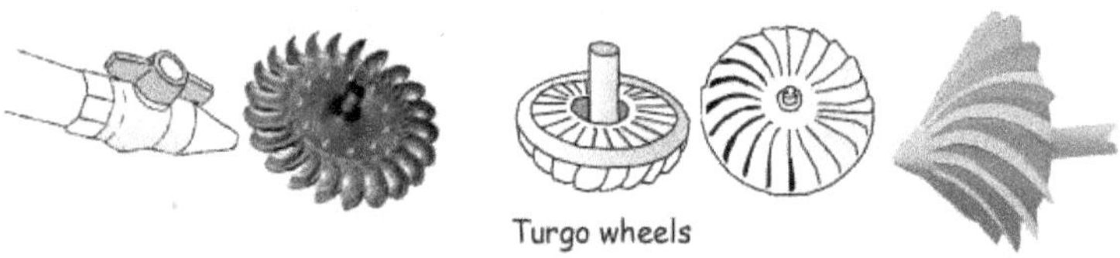

Turgo wheels

micro turbines can range from 3 to 4 inches in diameter – up to perhaps 12", and they can be purchased online from sites like Amazon and ebay for a pittance. They can also be purchased -attached to a small generator, if desired. Turgos and Peltons are designed to run at high speeds. A small uphill water source and enough penstock to reach it are the main requirements necessary to make one of these small impact turbines useful. Obviously not everyone is going to own a little piece of land or live near a water running water source in hilly or mountainous country. But for someone that does, a small Pelton or Turgo wheel of just a few inches in diameter, is capable of producing perhaps a useful 500 watts. In the absence of running streams, snow pack or abundant rainfall, an individual living in a mountainous area might still be able to collect up-slope groundwater, by using perforated drain pipes buried in boggy areas, from springs or from the drainage ditches alongside roads. A long run of water hose, polyethylene or polyvinyl chloride (PVC) pipe could conduct the water down slope, which would gain another pound per square inch of pressure for every 2.31 feet of drop. Water catchment from barn and house roofs could be redirected to holding cisterns and used by these little turbines when appropriate to augment other alternative off-GRID power systems, like solar panels. Sloped land and hydraulic head pressure (water head), are the main requirements for these impact turbines

> Windmills and wind turbines are generally differentiated by the reasoning that **windmills t**urn wind-power into <u>mechanical energy</u> whereas **'wind turbines'** convert wind-power into <u>electricity.</u> The creation of electricity involves the same considerations regardless whether <u>wind turbines or water turbines</u> are used.

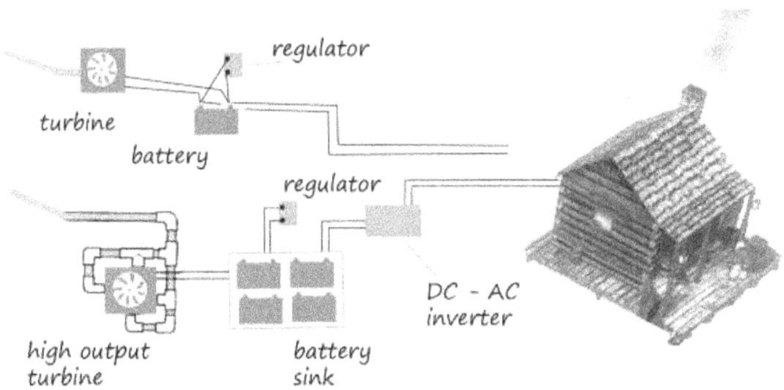

Whether an 'off-grid' power system is driven by sunlight, wind or moving water, the storage and usage parameters are about the same. The generated power is usually direct current (DC) because it can be stored in batteries for later use. Alternating current (AC) can not be stored, as is. It is easy to convert AC current to DC, but difficult and expensive to do the reverse satisfactorily, with an "**inverter**". That is why good inverters are expensive – they don't burn tools up.

AC current has benefits as well as drawbacks. For example AC can easily be "transformed" up to higher voltages, that can then travel great distances over wires, more efficiently than DC. A modern automobile's (AC) "**alternator**" will create more power than an old fashioned car's (DC) "**generator**" will; but that AC power still needs to be <u>converted</u>

into DC before the modern car can even begin to use it. As it is, most of the common household implements and tools manufactured, take AC power from the wall socket and convert that into DC voltage anyway, before it is used.

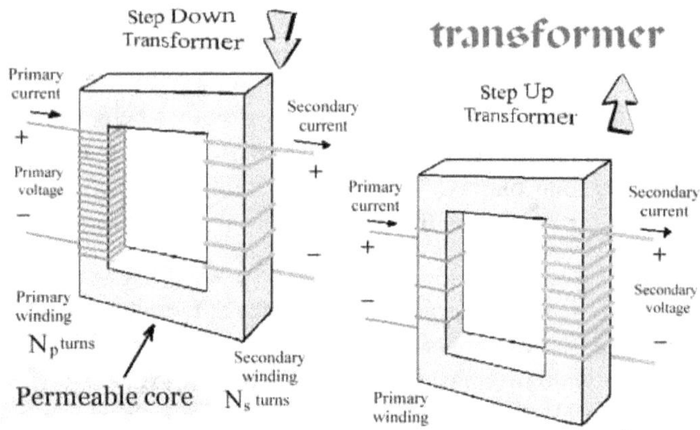

Iorn has good electromagnetic permeability - the ability of a material to support the formation of a magnetic field within itself.

Wind turbines, water turbines, Archimedes screws and centrifugal pumps in reverse, can be mounted by shaft or pulley belt to alternators or generators. Automotive generators (DC output) haven't been manufactured since the early 1960's. However most DC motors from tools, toys, treadmills, electric scooters, old printers, stepper motors, servos and motorcycle magnetos can be turned around and be made to produce DC current. Intrepid individuals might even take old AC motors from washing machines, ceiling fans or even automotive disc brake assemblies (especially as from a Volvo axle) and then re-wire them into permanent magnet alternators (magnetos).

(Some images that are tossed in here, might be reused later in a section that covers solar power. Subjects like charge controllers for battery protection, battery chemistry and deep cycle battery sinks for adequate off-grid electricity storage – need more room for explanation. For now the strategy is to maintain focus on wet things – water power).

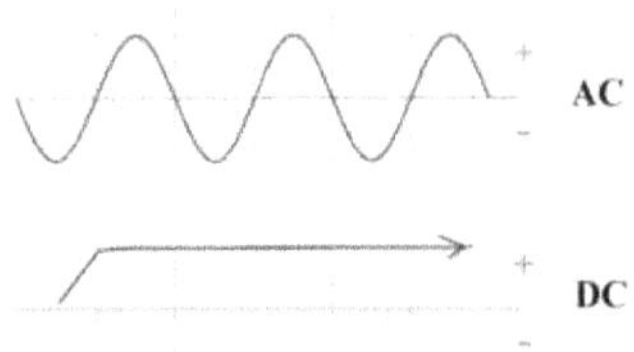

The 1893 Chicago World's Fair (also called the World's Columbian Exposition) was a grand affair which was themed upon the 400th anniversary of Christopher Columbus's arrival in the New World in 1492. This event was also the climax of a very unfriendly competition between protagonist of either DC, or AC power; called the "**War Of The Currents**". This was a downright media circus.

* Investors and inventors aplenty, on both sides of the Atlantic had from at least a decade before, vied for control over electrical technology, power transmission and service contracts.

Some celebrities or at least the most recognizable faces in this protracted disagreement were Thomas Edison and J.P. Morgan (representing DC), and their rivals Tesla and George Westinghouse (for AC). **Edison** formed 14 companies and held 1,093 patents. He preached DC, formed Edison Electric and encouraged demonstrations where AC was made to look dangerous – by electrocuting animals or by creating the first electric chair. But by 1889 he had lost control of his company when **Morgan** (a prominent Robber Baron banker) helped it merge with others to form General Electric. Brilliant Nikola **Tesla** immigrated to the U.S. in 1884 and even worked for Edison for a while – before the two fell out. Tesla perhaps gets too much attention for his avocation of AC development, because many others, especially in Europe, were doing the same things all along. **Westinghouse** dropped out of college after 3 months and then immediately received the first of his 100 patents and went on to found 60 companies; it is safe to say that he had talent.

By 1892(-93) Westinghouse Electric underbid General Electric (formerly Edison General Electric) to power the entire Exposition; which finally allowed a decent public demonstration of the adaptability, dependability and safety of the AC format. This success at the World's Fair also led to Westinghouse winning the lucrative contract to build a new hydroelectric power plant at Niagara Falls later that year.

But then: *"Water power was the most obvious choice for electric power since it had been used at mills for thousands of years. Good hydropower locations are limited, so it was necessary to transmit power over great distances in order to illuminate and power the great cities of the time".*[16]

(Photo 1904). *"Ten 5,000 HP Westinghouse generators at Edward Dean Adams Power Plant in Niagara Falls, the first large-scale, alternating current electric generating plant in the world, built in 1895".*

* But: 4 years before and thousands of miles away, a Tesla designed Westinghouse AC generator and identical motor were installed for the Gold King gold mill, at the Ames Power Station, outside of Telluride, CO. This generator was powered by penstock driving a Pelton turbine. The small power plant still in operation today.

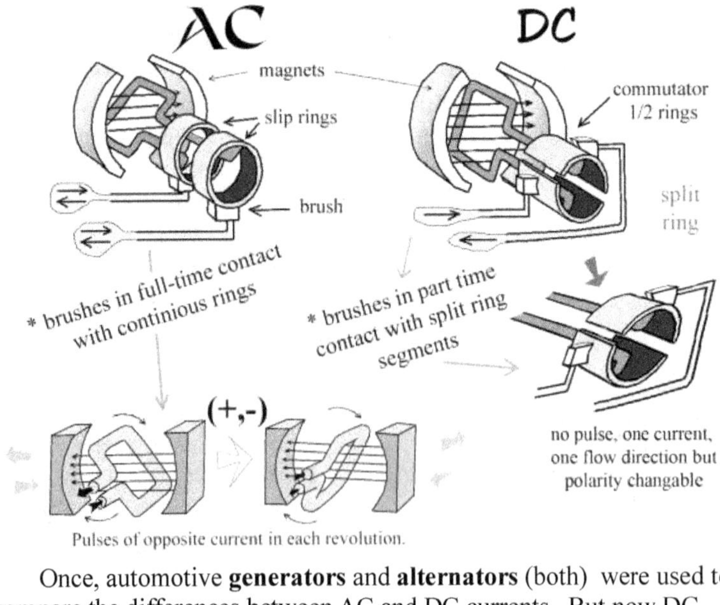

Once, automotive **generators** and **alternators** (both) were used to compare the differences between AC and DC currents. But now DC generators are rare and the last time one was installed on a car or truck

16 http://edisontechcenter.org/LauffenFrankfurt.html

was about 60 years ago.

The components in common modern automotive AC alternators are more complex, have miniaturized components , and are more unsuited for DYI experimentation. In the past for example, most car alternators were **externally regulated** (the voltage regulator was separate from the alternator and usually mounted on the firewall). For about the last 45 years though, alternators have been **internally regulated**, and use miniaturized electronics.

A "rectifer" converts AC current into DC current; easy to do.

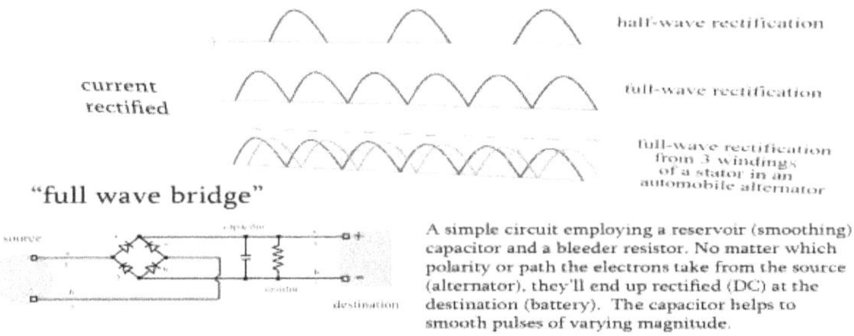

A simple circuit employing a reservoir (smoothing) capacitor and a bleeder resistor. No matter which polarity or path the electrons take from the source (alternator), they'll end up rectified (DC) at the destination (battery). The capacitor helps to smooth pulses of varying magnitude.

Automotive alternators succeeded generators, because they are more efficient and produce more power. Still there is a small price to be paid, as the alternator waste some energy as it establishes a regulatory field current.

An "inverter" converts DC current into AC current; hard to do.

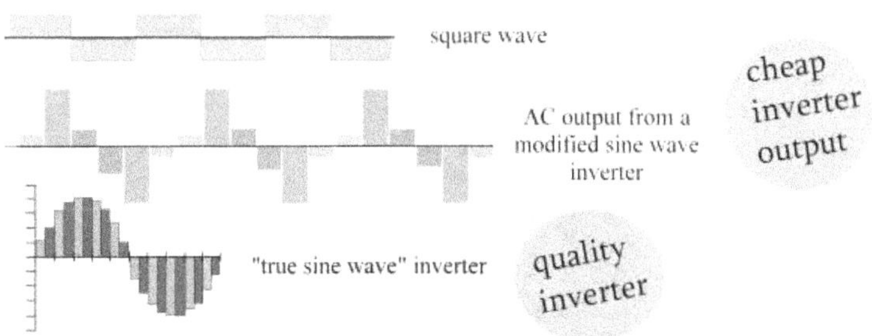

Modern car alternators are called “regular **field wound alternators**”. Up to 40 watts in a field-wound alternator (of the 1960's) was wasted to excite/energize the electromagnet field before any current was put out. Modern automobile alternators also require a high rate of revolution (or spin rate) of about 600 rpm, before they begin to produce voltage.

Most wind or water turbines don't want to turn that fast, so direct coupling to the alternator, is out.

Pumps / modern

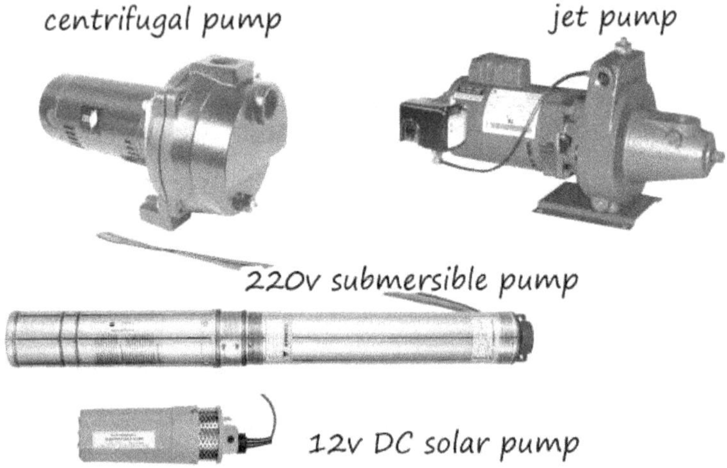

A centrifugal well pump and a jet well pump are basically the same – except for the jet. In essence the jet comes as part of, or is bolted onto a **centrifugal pump** body. Both type of pumps are mounted above ground. The ordinary centrifugal pump preforms well, where it is level with or not too much higher than the water it is trying to lift.

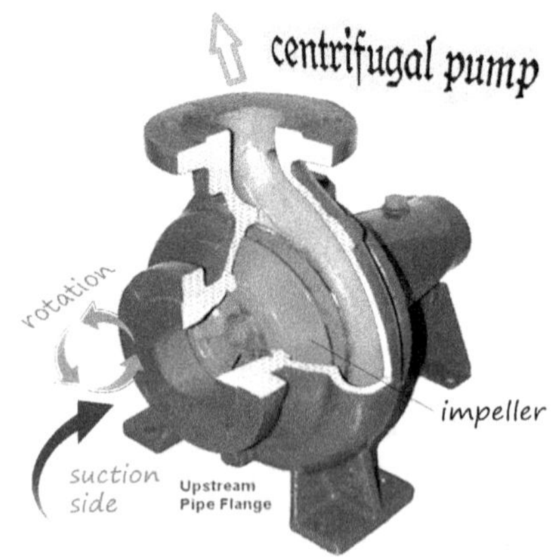

These ordinary pumps are good for moving surface water or for shallow wells that are no deeper than about 20 ft. (7 m). A "**jet pump**" at the surface can probably reach 10 ft. farther down and will have a better flow rate than a centrifugal pump. Placing the jet down in the well and using two pipes, greatly increases the draw.

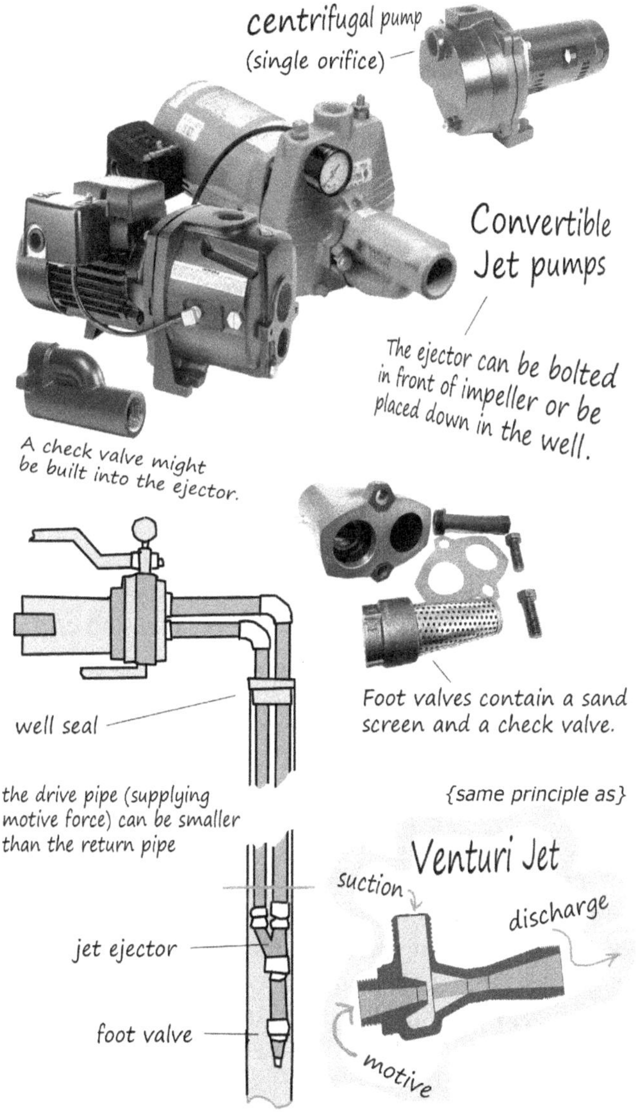

<u>Deep well</u>s are generally anything over 25'. The deepest normal depth that any <u>above ground</u> pump can draw from is about 165 ft. By using what is termed a "**foot valve ejector**". Instead of being bolted to the pump, the <u>jet ejector</u> is placed down at the end of the pipe.

Here a phenomenon similar to what is called a "**venturi effect**"is put to use. Inside what is called a "**jet ejector**" a motive force (a stream of water pumped in at high pressure) sucks along and entrains new water up from the well. The **motive** is at high pressure and sprays through a **nozzle**. [The same venturi principle works for a simple paint sprayer, where the motive force is compressed air].

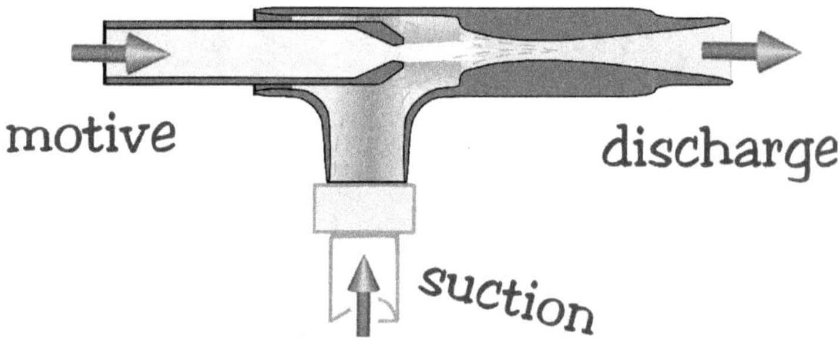

For this type of jet pump installation, a suction occurs where low pressure well water is drawn in, past a check valve or two (so that water can not back up) and into an expansion or discharge chamber. Two separate pipes are usually the favored installation for foot valve ejector type jet pumps, and the pipe that carries pressurized water down usually has a smaller diameter than the pipe that carries water up.

For double pipe, jet pump installations the two pipes connect to what looks like centrifugal pump, but has two orifices. One hole for input, the other for output. Again, an ejector mounted in front of the pump's impeller is only good for about 25' lift (a shallow well), but putting the ejector down in the well and then pumping water through it – can retrieve water from a greater depth (160'). However greater pumping depths are achieved by using **submersible pumps**.

When talking about submersible pumps and greater depths , then other factors come into play. First the **static water level** (the baseline depth of undisturbed water in the well) needs to be considered. The well's **capacity** needs to be estimated (how much water can be drawn without lowering the static level). The **horsepower** of the submersible needs to be determined. The deeper the pump setting, the more important it is to pay attention to the drop pipe's specification (**schedule**) and the type of couplings and check valves to be used.

^ For the deeper domestic water wells, **schedule 40 steel** drop pipe is favored; for safety, the valuable submersible itself is usually tethered by a separate braided steel cable. The weakest links in a deep well drop pipe, are the threaded connections. If possible, it might be wise to pump water into an above ground cistern, to reduce the cycling on and off of the submersible. A less expensive **booster pump** can take the abuse as it sends water to the house, from the cistern.

Pressure Tanks

A water pressure tank preforms for a house, the same function that an elevated tower does for a town. It provides backup water pressure when the pump is turned off. Back 50 years ago, household water pressure tanks had no **bladders**, but now most do. A larger tank is more desirable than a small one in most cases; more volume in reserve means less work for the pump.

Pressure Tanks

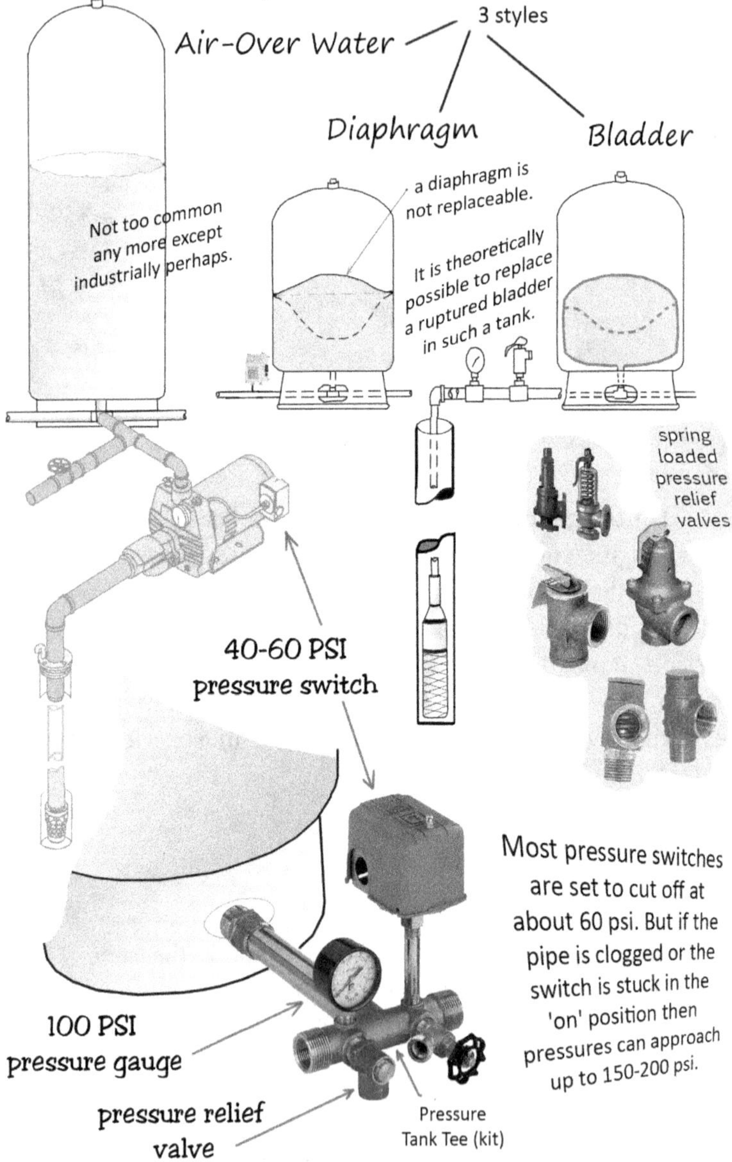

Hot Water

Water heater tanks store a reserve of fresh water too. These don't operate in the same way as a pressure tanks, but they require emergency pressure relief valves also. Specifically a T&P (Temperature and Pressure) relief valve that will be explored momentarily.

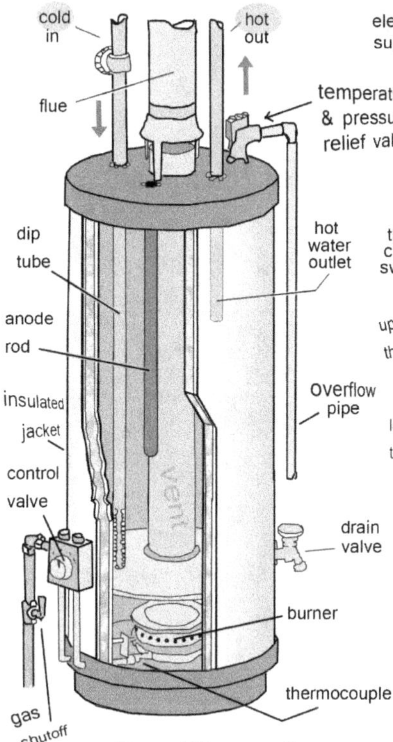

Gas Water Heater

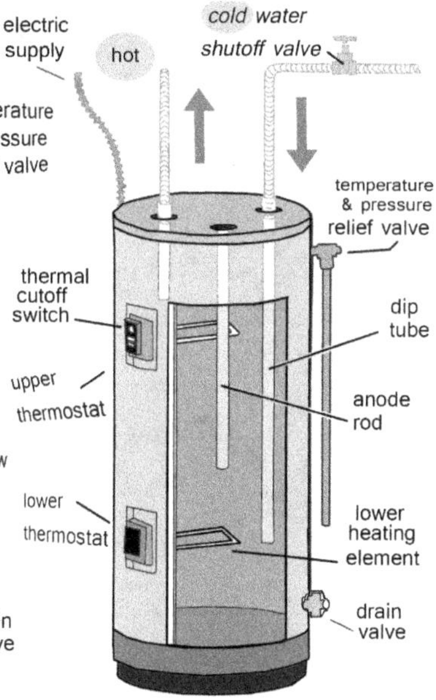

Electric Water Heater

There are several types of domestic water heaters. Some water heaters come with water tanks and some don't. Some run on electricity, some on natural gas and some only on LP (Liquid Petroleum).

The most expensive water heaters tend to be the <u>tankless types.</u> These "**flash heaters**" are thought to be more economical in the long run, because they only use energy (or fuel) when it is demanded, and therefore don't attempt to retain heated water. However, tankless water heaters are expensive to purchase, are normally complicated to install and are more problematic to maintain. Owners of tankless water heaters are often obliged to install water softeners or other treatment systems beforehand. Much depends upon the mineral content of the local water, but the typical upkeep regimes strongly suggest yearly maintenance inspections; where the unit is turned off, drained, filters emptied and mineral scale (if it exist) is flushed out with a vinegar solution for half an hour.

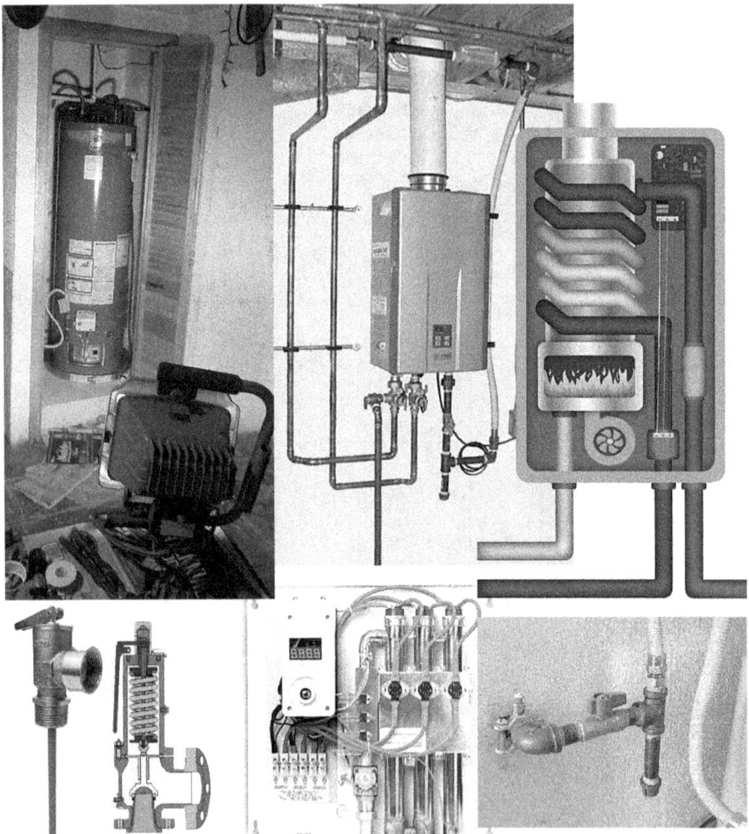

pressure relief valve

electric "point of use" "tankless heater"

sediment trap (for gas line)

At the right of the illustration above, the drawing shows a stylistic **heat exchanger** concept. Burning gas, this tankless water heater amounts to little more than a “fire-tube boiler”, as similar to what a Stanley Steamer race-about (a car) would have been using back in 1912. At the bottom of the same illustration is a picture if a small “**point of use**” tankless water heater. Rather than one big heater being located centrally, small electric units like this might be located under the kitchen and bathroom sinks. Copper heat exchangers are preferred in these units because of their high thermal conductivity, but heated copper tubes are also more susceptible to scale buildup. Lastly (for the same image) is a gas line **sediment trap**. This addition to required building codes, is intended to collect dirt, rust or scale that might flake off and bounce down the inside of a metallic gas line, possibly blocking an appliances gas control valve.

Electric water heater tanks, are the cheapest and easiest water heater tanks to install. Using electricity however is the most expensive way to heat water. Heating with LP is only marginally more efficient however. Natural gas when available, is usually by far the most economical way to heat house water.

Gas fired water heaters are less likely to run low of heated water during periods of high demand. At any rate they recuperate their warmth quicker than electrical heaters will. Many people prefer gas cooking ranges for similar reasons; quicker response and more precise control over the cooking heat.

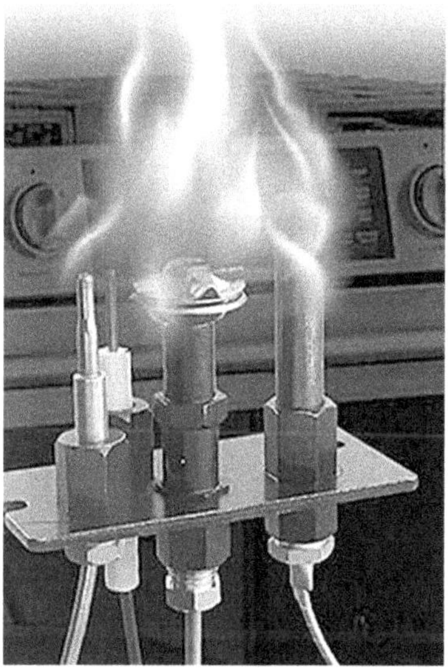

Natural gas (CH_4) and **propane** (C_3H_8) are two very different gases. Propane is the denser fuel and for the same volume, can produce twice the heat that natural gas can. One cubic foot of natural gas produces 1,030 BTUs (British Thermal Units) while 1 cu.ft. of propane creates 2,516 BTUs. 2.4 times as much).

Natural gas must be delivered by underground pipe. The main advantage of liquid propane (LP) is in it's portability. LP can provide remote isolated homes with crucial hot water and warm air. For someone living completely off-grid, LP gas alone can drive refrigerators, freezers, furnaces, ovens, range tops and power generators. One 200 year old device shown in the picture here, is particularly responsible for that means of independence. No electricity, not even a battery is needed. The picture is of a pilot light assembly, from possibly a gas fireplace or a LP gas water heater. From left to right would be a thermocouple, a piezoelectric igniter electrode, a hood on the pilot gas tubing and finally a thermopile.

The first **thermocouple** (which joins 2 dissimilar metals that create a tiny temperature dependent voltage; from 1821) detects whether the pilot flame is lit. If the flame goes out, the thermocouple stops generating voltage and the gas valve is closed.

Second from the left and towards the back in the picture is the igniter electrode. Using the same **piezoelectric** ceramic **perovskite** material that might be found in a butane cigarette lighter, pushing a start button on the heater's control panel will send a charge to the tip of the electrode and ignite the gas pilot. When the button is pressed, a little spring loaded hammer will strike a special crystal or PZT ceramic, causing a deformation which will create an electrical discharge.

** Conversely piezoelectric crystals and ceramics vibrate when excited with fluctuating electricity. A very tiny electrical signal is adequate to create sound in an old fashioned crystal earpiece. Here wires and a diaphragm of metal foil perhaps, are simply glued to a crystal.*

The last device, on the right in the previous pilot light picture, is a **thermopile**. A thermopile is made of many thermocouples that are bound together and wired in series – to increase the voltage output. A single thermocouple might make 30 mV; but its output is dependent upon temperature, which still makes it useful as a sensor. Thermocouples are used as sensors in diesel engines, gas turbines and ceramic kilns. But when combined in series as a thermopile, they can together create a useful (500 mV +) voltage that can be used to drive "**millivolt systems**" (with LEDs, solenoids and gas shutoff valves).

A standing pilot light that runs 24/7 will cost you an estimated $.60 per day, depending on the gas prices in your area.

Older gas appliances once made liberal use of thermocouples and pilot lights, but now most don't. Instead, modern natural gas furnaces, kitchen ranges and water-heaters have become IPI (**Intermittent Pilot Ignition**) systems. Which means they probably have control modules with integrated circuit boards, which on demand can open or close gas valves and prompt spark generation. Some such gas appliances can be remotely controlled. But these appliances have an 'Achilles heel'. Excepting specific models, these "gas appliances" won't work at all, without having dedicated electrical power service.

^ Natural gas is mainly **methane** but contains minor

amounts of other hydrocarbons. Methane is the lightest hydrocarbon, with just one carbon atom and four hydrogen atoms (CH_4).

When petroleum is fractionally distilled: methane, ethane, propane and butane are the first volatiles to boil off. At standard temperature and pressure (STP) they are gaseous. They are called **'the bottle gases'** because they can be liquefied with pressure and cryogenic temperatures. **LNG** (Liquid Natural Gas) is produced, but with much more difficulty than **LPG** (Liquid Petroleum Gas).

Ethane is primarily used in the petrochemical industry to make plastics. **Propane** has a lower boiling point and is easier to vaporize than ethane, so it makes a more versatile fuel. **Butane** is occasionally seen as a fuel in cigarette lighters and soldering torches. But in cold or freezing weather, the pressure of liquid butane in a lighter is reduced to where the device won't work. Normally bottled LP is about 30% propane and 70% butane. In colder months, the ratio of propane might be increased, so that good vaporization is maintained.

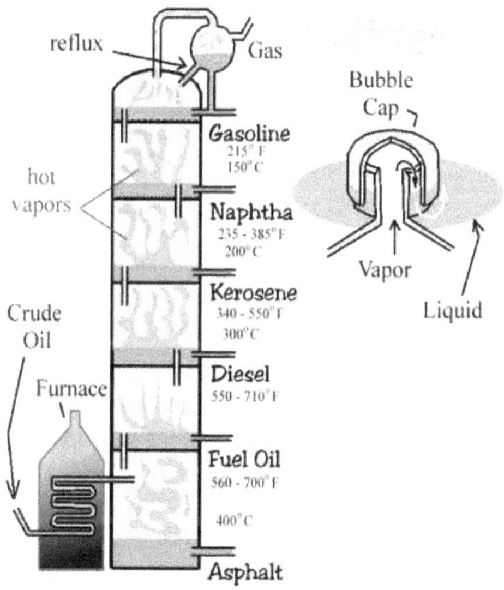

To the side is a representation of a "frac tower" (as in fractional distillation, preformed in an oil refinery).

The next commonly mentioned petroleum fractions to follow after the bottle gasses are mainly liquids: pentane, hexane, heptane, octane, petrol (gasoline), naptha, kerosene, diesel, residue fractions (fuel oil, lubricating oil, waxes) and bitumen.

Back to water heaters :

The workings of a pressure relief valve is of special interest because one could be useful in safely engineering a steam system – a tempting project for the next chapter. The type of thermostat used in a conventional automobile acts as a valve that controls the flow of coolant between the engine and the radiator. It stays closed when the engine is cold, helping it warm up quickly.

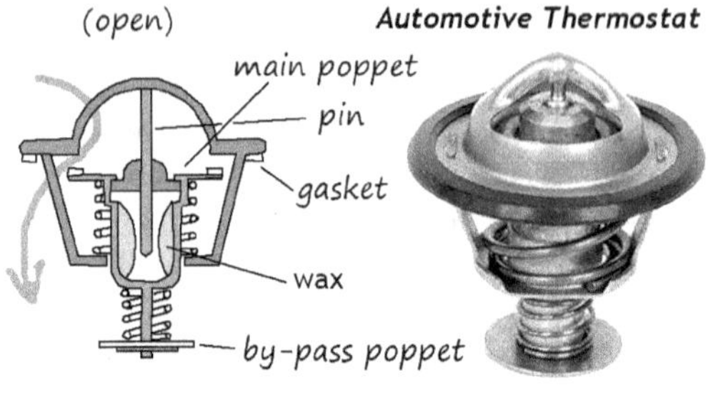

Once the coolant reaches a determined temperature, the thermostat opens to allow the flow of coolant, which protects the engine from damage. The mechanism that determines the set point temperature of a given thermostat is a specific **mixture of <u>wax</u>,** which is usually housed in a brass or copper capsule. When the wax (unique for different temperatures) melts and turns to liquid – it expands and pushes against a pin – which opens the poppet valve.

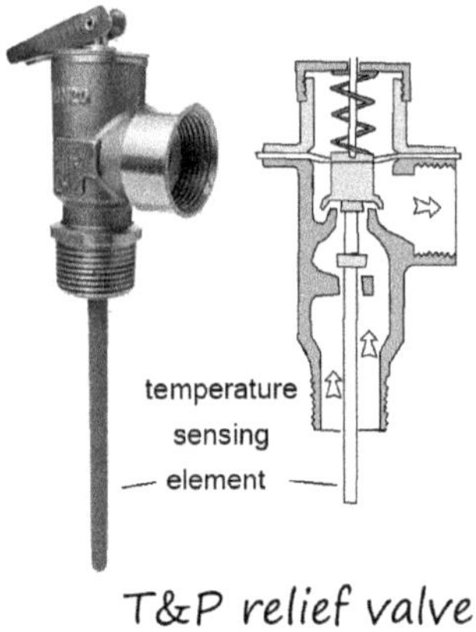

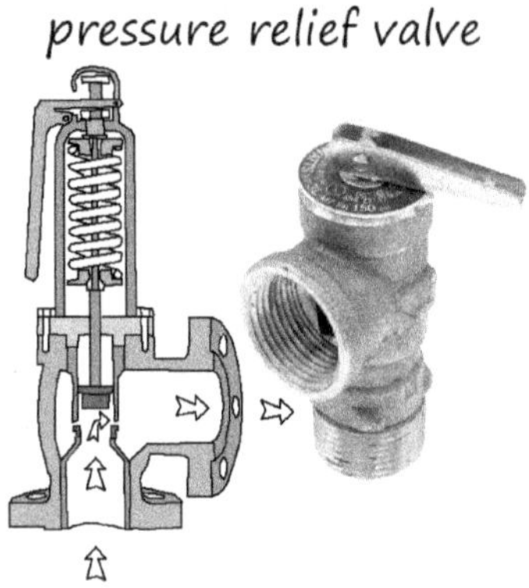

A regular **pressure relief valve** as used in a hot water or steam power application, have a calibrated internal spring that holds a disc or seal tightly against the valve seat. When a tank's pressure reaches the limit targeted, the force of the water pushes against the disc, overcoming the spring's resistance and forcing the valve open. A **temperature and pressure (T&P) relief valve** has an additional safety feature; that of being able to open because of temperature. Here a hollow thermal probe sticks down into the water at the top of the heating tank. The probe might be made of copper or stainless steel and it might contain a
specialized wax or other compound that expands rapidly when exposed to a certain heat. That expansion drives a piston rod to push against the spring-loaded disc, forcing the valve open to release overheated water

and steam. Most residential T&P relief valves are set to release pressure at 150 pounds per square inch (psi) and to relieve temperature at 210 degrees Fahrenheit (°F).

The boiling point of water is 212°F (100°C) at sea level. Standard air pressure @ sea level is 14.7 psi, 1 atmosphere, 1013.15 hectopascal (hPa) or 29.92 inches of mercury (inHG) on the dial of an old fashioned barometer.

(The temperature probe in a hot water T&P relief valve would prevent water from boiling in a steam boiler. However with the thermometer defeated, the spring loaded poppet valve part might prove to be very useful. While higher pressure valves can be purchased, 150 psi can still do a lot of work. Enormous ships once pushed themselves across the oceans with pressures 25 psi or less (over standard atmospheric pressure).

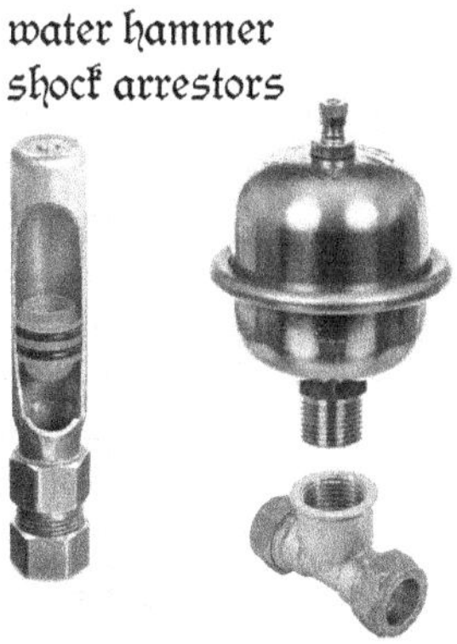

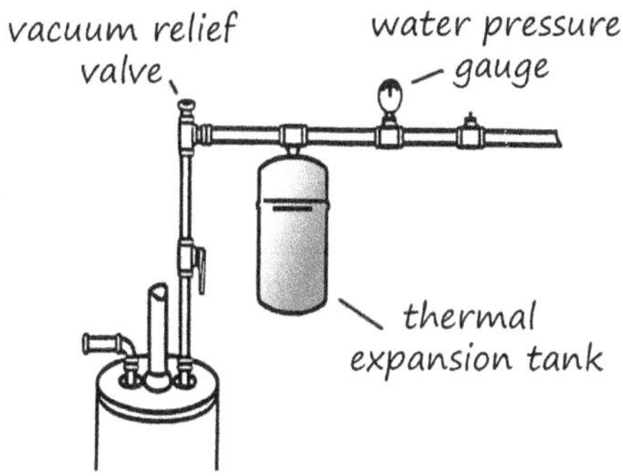

Water Hammer

Sometimes, depending upon the building, loud banging or knocking noises can be heard when a faucet or other valve is shut off. That's called a "water hammer" and it's a little shock-wave. A water hammer happens when water is rushing down a pipe, but then its momentum is suddenly stopped. Pressure abruptly surges up and vibrating pipes cause the banging noises. Water pipes will twist and swell against their constraints, which are often just holes drilled in wood boards. Water hammer might be more common in old homes,because time has allowed nails, screws and straps to work loose. Water hammer can cause leaks or even break water pipes. Devices like **expansion tanks** or water hammer **arresors** can be added to dampen or absorb these shocks.

The venerable Ram Pump

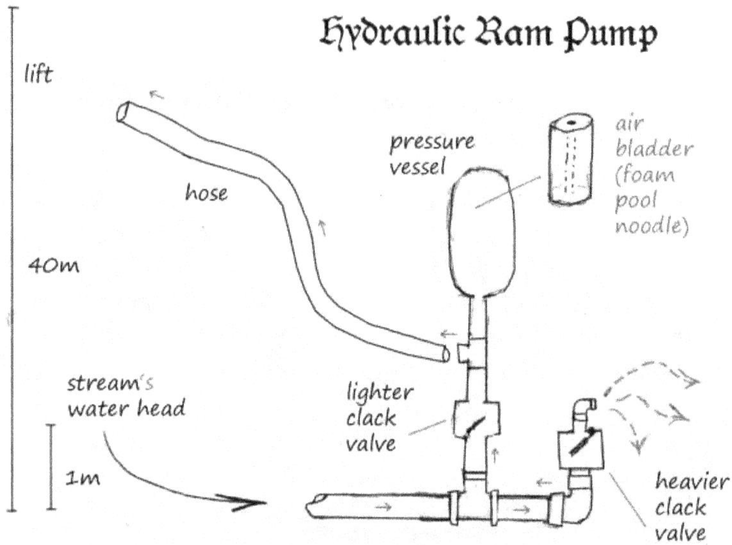

A device with origins that reach about 2 ½ centuries back, the ingenious "**hydraulic ram**" or "**ram pump**" can lift water to a higher elevation without using any power. In fact the apparatus relies only upon moving water and the water hammer effect to function! Its only moving parts are two check valves; one is a spring or weight loaded "waste" valve (or "clack" valve) and the other is called the "delivery" check valve. (**Check valves** permit fluid to flow in only one direction).

The equipment is set down in, or nearby a stream. A long pipe or hose runs upstream and a "water head" is built up. When the waste valve is opened, water rushes through and velocity increases. When the valve slams shut, pressure builds up and opens the second (delivery) valve. Then the pressure in the pump will fall back and the first valve will re-open to allow water to flow and build up momentum again. The cycle repeats. (A little manual stimulation might be necessary at first).

Most of the water (perhaps 90%) passes through and on down stream, but a small important percentage is forced to go up. That 10% or so can become significant after a 24 hour period. From a small 1/2" diameter hose one might expect delivery of about 300 gallons of water per day; from a 1" hose that output might rise to 1,000 gal/ day. Depending upon an assortment of factors, the typical lift of a ram pump might range anywhere between 10 and 300 meters. One suggested rule of thumb is that: "*for every 1 foot of water falling into the pump you can expect to get 7 feet of lift up from the delivery pipe*". A pressure vessel containing air, cushions the hydraulic pressure shock when the waste valve closes. Also an air chamber allows the delivery pipe to capture as much high-pressure water as possible during an impulse.

Although not mandatory, an **air chamber (pressure chamber)** greatly improves the pump's consistency and efficiency. One problem however is that if the air is not separated by some barrier or bladder, the gasses will slowly dissolve into the water and the cushion will eventually disappear. A remedy might be to insert a closed plastic bottle or a chunk of foam cell plastic (from a child's swimming pool noodle). Other considerations might improve the water hammer's effectiveness. For example the drive or source pipe should be of constant diameter and should be as straight as possible, leading to the pump. PVC pipe and fittings seem to work OK, but all metal counterparts would work better, because they will transfer the shock waves better.

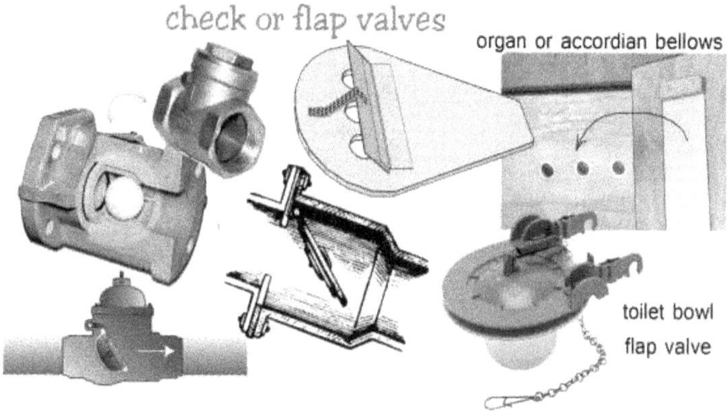

Humans have been using **check valves** for more than 23 centuries now. Or at least since Ctesibius of Alexandria (285 – 222 BC) used **flap valves** to make his first "pressure piston pump". Later his hydraulis (pipe organ) would exploit the compressed, air that this same flap valved water pump produced. A **reed valve** can consist of little more than a flap of leather covering a hole.

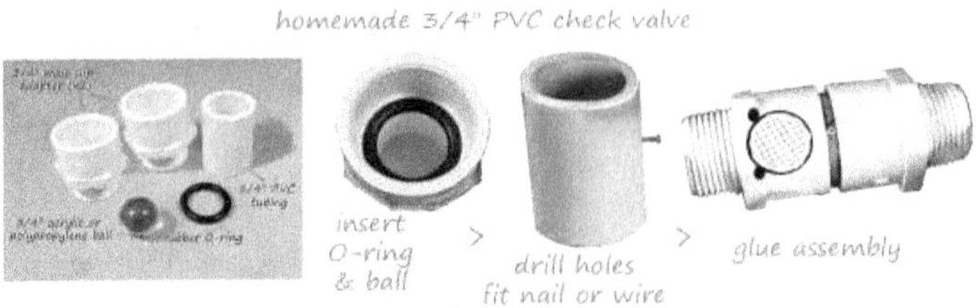

Check valves are relatively expensive. A way to make one cheaply, from 3/4" PVC fittings and a rubber ball is illustrated above.

Closing out topics related to water

In 1965 a character named Clayton Jacobson II, invented a stand up version of a "**pump jet**" or "**water jet**", recreational ski. Kawasaki soon used his idea to famously manufacture the"**Jet Ski**" (® Registered Trademark). This system which uses an impeller and a steerable nozzle, allows for better maneuverability and operation in shallow water, compared to traditional propellers. (A propeller is a type of impeller).

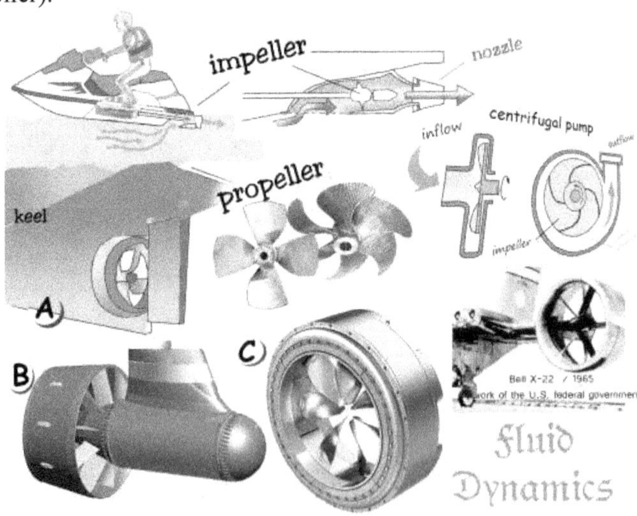

Impellers and turbines are opposites. Impellers impart energy to fluid flow. Turbines extract energy from fluid flow. "**Axial**" means; situated around, in the direction of, on, or along an axis. In the image above **(A)** represents a "**Kort nozzle**"; where the "duct" or " shroud" housing the propeller, does not rotate itself. A rudder behind the prop, controls the steerage. Shrouding a propeller can increase it's thrust and efficiency, while also reducing some of the unproductive vortices normally produced at the tips of propellers. Kort nozzles can be smaller diameter for the same amount of thrust produced by a standard propeller, be better adapted to use in shallow water and less vulnerability to debris. The design offers increased efficiency at lower speeds but, reduced efficiency at higher speeds.

(B) (previous page) is called an "**azimuth thruster**". This configuration places the electrically powered propeller(s) in a pod, and the pod itself rotates 180°. The huge RMS Queen Mary II (the only real ocean liner still in active service) uses four 260-tonne podded electric propulsors on its stern. Two are fixed and two swivel. The gigantic (148,528 gross ton) ship also has three transverse thrusters in its bow. The RMS QM2's combination of integrated electric propulsion and thruster layout would permit the leviathan to dock by itself in most ports, without the help of even one tugboat.

(C) In the previous image portrays an RPD (**Rim Driven Propeller**). There is no central shaft or hub in such a configuration. The rim amounts to the stator of an electric motor while the blades of the thruster are fastened to a ring that rotates within the electromagnetic rim. The potential benefits of this layout amount to increased efficiency, compactness and noise emission. The RDP is not a new idea, but one which is currently available despite its prospective complexities.

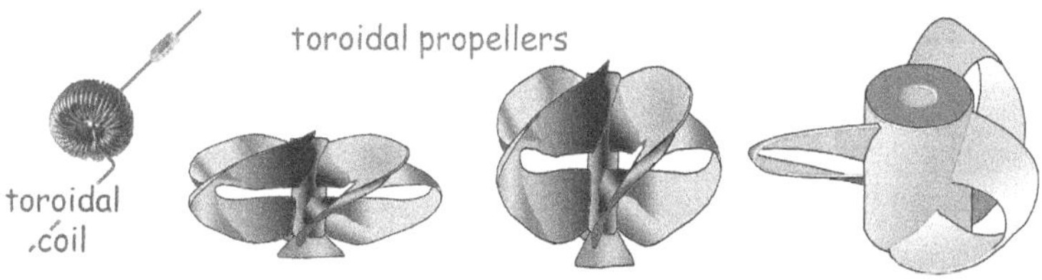

Propeller design is high science and crucially important in a field like submarine acoustics. Noisy turbulence caused by water flowing around a vessel's hull and by propeller **cavitation** especially, can give a submarine's position away. *(Cavitation is the formation of little low pressure bubbles – that make noise when they collapse)*. Stealth and quietness are paramount to submarine tactics. Above are some newfangled **toroidal shaped propellers.** A design still in its infancy that has already shown remarkable reduction in the noise and vibrations created by traditional motor boat props. Even aerial drones fly noticeably quieter when using toroidal propellers.

Sewage

There is a separate world of knowledge that could be investigated under this topic, but the space in this book is limited. Urban sewer systems must be well thought out if they are to preform adequately. For example the sewers under Paris, France are massive and famous.[17] There are 2,600 km (1,616 mi) of canals and pipes under the city. Using what the Roman's left behind, from 1370 the sewers have been "modernized" several times over the centuries. Parisians are proud of their sewers and they continue to offer guided tours to the public. (The tours began in 1867).

Sewage is less of an issue on the empty prairie, out in the woods or up in the mountains. But the denser the population, the more critical proper sewage becomes. In the past pioneers and settlers may have managed by just digging a deep hole and erecting an outhouse over it. Less odorous gray-water from kitchen sinks or laundry machines can be dealt with independently. While a composting toilet or outhouse might be sufficient for some situations today, it more likely might not be. The amount of acreage, composition of soil and disposition of the local planning commission will have a big say in what type of sewage system can be employed.

The expense and experience of creating a new septic system can range from simple to nightmarish. Sometimes constraints on when, where or if septic tanks can be installed on a property are extremely practical. You don't want to drink water from a well that taps into an aquifer that might be contaminated by your nearby neighbor's septic; and vice versa.

On another note, most states set rigorous requirements upon house plumbing, before they'll issue permit or allow a project to pass inspection. Most states do have licensing requirements for plumbers, but some states do not require a state-issued license for plumbing work. Plumbing is considered to be too complicated for the general public, or at least too critical a task to leave the work in an ambitious homeowner's hands.

17 https://stampmag.fr/the-hidden-depths-of-paris-exploring-the-citys-underground-sewers/

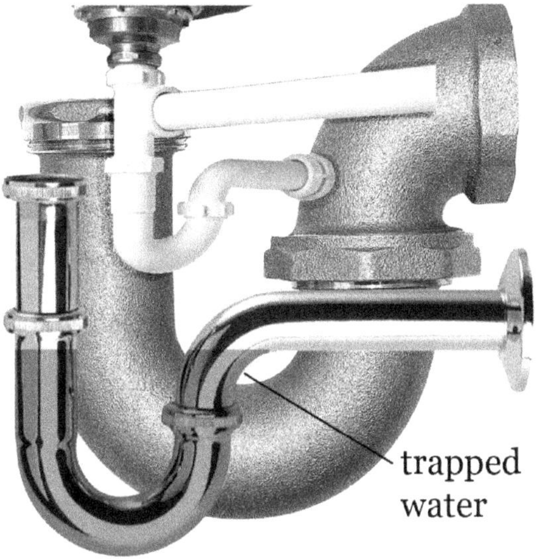

The task is not so daunting if one understands a few principles. Understanding drainage and ventilation is crucial; using the proper materials is important.

Typical plumbing code stipulates that a sewer line should be **sloped**; typically ¼ " per 4' of run. That's the minimum, throughout the house to where the pipe runs underground and into a sewer main or septic tank. All drains should have "p-traps" (**plumbing traps** / water traps**)**, which prevent sewer gases from backing-up and entering the dwelling.

Venting provides for the controlled escape of sewer gas, but also provides a path for air to enter the plumbing system. (When water flows down a drain, it creates negative pressure in the pipes). Without air from the vent to equalize that pressure then an "airlock" might occur and the effluent might not flow. Without adequate ventilation the p-traps might be siphoned empty of water, which would then lead to sewer gas entering the living spaces of the house.

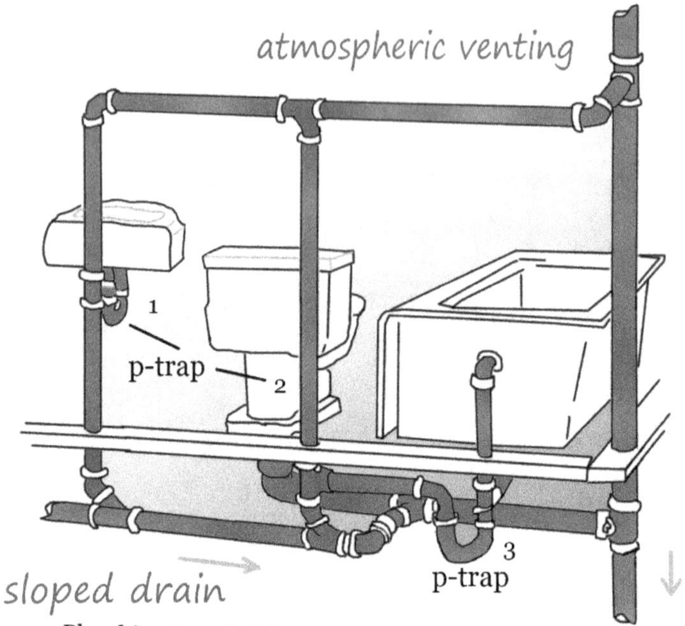

Plumbing traps hold water, and act to block sewage gas from backing up into a room or building. A toilet has a 'p-trap' built into its pedestal.

Shrewd planers often deliberately make plumbing walls – wider than normal. As in double framing the wall or bumping narrow 2"x 4" studs up to wide 2"x 6" studs in the wall. This gives the plumber more room to work and the pipes more badly need clearance.

Plumbing codes usually require stud protection plates (also known as nail plates), to protect pipes and wires within walls from being punctured by screws or nails.

Septic System

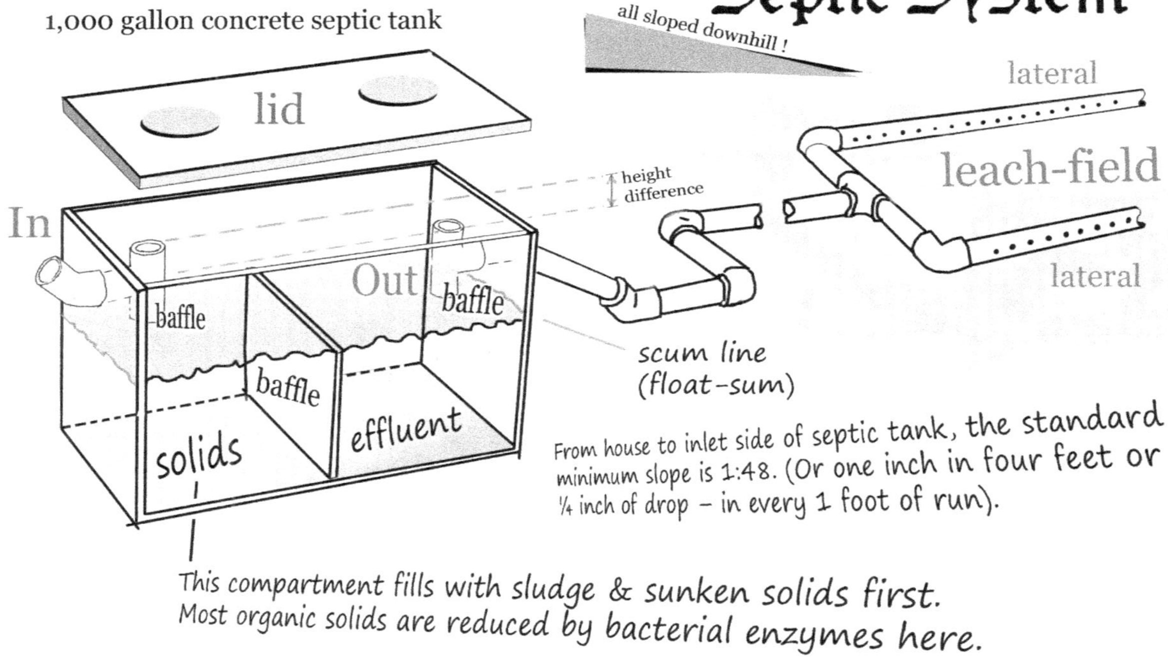

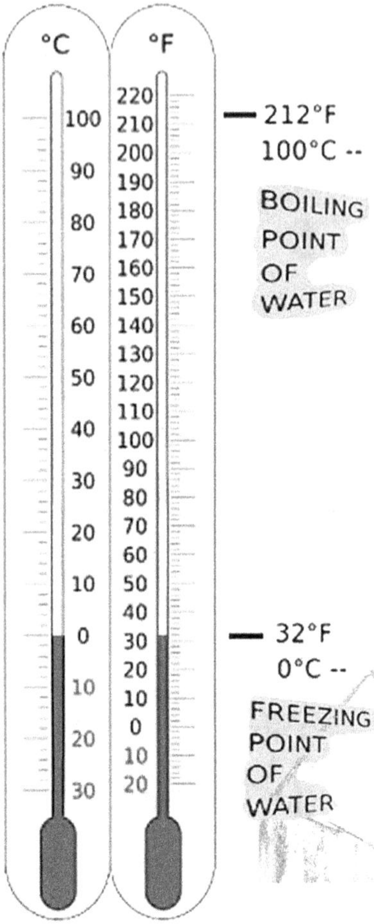

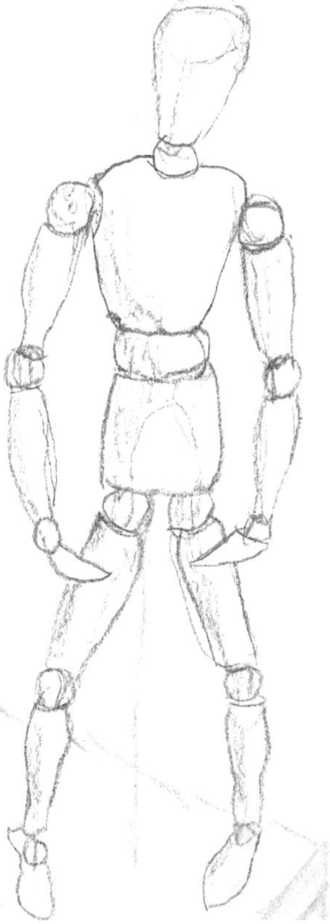

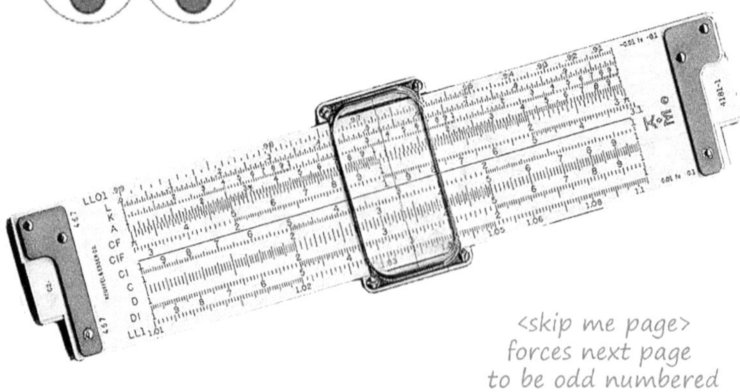

<skip me page>
forces next page
to be odd numbered
<a new chapter>

Chapter 3

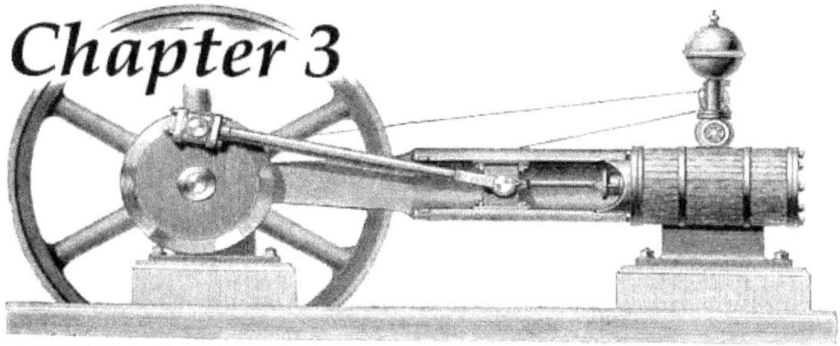

A "reciprocating", "single-acting", "horizontal", "simple" - steam engine.

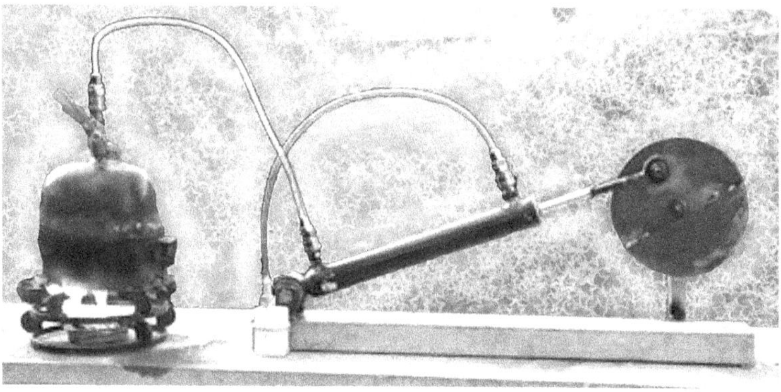

Steam Power

{How to turn teapots and scrap motor vehicle parts, into steam engines that can preform practical work or generate electricity. Some history to come first}.

Intro

Water is not specifically a liquid, but a molecule with three atoms and possessing three physical phases. **Vapor** and steam are not synonymous because many different substances can be turned into a gas and then be considered a vapor. "**Steam**" generally refers to heated water vapor. But the gaseous water vapor called steam can also be produced by simple evaporation or by the sublimation (transformation) of ice. Water vapor by itself, is invisible in the atmosphere. Fog or morning mist that results from water droplets suspended in the air is called an aerosol. When steam vapor becomes visible because of similar water droplets, it's then termed - **wet** saturated **steam**. But, **dry saturated steam** is invisible. At a specific temperature and pressure, dry saturated steam exists in a state of equilibrium where liquid water and water vapor coexist and no further evaporation or condensation can occur under the given conditions.

The **boiling point** (or **saturation temperature**) represents the temperature and pressure, at which liquid becomes vapor. At that point the vapor pressure of water equals the surrounding atmospheric pressure. The **critical point** of steam, defines the temperature and pressure beyond which water can no longer exist as a liquid.

Air molecules have mass; the atmosphere is pressing down upon us at all times. All of the air molecules from the earth's surface to the edge of space, directly above one little square inch patch of ground @ sea level, **weight 14.5 pounds (or 1 bar** or 105 Pascals) when added together.

The boiling point of a liquid increases as the atmospheric pressure increases and decreases as the pressure decreases.

- A camper at 14,000 ft. elevation can almost hold his finger in an open pot of boiling water, because it is not that hot.
- At sea level, dried pinto beans will cook easily in an open topped pot. At an altitude above 6,000 feet however, it takes forever to cook the same beans (without a lidded pressure cooker).

The atmosphere retains some water vapor at all times; typically about 1 to 2%. "**Relative humidity**" is the amount of moisture in the air, compared to what the air is capable of holding (at a specific temperature). 100% relative humidity simply means that the air has reached its water vapor capacity and can't accept any more. *"If more moisture comes along, or if the temperature drops, the air must shed water vapor in the form of water droplets"*. Rain.

Thermodynamics is the study of thermal energy and how it is transferred into other forms. There are only two thermodynamic laws of importance then trying to gain a superficial understanding of forces involving steam.

The First Law of Thermodynamics says that: ***energy cannot be destroyed, or created, it can only change form.*** Second Law of Thermodynamics states that: ***heat flows from hot to cold.***

- *(1st Law example): (using a coal burning power station):* chemical energy is changed into thermal energy, which is changed into mechanical energy and then power.
- Or; fuel > combustion > heat > steam > turbine > rotation > generator > electricity.

Also in thermodynamics, the term "**working fluid**" might be applied to a **gas**, rather than to a liquid. A working fluid is a gas or liquid that primarily transfers force, motion, or mechanical energy. In hydraulics the working fluid might be water or oil; but in pneumatics the working fluid is air or some other gas mixture. In "thermal engines" the working fluid must undergo a "**phase change**" (as from a gas to a liquid or a solid). In an air conditioner, a refrigerant undergoes a phase change in order to preform its work. Likewise injecting cold water into hot steam causes a phase change, where gas turns to liquid. The collapse in pressure causes a vacuum; an action that can be harnessed to preform work.

Efficiency

At first glance the efficiency rating of a typical steam engine might seem horrible. It may seem outrageous that the steam locomotives which moved freight across entire continents for a century, are generally rated at no better than 5% efficient.

Fuel economy (a result of thermal efficiency) is the ratio of effort (input) to work (output). It's the ratio of effort that converts chemical potential energy (contained in a carrier - fuel) into kinetic energy, or work.

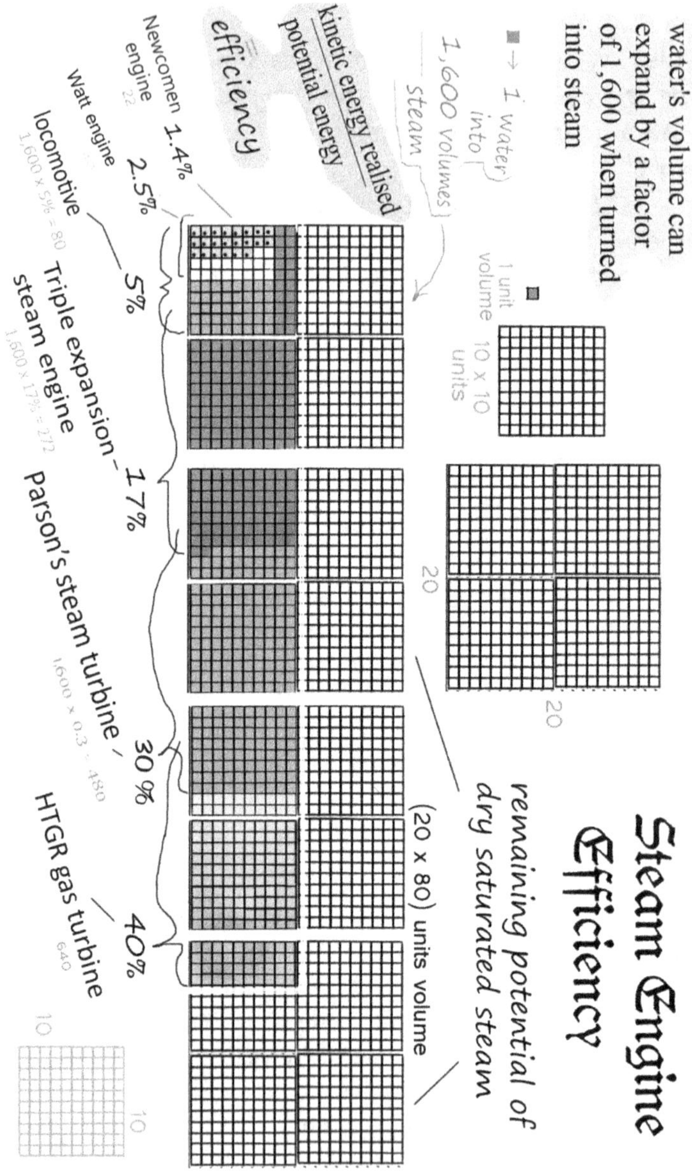

In the graphic above, a Watt's steam engine is rated 2.5% and a locomotive (train) steam engine 5%. Next a tipple expansion steam engine (for ships) 17%, a Parson's steam turbine 35% and a HTGR (High Temperature Gas-cooled Reactor) nuclear powered turbine at an electric utility plant 37%. Nearly identical but compact HTGR turbines on nuclear submarines 40%. These numbers represent the apparent maximums in terms of work, that can be extracted from the listed steam engines.

Not to get too tied up in in the notion and calculation of efficiencies, it should be noted that there are differing thermal engine types; which could sway such computations.

A steam engine with a piston is an **external combustion engine;** a reciprocating heat engine where the working fluid is heated by combustion, somewhere outside the engine. For example a steam turbine is an external combustion engine, because the working fluid is heated in a detached boiler and then piped into the turbine. Alternatively, in an **internal combustion engine** (like a car's petroleum motor) the working fluid is combusted within a chamber that is inside or integral to the engine. The metal of such an engine also retains some of the combustion heat.

So to mix efficiency ratings is a bit like comparing "apples to oranges". Gasoline and diesel engine thermal efficiencies are rated a little better than steam. Also the steam turbine is external combustion while a gas turbine is internal; (two different beast, thermodynamically speaking).

> Diesel and kerosene are more efficient fuels than gasoline, when it comes to bigger engines doing more work. The majority of modern automobiles on the road using gasoline, have an efficiency rating between about 12% and 30%.[18] In a car, anywhere between 68 -72% efficiency is lost by an engine in the form of friction or wasted heat. Another 3% to 25% of potential energy is lost by the drive-train, braking, rolling resistance and wind resistance. Electrical demands (from fuel pumps, seat warmers, climate control fans, headlights, etc.) subtract from the overall efficiency. Subtracts from the distance one unit of 'petrol' will push a car down the road.

While the efficiency of early steam locomotives and steam ships might seem abjectly wasteful today, these were once

18 https://www.fueleconomy.gov/feg/atv.shtml

civilization's prime movers, that managed to accomplish a tremendous amounts of work in their day.

The original Newcomen steam engines were as big two story houses (perhaps 9.5m tall x 9.5m long ~ (31' x 31'). Even at that size they produced only about 4.6 HP and were less than 1% efficient (0.8 %) [19]. But after years of slight improvements, that efficiency improved some. Primitive as they were, these Newcomen engines were capable of de-watering mines from more than 900 feet below the ground surface.

Fuels like coal, wood and oil were expensive back then; they are now. New, less wasteful steam engine designs were attempted. Eventually the "Watt triple expansion engine" and the "Parson steam turbine" would arrive, disrupt what was considered normal and then eventually become notable hallmarks in steam power – themselves. Steam is still very much an indispensable working fluid today. The engines in today's electrical power plants, aircraft carriers and nuclear submarines, all use steam. These all also emulate Parson's original steam turbine.

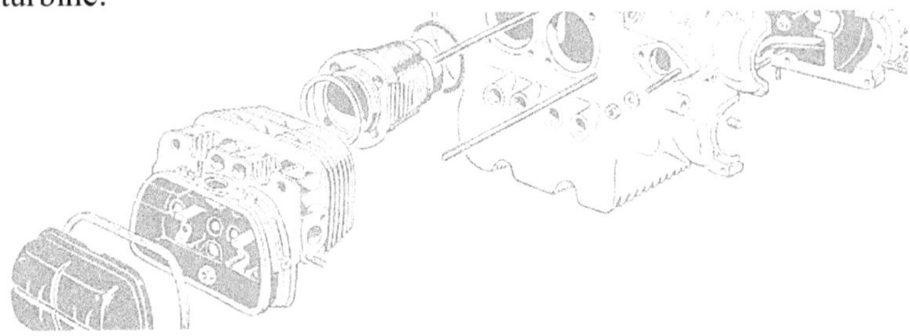

The only reason that this short study of steam power is presented in this book, is because the background knowledge might be of some usefulness to someone living "Off-GRID". If or when the GRID no longer exist and if auxiliary electrical power is no longer an option, then a little metalworking and some "know-how" might go a long, long way.

19 <https://peer.asee.org › the-newcomen-pumping-engine-a-capstone-design-project.pdf>

Hero's engine

In the Beginning

The study of hydraulics, pneumatics and steam power, all pretty much began with the Greeks. Some 23 centuries ago the great city of Alexandria (during the Ptolemaic Kingdom of Egypt) was founded and named after Alexander the Great. Alexandria quickly became a cosmopolitan city and the brightest intellectual hub.

The **Lighthouse of Alexandria** (one of the 'Seven Wonders of the Ancient World') was the world's tallest building for several centuries. Eventually the lighthouse was destroyed by multiple earthquakes over the centuries; its last stones gone to build other things, by 1480 AD. The **Library of Alexandria** was even more famous and largely responsible for the city being regarded as the world's capital for knowledge and learning during the **Hellenistic Period.** (*A **293 year period** between the deaths of Alexander the Great (323 BC) and Cleopatra VII 30 BC)*. Unfortunately the Library's influence was eventually marred by political purging of intellectuals, lack of funding and by fires caused by reckless Roman invaders. Most but not all of the Library's scrolls, which numbered in the hundreds of thousands – are lost to posterity. But once, many great minds flourished in or near that Ptolemaic library. One of those minds invented the **aeolipile** or **Hero steam engine** pictured above. Really a toy more than a useful machine; but nonetheless a steam engine created in the 2nd century AD.

Some off-topic history:

* After the death of Alexander, his empire (which stretched from Greece to India and encompassed some 2 million square miles) was divided up into some 20 satrapies or territories.

Some of his generals and officers were appointed as "satraps" or viceroys to govern those independent provinces.

* **Alexandria ad Aegyptum** (Alexandria by Egypt) then the capital of the Ptolemaic Kingdom, was to become the most influential of all the Hellenistic cities. **Athens** of course remained a cosmopolitan Greek center. **Ephesus** was a prosperous city along the Aegean coast (of modern Turkey), belonging to the Kingdom of Pergamon. **Antioch**, was the western capital of the much larger Seleucid Empire (located on the Mediterranean coast, in modern Turkey but at the Syrian border). **Seleucia** – "The Heart of the Seleucid Empire", was located on the west bank of the Tigris River near or in Baghdad, Iraq.

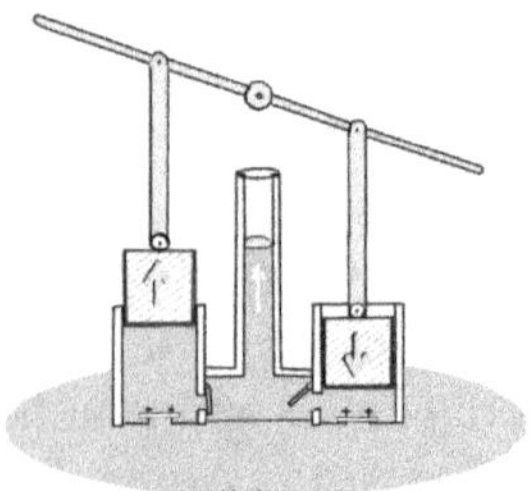

Ctesibius' piston pump

1st century AD Roman floor mosaic from Zliten, Libia; showing a hydraulis

- Ctesibius of Alexandria (285? -222 BC) a famous 3rd century inventor, mathematician and contemporary of Archimedes of Syracuse, is noted for the pressure piston pump, **hydraulis** (pipe organ) and improved clepsydra (water clock). He has been called "the father of **pneumatics**". Laws of pneumatics are practically pertinent to steam.

- Hero (or Heron) of Alexandria (around 10–85 AD.) was a notable mathematician, physicist and engineer of hydraulic and pneumatic devices. Several scientific works of Heron survived to be translated. Heron inherited many ideas from Ctesibius, a predecessor whom he could not have known. Aside from the <u>aeoliple</u> Hero also invented other items, like a wind driven version of Ctesibus' water pump pipe organ.

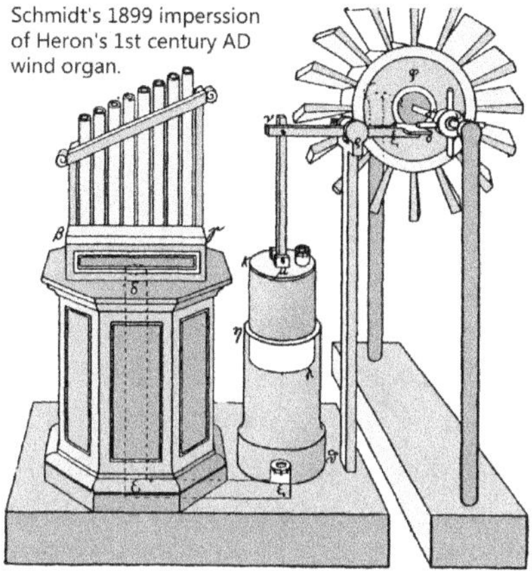

- Eratosthenes of Cyrene (276 -194 BC) also a contemporary and friend of Archimedes, was the first person to calculate the Earth's axial tilt and circumference of the Earth; both to a remarkable degree of accuracy. [He did that by analyzing the shadow of a sundial, and comparing that later to the sun's reflection while looking down into a well which was located some 526 miles to the south, 'as the crow flies'].

- Archimedes of Syracuse is considered the greatest mathematician of ancient history. His wealthy father in Syracuse sent him to Alexandria for a formal education. Beside math, Archenemies was also proficient in physics, engineering and astronomy. He (describing fulcrums and levers) is the one who said: "Give me a place to stand and I will move the earth". He is the one who jumped out of a bathtub and went running down the street naked, yelling "eureka", ("I've found it!"). Because he had determined how to tell if a king's crown was pure gold or not, by a buoyancy test in water. Archimedes however did not invent the Archimedes' screw named after him; the Egyptians did. Archimedes only described the screw with a litany of mathematics.

- Along with many others, these brainy Hellenistic Greeks were scattered across the Mediterranean ocean long before the

Romans came into power. Heron and Ctesibius may have been born in Alexandria, but Eratosthenes of Cyrene hailed from a Greek colony on the coast of Libya. Archimedes of Syracuse lived most of his life in the ancient Greek outpost, on the shoreline of Sicily.

- If you think that the names of Renaissance period European monarchs are repetitive and confusing, consider that a Greek dynasty governed Egypt for 293 years (305 -30 BC); giving rise to **15 Ptolemies** and **7 Cleopatras.** Alexander the Great's only full sibling was a sister, also named Cleopatra. "Cleopatra VII Thea Philopator" (69 -30 BC) was the famous one of Shakespearean and Hollywood drama. She had one child with Julius Caesar (reputedly) and three with Mark Antony. The last active ruler of the Ptolemaic dynasty, she was of full blooded Greek origin with no Egyptian heritage, and was the only Ptolemaic ruler known to have learned the Egyptian language. Egypt became a province of the Roman Empire after her suicide.

  {Ptolemaeus} from which Ptolemy is rooted, is an ancient name that was popular among the Macedonian upper class. Almost every pharaoh during the Hellenistic period used the name. Hollywood's Cleopatra (Cleopatra VII) had one father, two younger brothers, one son and one nephew – all named Ptolemy. Her siblings Ptolemy XIII Theos Philapator and the younger Ptolemy XIV were her consorts (husbands) but also rivals. Evidently she used "aconite" poison (see book #1) to dispatch the 14th Ptolemy.

- Claudius Ptolemy (or Ptolemy The Wise) {c 100 to 170 AD} lived after the Hellenistic period in Alexandria, Egypt. His extraction is unknown. He is famous for writing several formal treatises; the most important of which were the "Almagmest" and "Geography". His work named Geography was an atlas and treatise on cartography that pretty much summed up all the western knowledge on the subject at that time. The Almagmest was a famous astronomical compendium that used mathematics to analyze observations of planetary paths and motion of the stars. This important composition dominated thinking about the structure of the universe for well over a millennium.

Back to steam

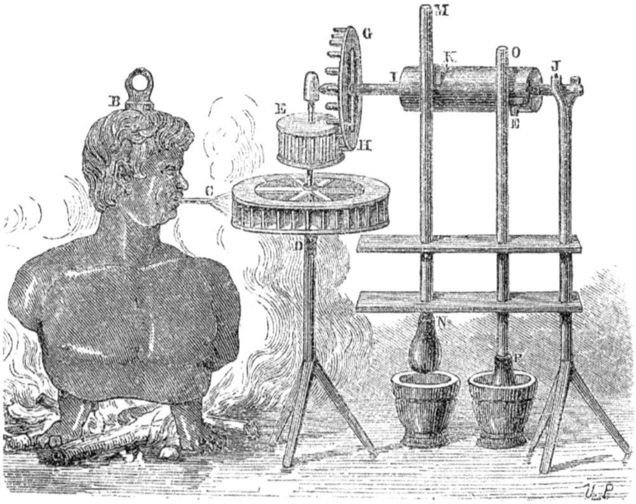

'Le Machine' - proposed by Italian engineer Giovanni Branca (1629) but illustrated two centurys later by a French civil engineer.

Coal was a crucial commodity during the **Industrial Revolution** (c 1760 onward). One thing desperately needed three centuries ago was a way to suck the water out of a mine shaft; from coal mines especially. Continuously too, because groundwater incessantly invades and floods such underground environs. Thomas Savery filed a patent back in 1698 for what is regarded as a pseudo-practical design for preforming the needed work. Savery defined his engine in a manuscript titled - *The Miner's Friend; Or, An Engine To Raise Water By Fire Described; And Of The Manner Of Fixing It In* Mines.

Savery's idea in the next illustration: providing that the four valves shown are one-way 'flap valves' or 'check valves' and that the globes are primed with water and are heated to make steam pressure: then fluid water should have no path to follow except up. Presumably. A modern drip coffee maker, works in a similar fashion.

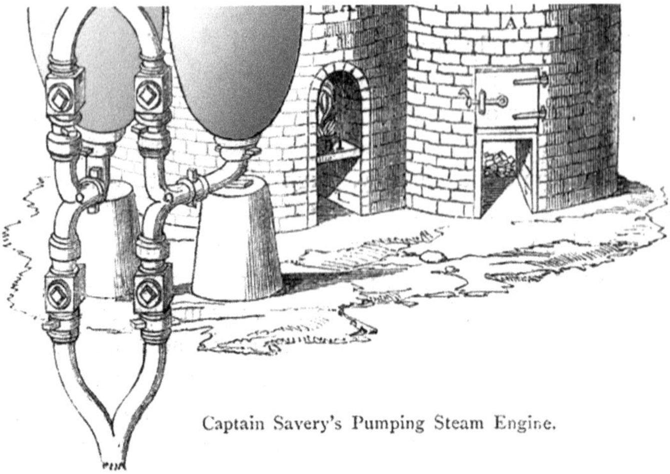

Captain Savery's Pumping Steam Engine.

* Steam is water in its invisible gas form. Usually the word "steam" refers to hot steam. But steam can be produced by evaporation at cold temperatures also.

* Water boils @ 100° C or 212° F (@ sea level). When the boiling point is reached, bubbles of steam begin to rise through the water and escape into the air. If open to the atmosphere and not confined, the water's temperature will remain the same until all the liquid has become gas.

* It takes 1 calorie - to raise 1g of water: 1°C. So it takes 100 calories of heat to raise 1 gram of water from 0° C (freezing point) to the boiling point. However it takes 540 calories of heat (without raising temperature) to turn that 1 gram of boiling water into steam. This heat is called steam's latent heat.

* At standard pressure - Saturated steam expands to roughly 1,600 times the volume of the original liquid. Superheated steam occupies even more space than that.

The next inventor to be mentioned here used cold water to shock treat hot steam into preforming work by creating a vacuum. So in a way, steam can be used to pull, too.

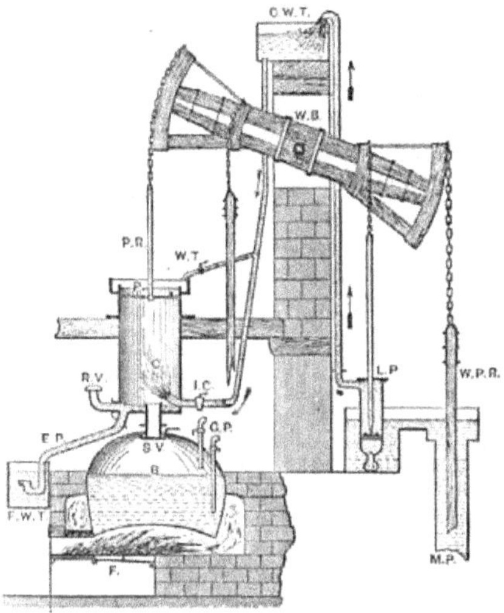

Thomas Newcomen [1664 – 1729] expanded upon Savery's idea of using a vacuum to harness atmospheric pressure. He converted thermal energy into mechanical energy; a **phase change**.

Newcomen's steam engine, was the first one that was practical for preforming work. Hundreds of his "beam engines" were eventually built all over Europe in the 18th century. These beam engines of his (where a pivoted overhead beam rocks to and fro) were also called **"atmospheric engines".**

A vacuum was quickly created below the piston when it reached "top dead center", by spraying cold water into the cylinder of hot steam. The steam collapsed and left in its place a vacuum, which assisted the weight of atmospheric air (14.5 psi) plus the weight of the piston, in returning the piston to the bottom of the cylinder. This down-stroke was the power stroke. Water from the condensed steam exited the cylinder during the down-stroke. Hot steam did little pushing up; the atmosphere did all the work by pushing down.

As the reciprocating motion of the beam swung back up, the piston was drawn up and the water pump's linkage (reaching down into the mine-shaft) went back down. A valve opened to admit new hot steam into the cylinder on the upswing.

Piston to cylinder tolerances were not tight. Being weak and inefficient, these engines needed to be massive, in order to be useful. At first they were used mainly to pump water out of coal mines. Despite the horrible inefficiency, Newcomen's engine was still capable of drawing water up from considerable distances.

By 1765 an engineer named John Page improved the efficiency (of "Newcomen" engines) up to 1.4%! Newcomen engines continued to be built and purchased until the 1820's, because they were simple and reliable. New, improved machines were expensive. Most Newcomen engines were eventually upgraded to a newer design. The last operational Newcomen engine was retired around 1930.

Realize that the 18th century coal mine owners had plenty of undesirable '**slack coal**' laying about on the ground. This poor quality coal that they could not sell, was still usable in their own boilers (for pumping out mine-shafts). There was little incentive for these coal companies to upgrade their machines initially. It would take sixty years before an engineer named Watt, would come along and make substantial improvements to Newcomen's design, and improve the efficiency.

Smaller experimental engines were being built all the while too. In 1783 (shortly before the French Revolution) a Frenchman named de Jouffroy created the first working steam ship, named the Pyroscaphe {Gk for "fire-boat"}. Paris was a hotbed of French scientific and engineering ingenuity at the time. Using a horizontal, double-acting steam engine the jury-rigged paddle-boat managed to work well for about 15 minutes; before it began to shake apart and sink from the stresses caused by the engine.

The picture on the left is titled "*"Plan of Mr. Fitch's Steam Boat", Columbian Magazine (December 1786), woodcut by James Trenchard"*. The engine sat in the keel and was not viewable from the side. An actual paddle (or oar) boat, the "*Perseverance*" staged a successful trial on the Delaware river in 1787. Few details of the engine have survived.

The 2nd boat, pictured on the right was launched by Fitch in 1790. This 60 foot long craft had several stern-mounted oars that paddled like the motion of a swimming duck's feet. During the summer of 1790 this boat carried as many as 30 passengers at a time, back and forth (on the Delaware River) between Philadelphia and Burlington, N.J. .

Fitch with the help of another watchmaker named Henry Voigt, independently constructed an uniquely American version of a marine steam engine. Although people traveled constantly across the Atlantic, governments were not so cordial. In 1790, the American Revolutionary War had only been over for a few years, while the War of 1812 was yet to happen.

Steam engines were state of the art technology - high value IP (Intellectual Property) and the British government would not deliberately share any important new engineering science with its former colony. American colonial metal foundries had historically been discouraged, for the same reasons.

The 1790's were tumultuous times. The Reign of Terror (the pre-Marxist communist experiment also known as the French Revolution) was also occurring. Between 1789 and 1799 the murderous French revolt witnessed some 16,594 sanctioned decapitations, 12,000 or so executions without trial and about 10,000 deaths of people abandoned in prisons.

(A preceding, more famous slaughter in France occurred about two centuries before, when some 30,000 Huguenots were murdered in just 48 hours. The St. Bartholomew's Day massacre).

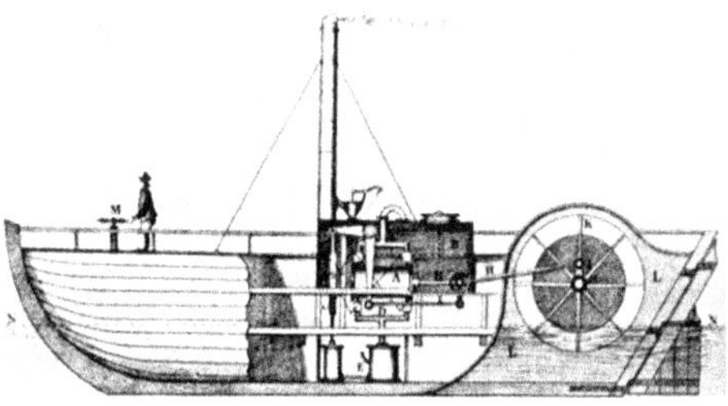

The Charlotte Dundas was a steamboat designed and built in Scotland. A prototype preformed successful trials in 1801; the paddle steamer shown was launched in 1803.

The next image to come is a model of Robert Fulton's financially successful and practical steamboat – the Clermont (in 1806). Fulton an American, spent some time learning and working in both England and France, before returning to America. He was actually commissioned in 1800 by Napoleon Bonaparte, to design a working submarine (called the Nautilus).

Fulton was apparently imparticular in terms of allegiance because he tried to sell his torpedo technology to everybody (the French, British and Americans all; while they were active enemies). The French tried his torpedoes against the British. The British tried to bribe him from enlightening Americans. He was unsuccessful at bargaining with President Jefferson, but had success selling a few torpedoes during Madison's administration - used against the Royal Navy during the War of 1812.

> *(His **torpedo** would be called an anti-ship mine today. Spar torpedoes were explosives on the end of a pole and Limpet mines were explosives intended stick to a ship's hull under the water line. All were called torpedoes back then).*

Once back in America, Fulton created the famous "North River Steamboat of Clermont"- a paddle-wheeler powered by an engine of his specification, but made and somehow exported from England by the Boulton and Watt company.

On its first trip (in 1807) Fulton's paddle-boat took 32 hours, to steam from New York City to Albany N.Y. . A 150 mile trip, upriver on the Hudson. From 1807 he and a financier, built more boats and established a profitable, weekly 300 mile round trip passenger service up the river and back.

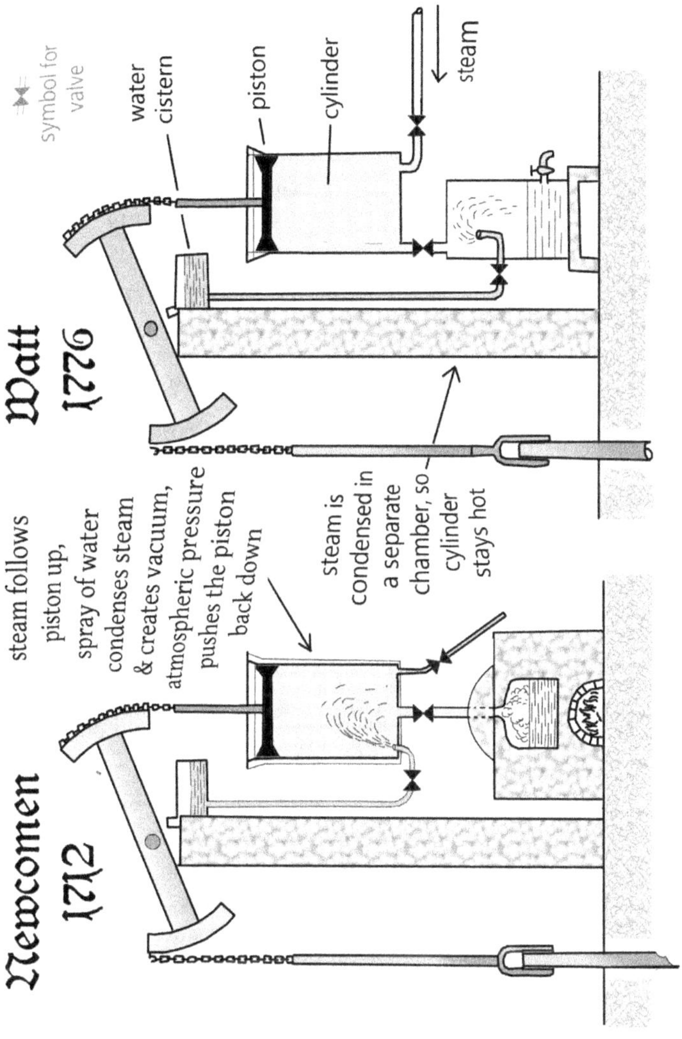

This simplified illustration does not show the linkages that control the valves which are critical to function, nor linkages that drive auxiliary pumps, which fill the elevated water cisterns. The cylinders of these **atmospheric engines** (Newcomen and then Watt's earliest versions) needed no closed top. Chains or cables rather than stiff rods, were mostly used for the linkages.

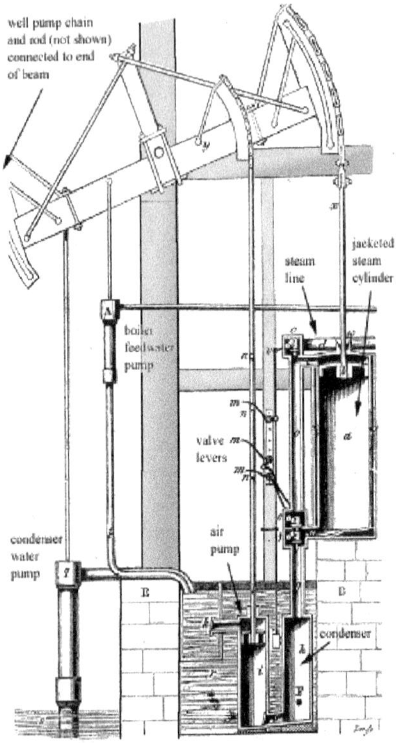

James Watt [1736 – 1819] comes along 64 years after Newcomen's debut and introduces an improved **condensing steam engine** in 1776. By adding a separated condenser, the cylinder itself would now stay hot; the atmospheric shock and cooling done in a different chamber. Watt would also eventually improve the piston's seal, enclose the top of the cylinder and insulate everything. These refinements led to a doubling of the engines efficiency.

Watt would receive a patent for his first engine in 1769, but did not build one under his own name until 1776 (with the help of another engineer and new partner Matthew Boulton). Watt's participation in steam research and development would continue for another 30 years, but he was only one out of a multitude, working constantly to improve steam engines. His initial design was a **single-acting** engine; meaning that steam entered on one side of the piston, powered one stroke and then was expelled to the atmosphere.

Jonathan Hornblower invented and patented the **compound steam engine** (steam expanded in 2 or more stages / cylinders) in 1781. Afterward James Watt again, developed the first **double-acting steam engine**; where steam is entered from <u>both sides of the piston</u>. Then two work strokes and twice the power are created. Watt also created a slick **parallel linkage** system for the beam engines, that kept the piston better vertically aligned with the cylinder. He also contributed the "fly-ball" **centrifugal governor**. An ingenious engine throttle control.

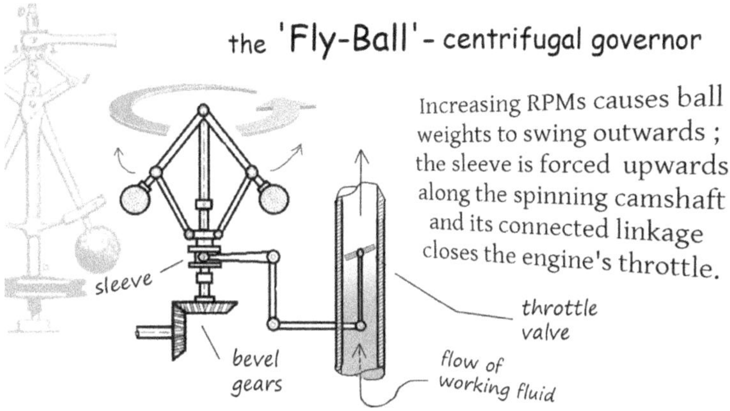

The double-acting (steam & beam) engine and parallel linkages together just begged for the power to be transformed into a more useful rotational motion that industry could interface with. That was accomplished by using a crank and flywheel.

A "**reciprocating** engine" might imply a beam's movement, and "**rotary** motion" might imply the presence of a flywheel. Watt engine's, of which there are many iterations, did both.

Spinning flywheels date back to Neolithic times. Leonardo da Vinci (1452 – 1519) toyed with flywheels too. Since someone else (another engineer and former partner of Watt's named James Pickard) held the patent for the flywheel-crank-rod on a steam engine idea, Watt and crew were initially obliged to use an **epicyclic sun and planet gear system** instead.

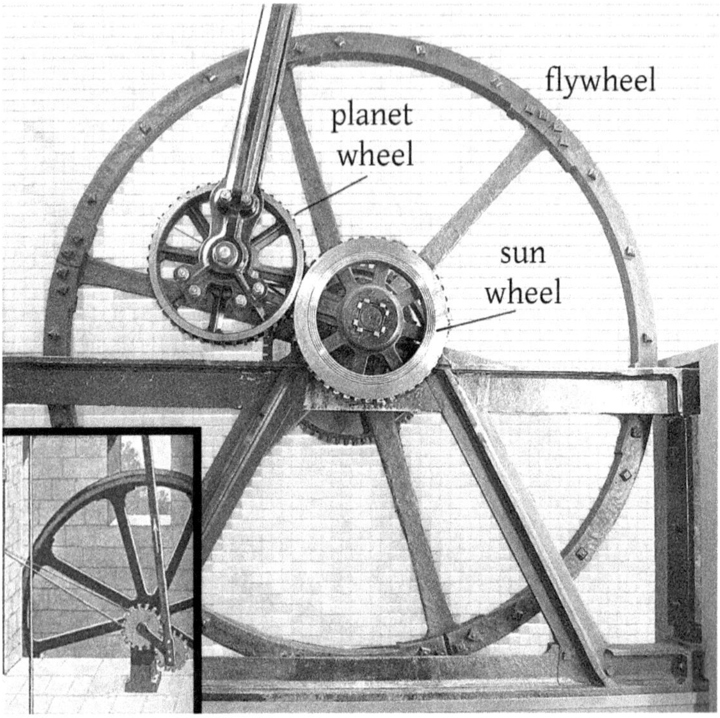

Inertia - is the property of a body that resists changes in its state of motion. A body at rest tends to stay at rest. A body in motion stays in motion unless it's inertia is affected by an external force.

> *"Flywheels store and release the energy of rotation, an inertial energy. The primary purpose of a flywheel is to regulate the speed and power of the machine. It does this through the amount of inertia contained in the flywheel, specifically the* ***mass moment of inertia****".*
>
> (The moment of inertia - is the ratio between the torque applied and the resulting angular acceleration about that axis).

Once set into motion, a **simplex** (single cylinder) steam

engine benefits from a flywheel, which helps maintain its motion. Smoothing the action of alternating strokes and making power more consistent. **Dulpex** (two cylinder) steam engines like those used later in the Stanley Steamer automobile, had more power and a more balanced motion to begin with, so didn't particularly require a flywheel.

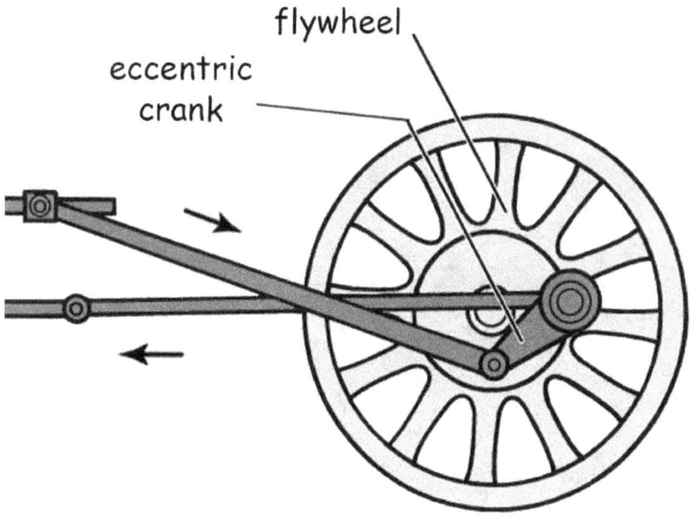

After the patent expired, every steam engine manufacturer could then use the favored rod, crank and wheel mechanism. Big flywheels like those used for many steam engines, were cast from iron and intentionally heavy. These stored energy, by conserving rotational inertia.

In physics, **angular momentum** (or also called rotational momentum) is the rotational analog of linear momentum. **Torque** is the rotational analogue of linear force. Angular momentum is a measurement of how much or how long in this instance – that a flywheel will continue to spin on its own, without any more torque being applied.

A museum display of a squat but very heavy flywheel.
https://en.wikipedia.org/wiki/Moment_of_inertia

A heavy flywheel attached to the crankshaft serves to prevent fluctuation of the engine and jerks from the crankshaft.

- **Rotational inertia** about a particular axis depends both on the mass and its distribution relative to the axis, increasing with mass & distance from the axis.
- A force distant from the fulcrum (or wheel's axis) stores more energy or is capable of more work, than an equal force that is closer.

The fact that belts and gears could be attached to the rotating crankshafts and camshafts of these engines was not insignificant. A great variety of machinery was developed to take advantage of the new spinning resource.

About nine decades of time separate the two images of textile factories (above and below). Both factories had steam driven, **overhead line-shaft** systems. The earlier factory used wooden drums, but the younger line shafts in the second picture had metal wheels and sheaves of differing diameters.

Coal mines and metal mines were the first customers for steam power. Soon though textile mills and then other factories adapted to the new power source. Matthew Boulton (mechanical engineer, silversmith and partner to James Watt) adapted the power of these machines to the task of striking coins. His machines stamped millions of coins, not just for Britain but for other countries as well.

Some historians suggest that the **Industrial Revolution** should be divided into two parts, the 1st being from 1760 to about 1840. This coincides nicely with sharp advancements in the use of water power, steam power and metallurgy. After a hiatus of three or four decades another uptick in intellectual achievement and industrial production would occur from about 1870 to 1914 (the beginning of WWI).

Obviously engineers in other parts of the world were trying to improve steam technology too, but in Britain the competition was particularly keen. It was steam that provided the necessary motive power during the industrial revolution (s). At the beginning of the 20th century, electrical power was still in its infancy; far from being an established commodity.

Back in the dark mine-shafts of western England, Cornish engineers would develop much better pumping engines for deep mines. Some of the best hard-rock miners in the world were Cornish. (The Cornish are a Celtic ethnic group of ancient Britain). The improvements of the so-called **Cornish engines** was centered around their usage of high pressure steam. Until this time the cylinders and boilers of engines had been necessarily large, because they had been designed to exploit low-pressure steam; (pressures little greater than atmospheric).

It was known that **higher pressure** could increase a steam engine's efficiency and allow for greater work. But boilers had an tendency to explode when pressed too far. Watt had earlier, cautiously avoided high pressure, because he rightfully thought it was too dangerous for the materials and technology at the time.

Technology caught up. When pressures were advanced up to 50-100 psi, then engines could be made much smaller. Cornish engineer Arthur Woolf (1766- 1837) would develop a high-pressure compound engine in 1804 *{but Jonathan Hornblower (1753 -1815) had already invented one in 1783. A patent held by Boulton & Watt company prevented its development though}*.

Woolf's new (Cornish type) engine was a double acting, high-pressure compound steam engine. It used the best features from older Boulton & Watt machines, but added another cylinder and higher pressure. It extracted more work from the same steam.

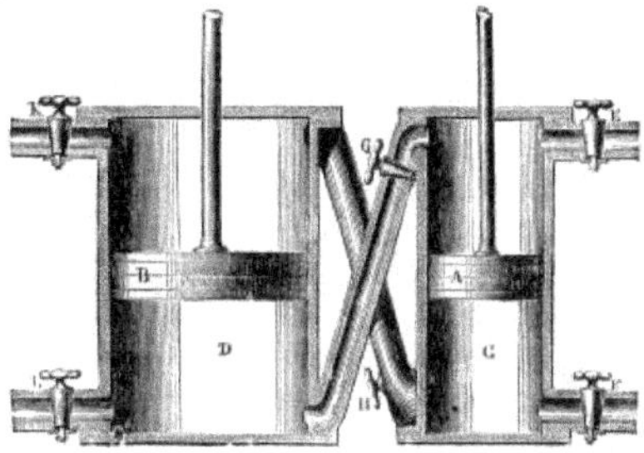

Woolf's double acting, *high-pressure*, compound steam engine *design*.

"**Double acting**" meant that steam pressure was applied to both sides of the piston. Once high pressure steam was used in the smaller cylinder, it was conducted to a physically larger but lower-pressure second cylinder, where the steam could continue to expand and still preform work. A "**condensing steam engine**" turns or reduces steam back into water, to reuse. "**Compound**" means multiple cylinders, and passing the exhaust from one to another; and extracting more work from the same steam.

Power Transmission

1st detour: **Pneumatic Power**

From 1866 in Paris, the French installed some 265 miles of underground pipes to be used in pneumatic message delivery systems. (The city's Morris code using electrical telegraphs were already overloaded with demand). Steam generated air pressure delivered telegram like messages called "***pneus***", to 130 locations about the city. Delivery boys dispatched the stamped, pre-payed uniform paper-stock pneus, onward from there.

An average 10 million 'pneu' messages per/yr were sent back and forth around the city during the 1930s. That number reached 30 million in 1945.[20] The "Poste pneumatique de Paris" operated for 117 years. Then in 1984 the Postal Ministry said it was doing

20 http://douglas-self.com/MUSEUM/COMMS/airclock/airclock.htm

away with the remarkably efficient service because it had been "overtaken by telephones and telex".

In 1880 (about 14 years after the initiation of the pneumatic telegraph) a Parisian compressed air network would also be installed underneath the city; using the same sprawling Paris sewer system to install the piping.

The Paris sewer system is old (begun in 1370) and massive. In places a semi-truck can be driven down the tunnels. Initially a pneumatic clock system would be installed; that required an extra 20 miles of large main pipe and almost 177 miles of branch and sub branch pipes – to power some 4,000 clocks around the city.[21] Air clocks were placed in hotels, train stations, houses, schools and public streets. These pneumatic "**Popp clocks**" were named after their inventor, an Austrian engineer named Victor Popp. Electrical power was very much in it's infancy in a period when accurate time keeping was becoming more fashionable.

21 https://www.resilience.org/stories/2018-05-16/history-and-future-of-the-compressed-air-economy/

PNEUMATIC CLOCK IN PARIS.

The clocks were all synchronized to the main city clock, with puffs of air. *'An automatic timing mechanism opened a valve to release a 20-second pulse of air every minute and then the valve closed for 40 seconds".*

On top of the pneumatic mail, the successful clock system (retired in 1927) helped lead to a city-wide power distribution network (which would also last for over a century - (from 1881 to 1994)). The compressed air network proved to be highly successful in small industrial and service establishments.

> As F.E. Idell wrote in 1892: *"among the smaller industrial purposes for which the air motors are used in Paris, I find the driving of lathes for metal and wood, of circular saws, drills, polishing machines, and many others. They are also used in the workshops of carpenters, joiners and cabinet-makers, of smiths, of umbrella makers, of collar-makers, of bookbinders, and naturally in a great many places where sewing machines are used, both by dressmakers, tailors, and shoemakers, from the smallest to the largest scale."*[22]

Commercial and domestic use of compressed air eventually decreased but industrial consumption of compressed air kept

22 "Compressed air: experiments upon the transmission of power by compressed air in Paris (Popp's system)", F.E. Idell, 1892

growing rite up till the end.

Above is a typical air compressor of the 19th century; driven by coal -fired steam. Paris had large buildings with fleets of engines like this, and air accumulators (air pressure tanks) too. Paris was not the only city to exploit compressed air, but it was the only one to develop and distribute that power so widely; due largely to easy installment in the city's unique sewer system. London, Washington DC and New York City all had large underground pneumatic tube systems too. NYC had 27 miles of tubes that from 1900ish to 1957, distributed about 30% of all its mail, between 23 separate post offices.

New pneumatic tube system (for cash) in Marshal Fields department store, Chicago,Illinios. Nov 26, 1947

Alternately, in the 19th century other cities like Antwerp, Geneva, Melbourne, Sydney, Buenos Aires, London, Liverpook and Glasgow would develop hydraulic power networks. These would use the same or similar coal burning **steam engines** to create a motive force that was converted into mechanical energy.

You can't pipe hot steam a great distance, because it cools and condenses back into water. loosing it's oomph. But you can compress air and push that a long way with out loosing much energy. Likewise you can press and push against water to transfer power a great **distance**.

The next image shows an early, primitive electrical power plant that uses a few steam engines and leather belts to spin some low-voltage DC generators. The puny output might have powered a few dozen light bulbs, but that would hardly be adequate to illuminate a large city. DC power is inefficient to transmit over a distance, so its generation needs to be preformed - near the work. This was the problem that limited the widespread adoption of Edison Electrics' DC power scheme.

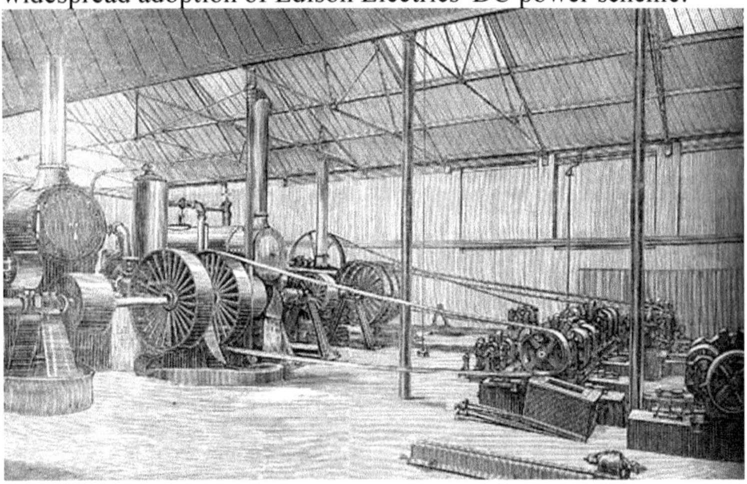

In the early days compressed air and hydraulic systems, both had good long distance transmission efficiency when compared to DC current. Of the two: compressed air has the advantage over water pressure, because it can be found and then exhausted anywhere, while water needs to be supplied from a source (is not omnipresent) and is harder to dispose of after being used.

2nd detour:
Hydraulic Power Networks

There was a period beginning in the second half of the 19th century, when small "water motors" were used to power forerunners of machinery that would require electricity today. Small household water motors were mass produced and sold in cities that could provide strong water pressure from a faucet. Sink top water motors (in the U.S.) were most often 12" diameter Pelton impact turbines, which were enclosed in a cast metal housing. They drained from the bottom and had a shaft or belt pulley drum on the side. These small motors were surprisingly powerful when water pressure was good, but very wasteful of tap water in general. They could be found powering sewing machines, fans, food grinders, air compressors or even small electric generators. Larger models were offered, that might have been used in some factories, or have been tailored for use in more remote locations where water from independent penstocks could be harnessed.

PELTON WATER MOTOR FOR SMALL WORK

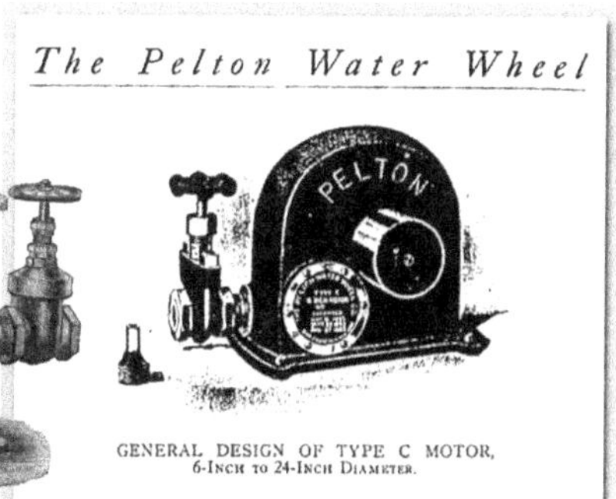

GENERAL DESIGN OF TYPE C MOTOR, 6-INCH TO 24-INCH DIAMETER.

In the US these water motors faded away by the early 20th century, but in Europe the exploitation of hydraulic power was carried further. In cities like London, special "power water" networks were built, which distributed water under pressure for motive power purposes, only. The switch to a much higher and more regular water pressure, was made possible by the invention of the hydraulic accumulator.

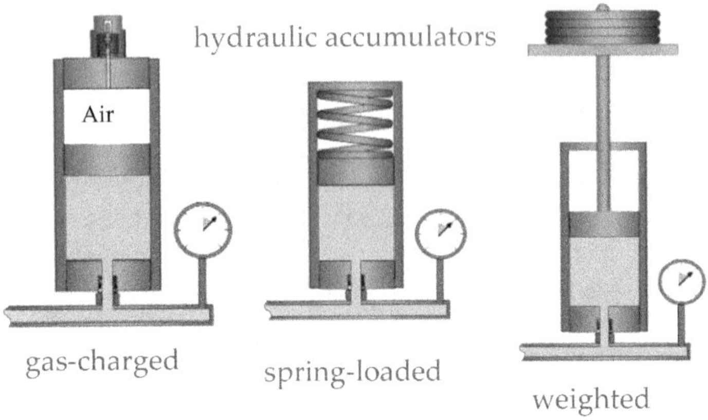

Hydraulic accumulators store energy as pressurized fluid. Accumulators apply pressure to stored fluid networks (pipes and tanks) and they can be made in several ways. City water towers and bladder pressure tanks for household wells, work the same way as hydraulic accumulators, but at lower pressures. **Hydraulic jacks** share the same principles and abide by Pascal's principle of **fluid mechanics**. Blaise Pascal (a French mathematician in the 1650s) proclaimed that - *a pressure change at any point in a confined incompressible fluid is transmitted throughout the fluid such that the same change occurs everywhere.*

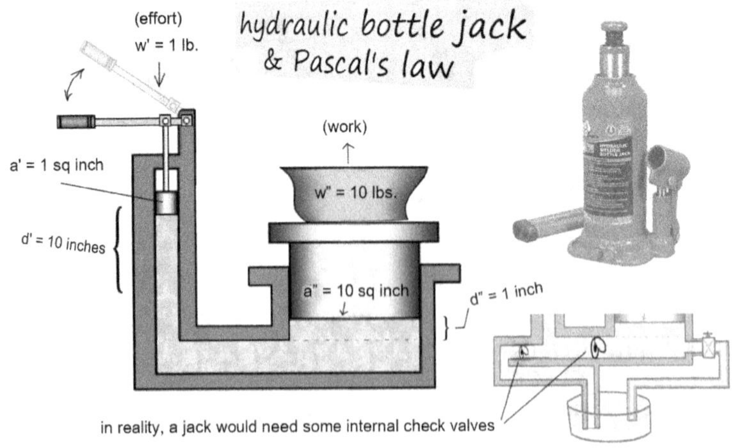

In the illustration: 1 pound of force applied to the small piston with 1 square inch surface area, will lift 10 pounds of weight pressing down on the larger piston (provided that its surface area is 10 square inches). Pascal's law stipulates that all points in the liquid share the same pressure, therefore in this instance, 10x the force is pushing up on the larger piston. However the distance traveled by the larger mass would be only 1/10th that of the smaller piston's travel. (What is gained in mechanical advantage is lost in velocity ratio). In a jack, some check valves are needed to maintain pressure and to keep the fluid moving in one direction.

The first **hydraulic press** was invented by a carpenter/ locksmith, in 1796.

The hydraulic accumulator's history is tied to shipping docks. Shipping docks, wharfs or quays are necessary for seagoing ships to approach and tie down to, in order to embark or disembark passengers or to load or unload cargo. Cargo ships are the most crucial type of ship to a nation's economy. Originally multipurpose by design, cargo ships might have carried the mail and a few passengers also. Historically the major reason of maintaining warships and therefore any Navy at all, was to protect a nation's commercial cargo fleet on the high seas. Originally cargo ships were loaded by hand. Later freight or cargo was bundled and lifted by ropes and pulleys. A ship with a pulley rigged in the mast above head could use a man-powered **capstan winch** and sailors, to load and unload cargo (the same winch was used to lift heavy anchors and mainsails).

German sailors, operating capstan, aboard the MAGDALENE VINNEN, Sidney Harbor, 1933

Eventually docks and harbors themselves began supply their own winches and cranes. With the Industrial Revolution and capitalism came bigger ships, bigger cargoes and new ideas.

Starting from the 1830s investors began to making iron hulled cargo ships. As these grew in size and capacity, the need for stronger lifting power grew too. From the 1850s onward, steam powered cranes and maritime winches were installed on ships and on wharfs both.

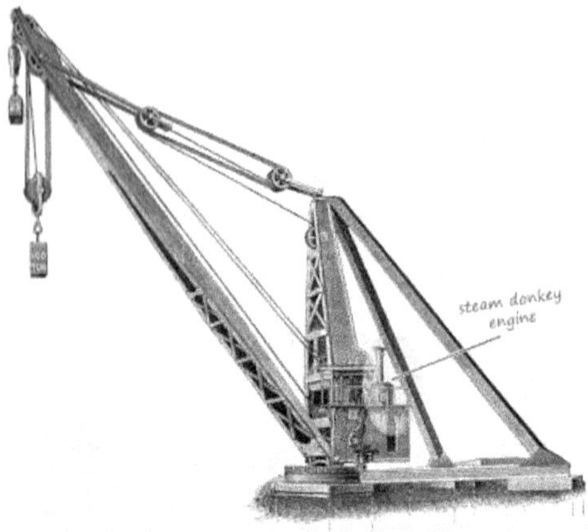

These steam powered cranes and winches used relatively small "**hoisting engines**" to supply their power independently. While some of the cranes were huge, it took only a small engine to power them. (Donkey Engines will be addressed in more detail later).

But to better fulfill a persistent need to hoist supplies to and fro in the larger, busier dockyards, an engineer named William Armstrong, began building and selling hydraulic cranes in the 1840s. These cranes worked on the water stored in a network of pipes and tanks. Hydraulic accumulators were used to store energy and maintain pressure. Like the hydraulic bottle jack, a pressure from a small force can be made to move a much larger force (weight). And this same pressure can be transferred miles down a high pressure pipe to preform the same work. The water cranes were quick to work and effective. A fire for a steam crane's boiler did not need to be lit every-time that a ship rolled up to the dock.

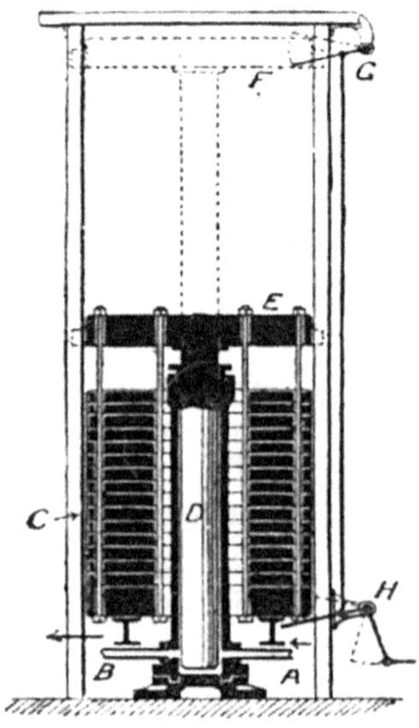

FIG. 19.—Hydraulic Accumulator.

Yes, coal burning steam engines somewhere in a centralized warehouse, had to power the water pumps that provided the initial accumulated pressure. But depending upon the size of the hydraulic network and the number of accumulators storing energy in the system, the water pressure was ready to go in an instant's notice, and easy to maintain. Such a hydraulic power network was built and it was spread all over central London.

The pressure and supply from city water mains and water towers was impractical for Armstrong's needs in 1840, so with financing he created an independent hydraulic network. His pipelines maintained a pressure in excess of 700 psi; his accumulators were little more than a piston in a long vertical pipe, with a heavy stack of bricks on top.

The motive force later found many new customers and high pressure water was adapted to a variety of industrial applications; like handling freight and rotating turntables in railway yards, rotating stages in theaters, powering elevators, or forging, stamping, shearing, punching and riveting in iron-works factories. Aside from London, similar hydraulic power networks were installed in Liverpool, Birmingham, Manchester, Glasgow, Antwerp, Belgium, Buenos Aires, Melbourne and Sydney.

By act of Parliament The London Hydraulic Power Company was established in 1883, and it merged with at least four private hydraulic companies. The hydraulic network was finally replaced by electricity, in 1977. Its pipes and accumulators veritably maintained a nominal pressure of 55 BAR (800 psi) for 94 years.

"The most extensive hydraulic power network was built in London, operated by the "London Hydraulic Company". At the company's peak in 1917, five interconnected central power stations pumped high pressure water in about a dozen hydraulic accumulators and almost 300 km of supply mains, powering more than 8,000 machines and serving most of the city".[23]

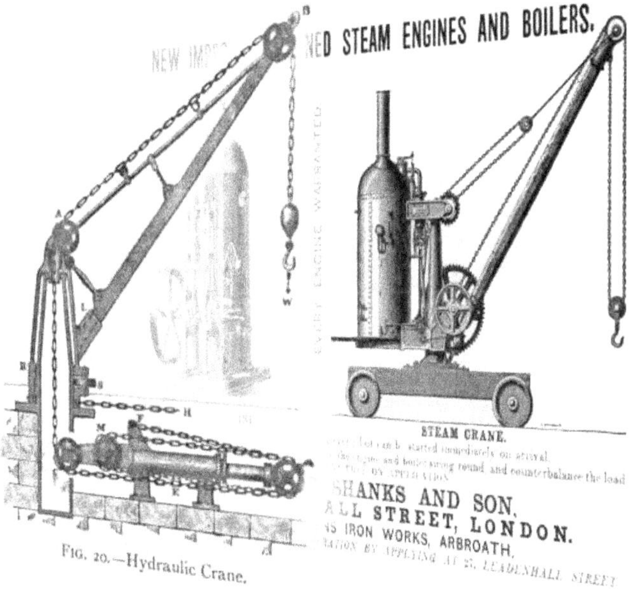

Fig. 20.—Hydraulic Crane.

23 https://solar.lowtechmagazine.com/2016/03/power-water-networks/

Energy Conversion

Useful power is lost each time energy is transformed from one type, to another. For example, the mechanical energy from a water wheel when used to compress air, is one energy conversion (mechanical energy > into > potential energy). But during the conversion some energy is lost and efficiency is reduced. Consider that perhaps only 75% of the original energy remains after this conversion (or any other single conversion). Then using this new compressed air in a tool like a pneumatic hammer drill then, represents a second (back to mechanical) conversion of energy. So now two sources of energy loss occur and the total efficiency drops to something like 56%.

Take a modern factory as an example, that uses electricity to compress air (like a food packaging factory, or a blast furnace that injects air into molten iron to make it hotter). Here; electricity must first be generated (mechanical energy converted to electricity, or M>E), then an electrical motor converts the power back into mechanical energy (E>M), then an air compressor converts mechanical energy into air pressure (or potential energy, M>P), then a tool usually converts that potential back into mechanical work as the gas is expanded (P>M). So that is four separate energy conversions [M>E, E>M, M>P, P>M] and a poor system efficiency of 30%. (75 x .75 x .75 x.75 = 31.6).

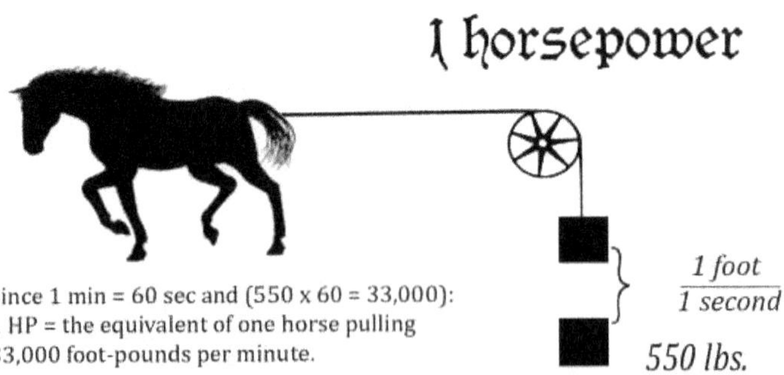

Since 1 min = 60 sec and (550 x 60 = 33,000):
1 HP = the equivalent of one horse pulling 33,000 foot-pounds per minute.

-Or 330 lbs lifted 100 ft/min
-Or 33 lbs lifted 1,000 ft/min.
-Or 1,000 lbs lifted 33 ft/min.
-Or 735.5 watts (in SI)

* James Watt created the horsepower unit (before 1776) to express large forces in terms that other people could relate to. SI named its equivalent unit of force after him in 1960.

Back to Steam

About a dozen pages ago before the subjects of pneumatics and hydraulics were injected, the topic was "**Cornish engines**". The term does not identify any specific type of engine, rather it might just imply that an engine works at higher pressure than normal. Before, engines had worked at atmospheric pressure, or just a little above. What quickly became known as Cornish engines however, qualified because they worked at about twice that pressure, or were compound engines or were were designed by Cornish engineers. (25 psi – the working pressure of a modern day kitchen pressure cooker, was considered high industrial working pressure at the time). Woolf' was the early Cornish engineer who held a patent on the high pressure compound engine in 1805 (mentioned right before detour into pneumatics & hydraulics). But he had many capable and competitive contemporaries. Also, improvements in industrial machining and tooling were continually being made.

London Steam Carriage (1803)

Oruktor Amphibolos (1805)

Another Cornish engineer **Richard Trevithick** (1771 – 1833) and an American named Oliver Evans (1755-1819), both made good high-pressure, non-condensing engines about the year 1802. These had small, well-machined cylinders, that could fit contraptions or work spaces smaller than ships or entire factory buildings.

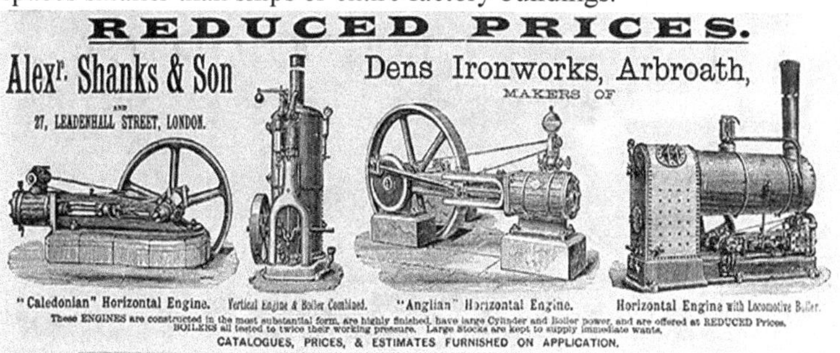

However, some Cornish engines were still humongous.

(A Cornish engine working in Virginia City, Nevada; 1895)

This looks like the same Cornish engine as in the previous image, and shows a low slung, space saving beam. Representing a **side-lever engine** configuration.

(The crank but not the wheel is shown in this image of a high-pressure, overhead beam - **Cornish** steam engine).

Trevithick would go on to invent a (non rail) steam road locomotive called the "Puffing Devil", then a wheeled contraption named the "London Steam Carriage" and the world's first working railway steam locomotive.

On the other side of the Atlantic, one of Oliver Evans' creations was the world's first but not quite successful amphibious automobile, called the "Oruktor Amphibolos".

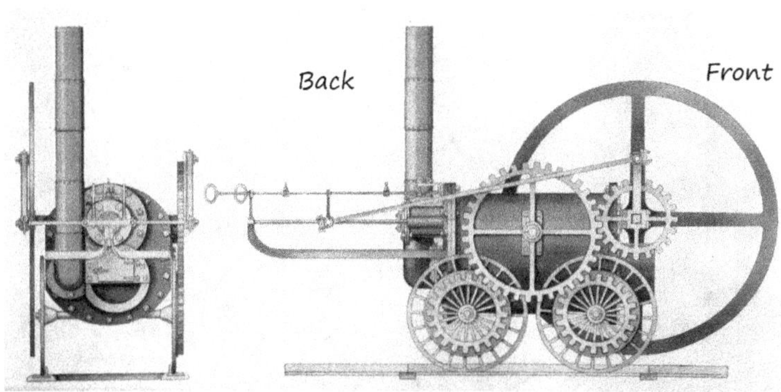

This 1803 illustration is the only surviving evidence of Trevithick"s earliest attempt to build a working railway locomotive. The financing ironworks company dropped Trevithick's railroad project after a worker was killed at the engine. Both piston and cylinder were horizontal and enclosed within the high pressure boiler. A year later, Trevithick's slightly improved 1804 locomotive however, was a resounding success.

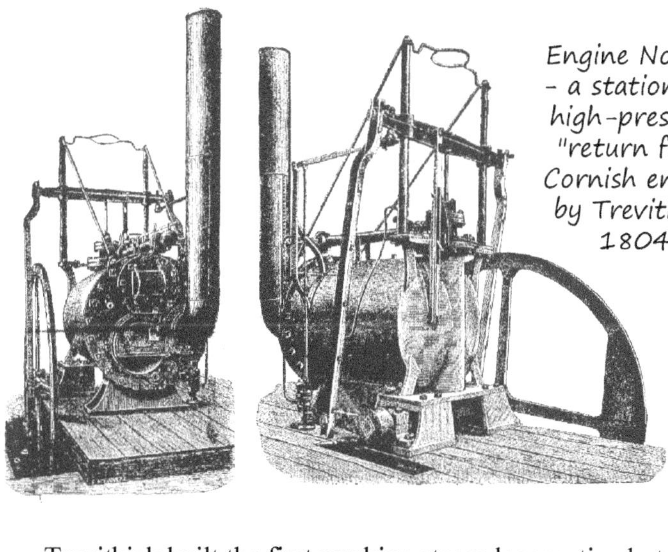

Trevithick built the first working steam locomotive but his tracks didn't go anywhere. Later, by receiving easements granted by Parliament, the Middelton Railway became the first commercial railway to use steam locomotives successfully; in 1812. But only for the purpose of ferrying coal.

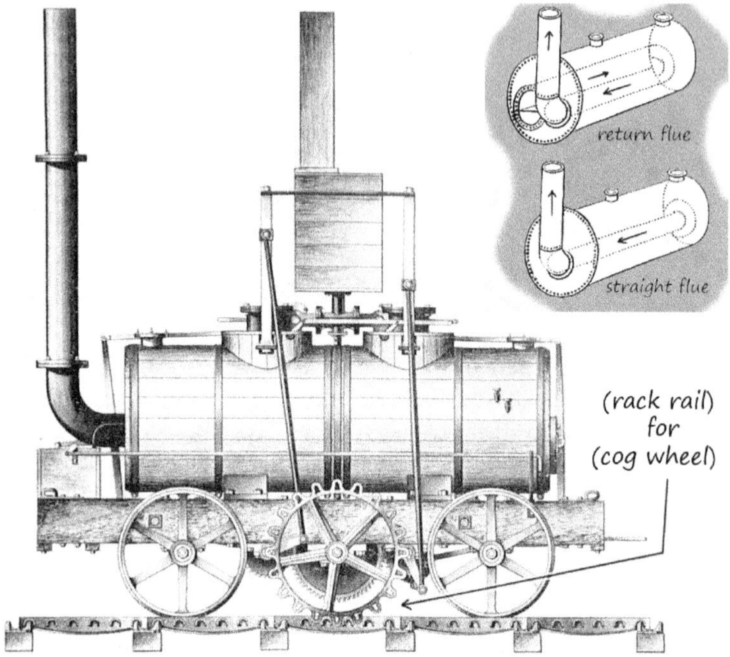

Blenkinsop's 1812 rack locomotive - with straight flue boiler

An 1814 aquatint engraved from a painting named "The Collier". Showing a new "rack locomotive" pulling cars from a coal mine. The mine had its own steam power - to drain water, provide fresh air and perhaps power a practical device called a “man ladder”, down in the shaft.

But it was George (and son Robert) **Stephenson** who built the first railways to become profitable by transporting passengers alone.

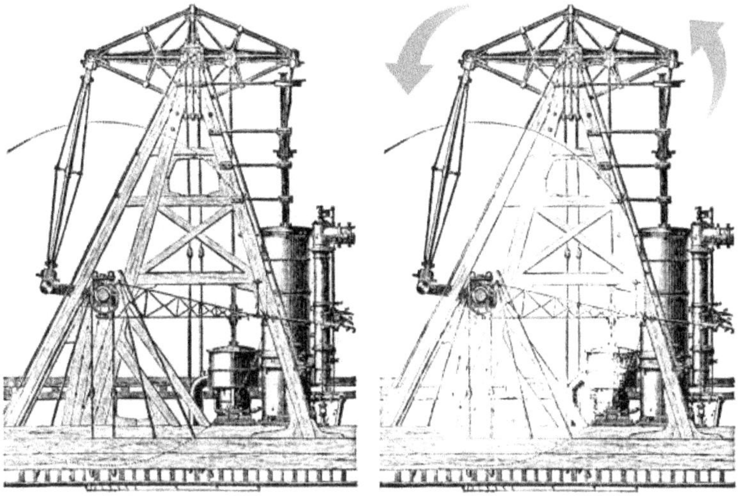

Walking beam engines were used to power many American paddle steamers in the 19th century. The paddle wheel, like flywheel, could be turned by the crank and rod (or secondary beam). From a distance and with the right perspective, the rocking motion of the tall beam was reminiscent of a giant walking skeleton – hence the name "walking beam".

The walking beam steam engine paddle-boat was popular and became a common sight on America's eastern Intracoastal Waterway. Walking beam engines were still being occasionally manufactured as late as the 1940s. On the right in the previous image is a **steeple engine**, also sometimes referred to as a "**crosshead**" engine. It was an early attempt to make engines more compact, by breaking away from the overhead beam. The tandem compound steam engine shown was built to stand vertically; it looked tall and pointed like a church steeple.

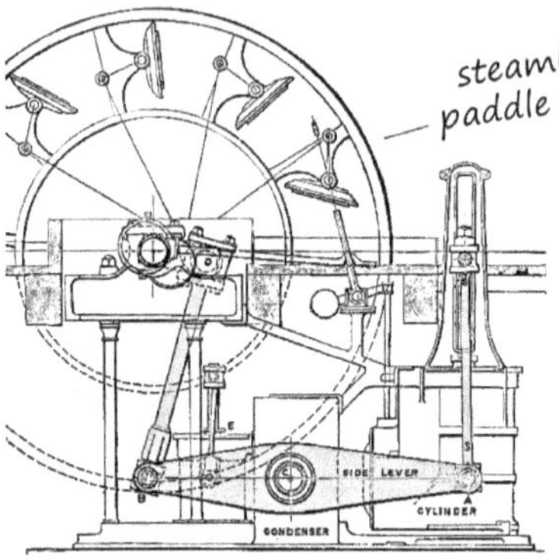

FIG. 92.—The Side-Lever Engine, 1849.

The side-lever type steam engine which was developed in the 1840's or 50's, placed the beam down low, to the side of the boiler and cylinder. Thereby making an engine more compact.

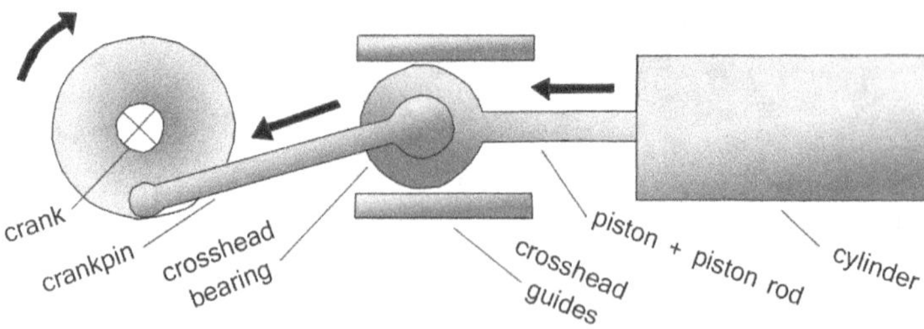

A "**crosshead engine"** implies that a crosshead bearing is being used between the piston rod and crank bearing. The design feature allows reduction of engine size and prevents the piston's rod from striking the edge of the cylinder at the end of a stroke.

This type of bearing slides along a rail or inside a groove.

Some of the next engines are "direct acting" (not the same as double acting) which means that they are not beam engines and are instead attached directly to crankshaft by piston or connecting rod. These designs became common configurations seen in the 1840s to the 1870s; usually intended to work in confined spaces.

The "vibrating-lever" type engine for example was a variation of the trunk engine; like used in the USS Monitor and other Federal ironclads during the Civil War.

- Probably named "trunk" because of its looks. A trunk was a large but stack-able piece of luggage. A trunk engine was but one type of 'vibrating lever' engine.
- There were many different configurations of steam engines; just a few examples are shown here.

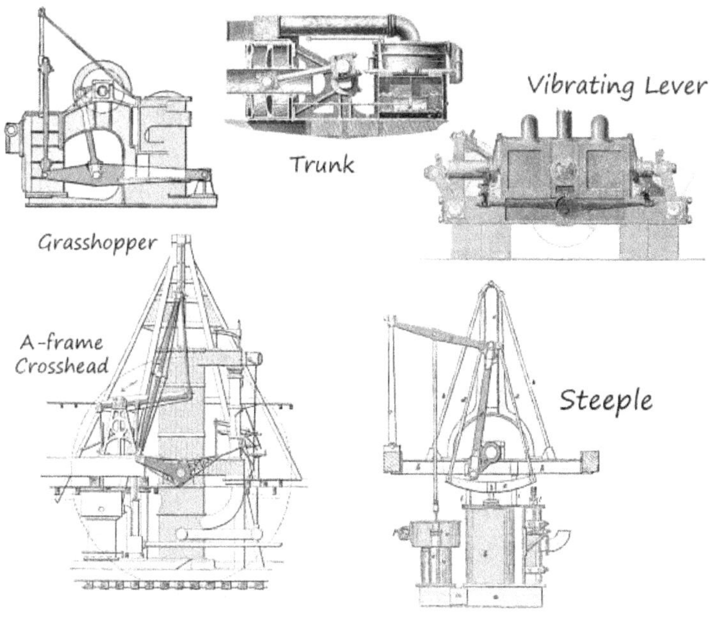

When the American Civil War ended, the US Navy had almost 700 warships. Some 65 of those were ironclads. Many of those ironclads were nicknamed - "**Monitors**" because of their resemblance to John Ericsson's original USS Monitor (1862).

The USS Catawba shown below, was completed shortly after the end of the war (1868) and was never commissioned. Instead, she was sold to the Peruvian Navy and renamed - "BAP Atahualpa". This ship used (4) fire-tube boilers and (1) Ericsson vibrating-lever steam engine to power (1) propeller.

The trunk engine, (again a type of direct-acting engine), was originally developed as a means of reducing an engine's height while retaining a long stroke. The trunk design was popular in monitors because of its low profile, which kept it below the waterline. Operation was smooth at normal pressure (20 -25psi). It was used in Sweden a bit also; but not popular elsewhere.

On her maiden voyage (in 1819) the American made SS Savannah was the first steamship to cross the Atlantic. She used mostly sail power to do that, but it would take another 30 years before any other steamship (meaning steam/sail hybrid here) did it again. For sensible reasons, steam ships would retain their masts and sails - as backup contingency, for the next century.

The engine of the Savannah is described as being single cylinder-ed, inclined and of the direct-acting type; and it supposedly produced 90 hp or 67kw. (Direct-acting in this instance merely means probably, that it was not a waking beam or side lever type engine).

SS Sirius

SS Great Western

In 1838 at least three shipping companies were in close competition to be the first to establish a transatlantic service with new, fast wooden hulled paddle-wheelers. At the time it took a normal "sailing packet" 40 days to make a westbound passage across the Atlantic ocean. (In the 18th and 19th centuries a **packet ship** or boat was a medium sized vessel that delivered mail, passenger, and freight transportation, on a regular basis). The SS Sirius pictured above hurriedly left Cork, Ireland and was the first to arrive in New York, 18 days later and without any remaining fuel. The much larger SS Great Western started from Bristol, England 4 days later - and arrived one day after the SS Sirius. The Great Western had enough fuel in reserve to go much farther. A British engineering prodigy named Isambard Kingdom Brunel designed and launched the SS Great Western in 1837, which was named after the railroad he had previously engineered and supervised. *('The Great Western Railway' was a project founded and financed by Parliament in 1833. The state-owned company was subsumed by British Rail in 1948).* The largest passenger ship in the world for a couple years, the Great Western used a 2-cylinder, side lever engine that ran at 12-15 rpm. For the next two years the Great Western would average 16 days for a westward passage to NY, and 13 days to return on the eastern return trip.

There was fierce, continual competition between companies

and engineers to build the next great steam engine; be it for ship, railway, factory or mining equipment.

Isambard helped revolutionize public transportation in general. He built bridges, tunnels, dockyards and railways. Although his first ship was made of wood, he then built the first iron hulled, purpose built transatlantic steamship to cross the ocean and later, an even bigger iron hulled steam powered ocean-liner that could transport 4,000 passengers.

1843 launch of SS Great Britian -painted by Joseph Walter

When Brunel's **SS Great Britain** was launched in 1843, she was by far the largest ship afloat, and the first to be simultaneously iron hulled and screw propelled. In 1845 she made the first iron-steamer crossing of the Atlantic, taking 14 days. She had a 2–cylinder, inclined, direct acting steam engine that operated at 18 rpm and produced 500 hp.

Deliberately, Brunel strengthened the hull with massive longitudinal iron girders, thick iron ribs, keel plates, a double bottom and several watertight bulkheads. Fortunate, because the well endowed hull was to endure some epic abuses.

In her second year of life and after two trips to New York and back, the SS Great Britain was accidentally run aground off the coast of Ireland. She spent almost a whole year (1846+) beached and neglected; until (at great cost) they managed to float the huge ship free.

The propeller on the model shows the relative position of the propeller shaft; but the engine was amidships while the propeller was far away at the stern.

Once retrieved and refitted however, the owners were bankrupted and she was sold at auction. Her new owners put her to work for the next decade, transporting thousands of emigrants to Australia. Then in 1863 she was wrecked again somehow; at Santiago, a Cape Verde Island (Portuguese). In 1874 she collided with another ship in a British dock and sustained hull damage.

In 1882 they yanked her engines and converted the Great Britain into a mere collier (a coal carrying cargo ship). In 1886 while ferrying coal, she caught fire and was run aground in Port Stanley, Falkland Islands. Damaged beyond repair, she acted as a storage hulk for the next 50 years. Then in 1937 the hull was towed away, scuttled and abandoned.

But then in 1970 (after 33years underwater) the hull was retrieved, patched, and hauled back to her home port (Bristol) where she was restored and turned into a museum ship! Very few ships can boast such an eventful life.

Isambard Brunel's next ship would be twice as big. When launched (1858) after 4 years of construction the behemoth named the **SS Great Eastern** broke all records for size and capacity. With a displacement of 32,160 tons and a respectable length of almost 700 feet, the SS Great Eastern could ferry 4,000 passengers + 418 crew and 3,000 tons of cargo - from England to Australia without refueling.

The Great Eastern had three sources of propulsion: sail, paddle and screw. An enormous 15,000 tons of coal was bunkered around and over the top of her 10 boilers, that supplied two widely separated engines. One drove the propeller, the other the paddle-wheels. The state of the art engines were unalike to each other, weighed about 90 tons each and were supplied by James Watt & Co. Each produced about twice the power (@ 3,150 hp each) of any other ship engine afloat at the time. \

The engine that propelled the 56 ft diameter paddle-wheels had inverted cylinders which possessed a bore of 74" and a stroke of 14' (1.87 m x 4.3 m). The engine for the screw had an "opposed- four configuration. Steam pressure was regulated at 25 psi (172 kPa).

{Next images from The Illustrated London News -1859}

THE PADDLE-ENGINE ROOM OF THE "GREAT EASTERN."—SEE FACING PAGE.

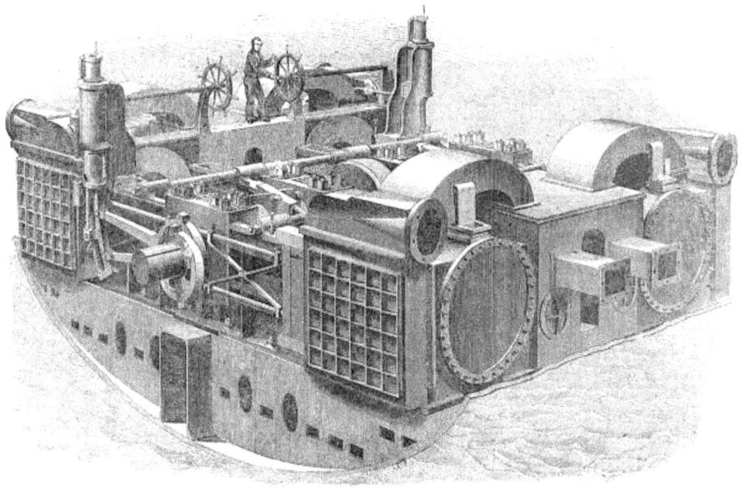

THE SCREW ENGINES OF THE "GREAT EASTERN" STEAM-SHIP, BY JAMES WATT AND CO.—[illegible]

The Great Eastern operated as a passenger liner between Britain and America, from her launching in 1858 and through the American Civil War. Not all her passages were smooth though. In 1861 she was trapped in a major hurricane for three days. She pitched and rolled violently where people, pianos and furniture were tumbled and across the floor and slammed into the walls of the Grand Salon.

She was damaged badly, her paddle-wheels and rudder torn of; but she limped on to her destination, while a lesser ship might have failed. The Great Eastern being unprofitable, was auctioned off a few times. She was not a particularly lucky ship and experienced many setbacks during her existence. Eventually she was put to work laying the underwater Transatlantic telegraph cable. When the Suez Canal was opened in 1869, her 30 ft draught was too deep to navigate the shallow channel.

Side view of the great ship, circles showing the locations of the two engines.

She sat moored and unused for 11 years at Milford (a waterway and natural port in Wales). Later she was used as an advertisement barge, showboat, a floating palace/concert hall and gymnasium for a few years. Sold as scrap in 1888, the Great Eastern's story was not quite done. Isambard's ship being double hulled, was so well built that it took a year and a half to break it apart. Salvage crews and laborers were normally paid by the ton, for scrap metal. When they discovered that they weren't making any money because of the stubbornness of its construction, they became frustrated and actually went on strike.

Below: a French-built oscillating engine, dated in the 1860's,

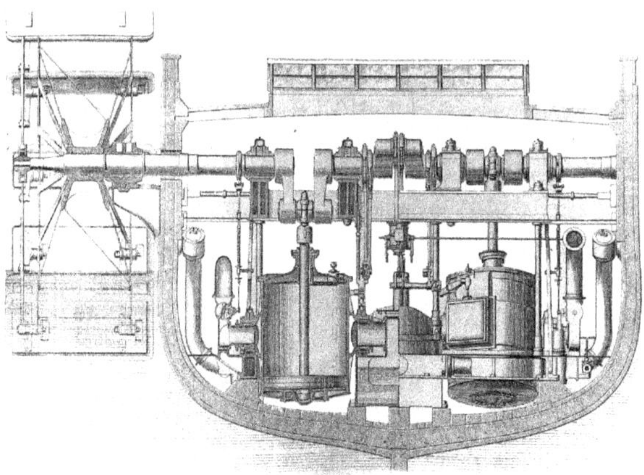

2.25 *French-built oscillating engine; constructed in early 1860s by Mazeline & Co. for the yacht* L'Aigle

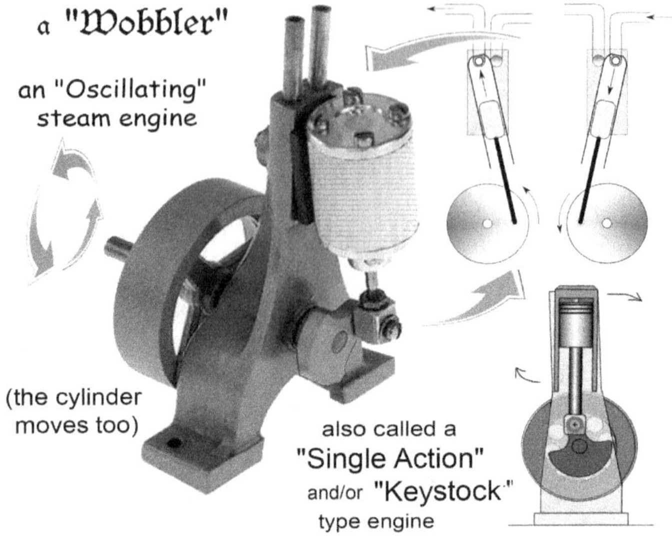

The working toys above and below, are both "oscillating steam engines". As well as pistons moving, the cylinders themselves in this type of engine move or wobble back and forth. ("Keystock" is simply a name for round or rectangular, uniform metal fabrication stock, used by machinists).

It was the vertical, reciprocating (piston & crank), triple expansion - compound engines that soon became favored for big ships.

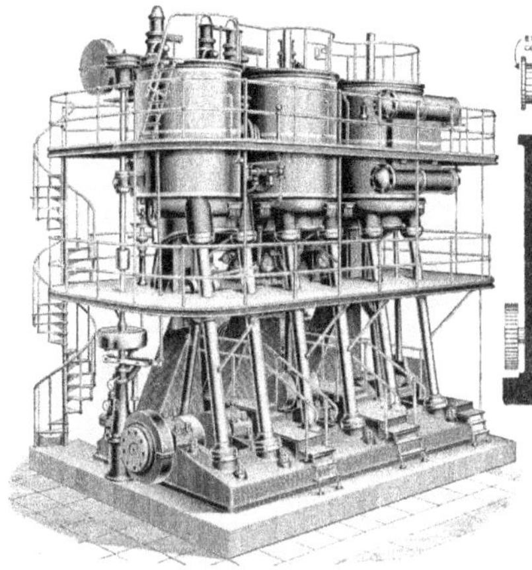

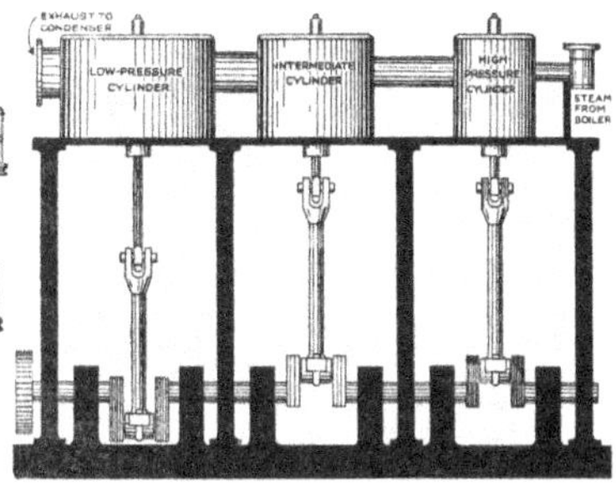

Standing Tripple Expansion
- Compound steam engine

Huge three and four cylinder reciprocating (piston & cylinder), direct acting engines were built for the day's most important ships. “Tipple expansion” means that the steam is used three times or in three stages. The highest pressure is released into the smallest cylinder first. The exhaust from that was then passed to the next biggest cylinder, and so on. Quadruple cylinder-ed versions were common too.

Cutaway of triple expansion compound steam engine, from an 1888 Austrian torpedo boat.

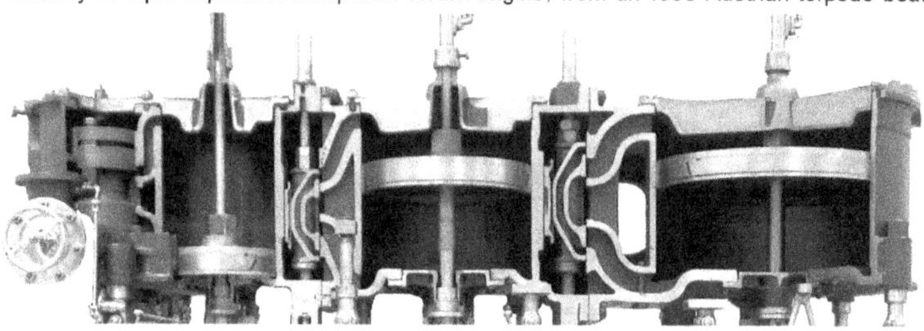

{A museum placard directs that this “cutaway” from an engine that was built by Eibinger Schichau-Werft in 1888, produced a power of 660kW, and a ship's speed of 17 - 22 kt @ 280 - 370 rpm}. That's about 884 hp for a 'small' marine engine; roughly 12' high and 12' long; necessary for a slender 19th century torpedo boat. A high-pressure boiler had to fit inside the canoe too.

Note: Horsepower > Watts

1hp = 746 W **1hp = 0.746 kW**

The largest number ever of any one ship design to be constructed, was the Liberty Ship of WWII. Between 1941 and 1945, eighteen American shipyards completed an average of three ships every two days. Totaling 2,710 nearly identical ships that used the same engine, pictured above. The engine was a fairly obsolete triple-expansion design that was about 19 feet tall and 21 feet long. It was configured to rotate at 76 rpm and push the cargo ships at about 11 knots (13mph / 20 km/h). This engine was chosen because it was rugged, well understood, easily maintained and economical (figuratively speaking).

The previous page's 2011 Flicker image, is of the world's largest operating triple expansion steam engine remaining. Actually, the Kempton Steam Museum in London has two. Each engine (1,000 tons heavy) was capable of pumping 19 million gallons of water daily and ran 24 hrs a day, seven days a week to supply North London with drinking water. The pair began working in 1929 and were retired in 1980. The restored one (shown) is started occasionally.

A pair of similar sized engines this big, went down with the RMS Titanic in 1912. Just like it's equally large sisters (RMS Olympic and HMHS Britannic), Titanic's wing propellers were driven by two huge reciprocating engines and a third, central propeller was driven by a low pressure steam turbine.

* From experience, an ocean liner with all turbine propulsion was fast, but fuel costly and annoying. Because turbine caused vibrations carried throughout the ship. Combining reciprocating engines with a turbine though was not only more comfortable, but more economically practical too. (Turbines are fuel hungry, or at least they can demand a lot of steam. When it comes to the economy of a steam engine system though, the limiting factor is often in the boiler(s) themselves. Trippe expansion engines are more economical at lower speeds and RPMs. Turbines only become economically practical, at higher speeds).

** The Olympic-class ocean liners and some of their competition were simply too large to dock in certain harbors So they anchored offshore and were serviced by 'smaller "tending-ships"'. A tender like the SS Nomadic could still carry 1,000 passengers with luggage and other supplies, back and forth from ships to constrictive harbors like (Cherbourg Harbour, Normandy, France).*

** The Titanic which was 852' long and had a displacement of 52,310 tonnes, possessed 24 massive double-ended and 5 single ended Scotch boilers. The ship carried **7,700 tonnes** of coal. (That weighs as much as an entire, modern Naval Frigate).*

- *Feeding 53 separate boiler fires (24 x 2)+5) was a nightmare. It took the combined effort of 176 firemen shoveling in shifts around the clock, to keep the ocean-liner moving at pace. About 600 tonnes per day were required.*

Boilers

Isambard Kingdom Brunel and two other engineers at the International Exhibition London 1862, standing in front of three typical Lancashire boilers. This type of boiler predated the shorter and more popular "Scotch" boiler. Aside from the length, another difference is in a number of small fire-tubes that the Scotch boiler has but the Lancashire lacks. (Numerous tubes ensured more surface area and better heat exchange between hot gasses and water).

This double ended Scotch boiler and 23 others just like it went into HMHS Britannic (*His Majesty's Hospital Ship / sister ship to RMS Titanic).*

*(**RMS** = Royal Mail Ship; a prestigious designation).*

Britannic struck a naval mine and sank in the Aegean Sea (1916). This image of a boiler in the air, shows three furnaces on this end. Three more furnace openings are on the other end. Coal fired boilers and Scotch Marine boilers like this come in a few slightly different configurations and are still built today for industrial customers that need them.

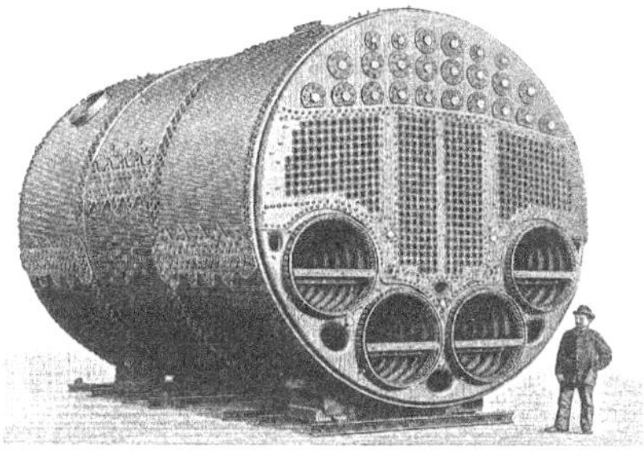

A “dry-back” Scotch boiler has a heat resistant brick refractory lining at its end that reflects the heat back, and the “wet-back” version has a water reservoir at its back end rather than refractory. A “double ended boiler” is basically just two single ended boilers back to back, which share the same exhaust.

Watercolor of the North Atlantic Squadron (by Fred S. Cozzens - 1892)

Sails were not shown in many images of steamships at the end of the 19th century. Such ships were usually very proud of their new self propulsion. Seldom scene are images of these hybrid ships being under sail. But these coal burning ships were inefficient and had 'short legs' so to speak. So out on the high seas, whenever the winds were favorable, they used their sails to conserve fuel.

The Spanish American War (April - December, 1898) saw the U.S. acquisitions of Spanish holdings in Puerto Rico, Guam, the Philippines, and Cuba. Cuba would remain a protectorate of the U.S. until it gained its independence in 1902. France began construction on the **Panama Canal** in 1881 but quit; the U.S. stepped in, only to complete the project by 1914 (WWI). Taking about 33 miserable years to complete, the canal project cost some 25,000 lives. Most of the deaths were attributable to **malaria** or **yellow fever**; two swamp diseases carried by two different species of mosquito.

The Great White Fleet 'en route'... (1907)?

Before the canal's completion its loudest protagonist, the 25th president Theodore Roosevelt sent a fleet of warships on a world tour to project American prestige abroad. From 1907, it took the **Great White Fleet** as it was called, a year and two months to complete the voyage. The fleet consisted of 16 battleships, various small escorts and 14,500 officers and sailors. The battleship's superstructures were painted a yellowish, ochre buff, but the hulls were painted white – hence the name. The ships in this fleet were already obsolete when compared to the new (1906) HMS Dreadnought, but put together the 16 battleships made an impression everywhere they went, and comprised the most powerful fleet to have ever circled the globe.

An underappreciated consequence of the voyage was that the American admiralty learned the hard way, what the British admiralty already knew. In order to navigate the blue waters of the world's oceans with a coal burning steamship, you need lots and lots of coal. The U.S. Navy was incapable of resupplying its own ships with adequate fuel on the high seas. They had to rely heavily upon colliers of other nations (including Britain, a naval rival) to supply the fuel. At times they had to make long, out of the way detours to dock and purchase fuel. In retrospect this debacle undoubtedly impressed upon the American admiralty – the strategic need for better logistics.

During this period the arms race between European countries was fierce, and continual improvements were being made at a breakneck pace. Compared to the needs of other military services, Naval ships combined the most numerous technologies and required the highest expenses. This was the Industrial Revolution after all. But don't forget Japan. In 1905 a large Russian fleet sailed 18,000 miles, from the Baltic Sea to reach Vladivostok. Only to be humiliated and almost wiped out by the Japanese Navy at the decisive Battle of Tsushima. At the time the Japanese fleet was composed of ships designed and made primarily in Britain, Italy and the U.S. Japan might have lacked the industrial facilities needed to build its own large warships, but that would soon change. The Japanese were quick to learn.

A cutaway of a Babcock & Wilcox boiler (1913). These were the favored coal burning, water-tube boilers at the turn of the 20th century; both for capital naval ships and for ocean liners, which were looking for maximum steam and the fastest speeds. In the picture, almost all the boiler's encapsulation has been removed – so that we can see inside. Only some of the refractory firebricks are visible at its base. Hot water at high pressure filled all the tubes, and half the "steam drum" on top.

The two battleships about to be mentioned, illustrate the important transition from coal burning- to oil burning technology. The revolutionary and iconic **HMS Dreadnought** (launched 1906) had 18 individual (UK made) Babcock & Wilcox boilers. A later and similar battleship, the **USS Texas** (BB35) was built with 14 (U.S. made) Babcock & Wilcox boilers. The original American company (est. 1867) built a branch company in the UK (1891). Their idea for a safer boiler, dispersed water into many separate tubes.

Today the USS Texas is a museum ship that was launched over a century ago (1912). The Texas is the last remaining battleship from the brief Dreadnought era. The two ships (Texas & Dreadnought) shared many similarities; one being that they originally used similar boilers. The Texas was also the last American battleship to be built with coal burning boilers. But in 1925 those B&W coal burners were swapped out.

Due to the **1922 Washington Naval Treaty** (an agreement between nations to limit the size of their navies), the Texas was upgraded with 6 new "**express**" oil fired, water tube boilers from a newer upcoming American battleship that was canceled. These were Yarrow-styled A-frame 3-drum boilers of the period; a variation designed and named after Rear Admiral Charles W. Dyson, U.S.N. *(Express in this instance meant that the ship could accelerate quicker, due specifically to the boiler design)*.

An "A-type", 3-drum, small water tube "express" boiler.

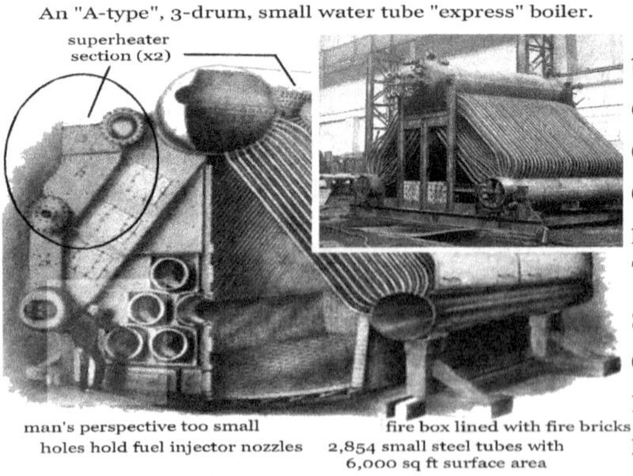

The Texas was to retain her two original triple-expansion steam engines. These required lubrication. The steam in these ships was condensed back into fresh water and reused. Seawater was undesirable. Injecting oil for lubrication and re-circulating that through the boiler tubes was out too, because that fowled the works. The solution was to lubricate with the fresh water droplets (no oil) in regular wet steam. So on the USS Texas, the new boiler's superheaters were removed. **(Only turbine engines thrive on dry saturated steam).**

Photo # NH 60219 Coaling USS Georgia, circa 1919

There are a couple of good reasons why fuel oil should be preferred over coal, as a ship's fuel. First, fuel oil (a low fraction distilled from crude oil) is a denser form of energy than coal. Coal has more carbon and produces only ½ to ¾ the heat of fuel oil when burned. A second and more compelling reason to use fuel oil is in the storage and supply logistics. It is relatively easy to transfer oil between ships using pumps and pipes. On the other hand it was a nightmare to refuel a warship with coal. A typical capital ship of the WWI era might have had a coal carrying capacity of 3,000 tons and an appetite that could use that up in a short time. The ship's compliment (crew) for a large ship might have been 2,000 sailors and 120 officers. Woe to the sailors on "coaling day".

> Aside from being the first battleship to install **turbine engines**, the Dreadnought carried some 1,140 tons of fuel oil (in addition to twice that weight of coal) which was to be used to spray over the coals; to increase the burn-rate in her boilers in case she needed to sprint. Her top speed of 21 knots, was faster than any other battleship in that day.

If a random capital ship needed to move at maximum speed, it might have burned through 1,000 pounds (½ ton) of coal per minute. The switch to oil-fired boilers changed maritime design dramatically, and also contributed to the development of bigger warships.

The next three images are of the same ship's hull. The USS Jupiter (AC-3) was one of the few purpose built colliers in the Navy. But she was converted some 10 years after her launch, into the 1st American aircraft carrier - USS Langley (CV-1) in 1920-22. Possibly because the Jupiter also happened to be the US Navy's first ship to use turbo-electric-powered propulsion.

Photo # NH 81339 USS Jupiter fitting out at Mare Island Navy Yard, Dec. 1912

Photo No. USN-185915 USS Langley (CV-1) in 1930

Returning to the cutaway boiler image a few pages back and to come, fuel oil could be swirled around, mixed with air and atomized before being injected into the hot firebox. This made intense flash heat that could be quickly transferred to the water flowing inside thousands of small diameter steel tubes (representing a tremendous surface area). The front of a boiler might have had a dozen apertures for fuel injector nozzles. For normal speeds, only a few of the injectors were ignited though.

Above are some Yarrow type boilers with their water tubes exposed and below are the backside of a cluster of similar boilers inside their shells or housings and sharing an exhaust.

The RMS Queen Mary carried 24 of these Yarrow, oil fired boilers to power 4 Parsons single-reduction geared steam turbines. With adequate steam those engines could make 200,000 shp (150,000 kW). Built as an ocean liner for Cunard

Line she carried passengers across the North Atlantic Ocean from 1936 to 1967. During WWII she was annexed to ferry troops across the ocean. At a huge 81,237 gross tons, she was still so fast (over 30 knots) that she made those crossings alone and without escorts to protect against enemy submarines. Speed was her best defense, and no escort ship could have matched her pace anyway.

Naval Petroleum Reserve

Following the lessons learned by the Great White Fleet's world tour in 1907 -08, and combined with new discoveries of immense oil fields in the American West, the American Navy decided to convert from coal to oil fueled ships. On July 2, 1910, President William Howard Taft (authorized by the 1910 "Picket Act") set in motion the protection of three Naval Petroleum Reserves. (Two of these were located in California, the third in Wyoming). Later, President Warren G. Harding created Naval Petroleum Reserve Number 4; better known today as: National Petroleum Reserve in Alaska (NPRA).

Not for automobiles, the nation's oil reserves were meant for ships; to insulate the United States from foreign dependency on oil during times of war. In truth, from 1900 onward, oil began to replace coal as the preferred fuel source in railroad locomotives too. These Reserves remained mostly undeveloped until 1976, when Congress passed the Naval Petroleum Reserves Production Act, authorizing full commercial development of the Reserves. *"The crude oil, natural gas, and liquid products produced from the Reserves were sold by DOE at market rates. Revenues were deposited to the U.S. Treasury".* [24] Since that time those first three federal oilfields have been largely privatized and depleted.

The NPRA (Alaska) is largely untouched, but recent studies suggest that original estimates of the reserve's capacity were grossly overestimated 20 years ago.

24 https://www.energy.gov/ceser/naval-petroleum-reserves

This section is suppose to be about steam producing boilers, but some of these big ship pictures are getting in the way. As mentioned the Queen Mary (both images above) used 24 steam boilers. But there is little information on how much strain, how much pressure those boilers worked under. So the ship's designers would have dictated a comfortable compromise between speed (a function of pressure) and safety.

Before the introduction of nuclear aircraft carriers -which can be very fast, some of the speediest big naval ships (ignoring cruisers) in the oceans were America's Iowa class battleships. These were uniquely designed capital ships built with an emphasis on speed. (The only other contemporary WW2 era battleships remotely as fast – were the two uncompleted French battleships: Richelieu and Jean Bart).

Each of the 4 ships in the Iowa's class were about 53,900 tons heavy, could produce about 212,000 HP and could move at a rate of 32.5 knots (37.4 mph). At average speeds their 8,841 long tons of stored fuel oil gave them an approximate 18,300 mi (29,400 km) range too.

The Iowas had 4 engine rooms and 4 boiler rooms balanced throughout the lower ship decks. Each boiler room held two - Babcock & Wilcox M-type water-tube boilers producing steam at **600 psi @ 850 °F.** Boilers like these would still be considered 'state of the art fuel oil boilers' even today, some 80 years later; because oil boiler development changed following the war. U.S. Naval engineers turned to nuclear energy to make steam. All commissioned U.S. Navy submarines and supercarriers built since 1975 are nuclear powered. (Some Virginia class guided-missile cruisers were nuclear powered. But in the face of budget cuts these were deemed to be too expensive to operate, and were all decommissioned and recycled by 1998).

Creative Commons Attribution- to author- Stasqd

Both fire-tube boilers and water-tube boilers are being made today. Again, within a **fire-tube boiler** - the pipes hold the moving hot gasses and water surrounds the outside of the pipes. The situation is exactly reversed in a **water-tube** design. Fire-tube "**package boilers**"are used in most large manufacturing industries. And these low-pressure boilers are commonly used to heat large buildings.

The Refining industry, the Paper & Pulp industry, the Petrochemical industry, Textile industry, Food Processing industry and the Pharmaceutical industry; all can be found to be operating low-pressure, fire tube boilers like these shown.

Industries might employ water tube boilers too, however these use higher pressures that are more often applied to working conditions that require pure horsepower. Water-tube boilers are predominately used in power generation plants, in ships or other instances that use steam turbines.

Water tube designs are more efficient than fire-tube designs, and can generate more steam for the fuel used. Oftentimes they dissipate heat better, prevent overheating and reduce the risk of explosions. But these high pressure boilers are not free from having their own set of complications.

An average fire-tube boiler's maximum operating pressure might be about **25 bar** (362 psi), while a water-tube counterpart might work at **200 bar** (2,900 psi).

There have been many, many patents and plans for steam boilers over the years. The individual differences might not interest a reader, but note that even within the set of high pressure water-tube boilers, there were once prominent and recognizable manufacturer names (like Babcock and Wilcox, Benson, Lamont, Loeffler, Stirling and Yarrow). The 3-drum Yarrow design and its copycats, was singled out mainly because it was a fast, efficient and compact unit, with high water capacity and high evaporative power. For these reasons the Yarrow styled boilers were frequently found installed in ships; but seldom ever in land-based, stationary installations.

The Benson boiler (invented in 1922, but won't be explained) is a supercritical boiler that forms no bubbles when it directly converts water to steam. A Benson boiler regularly operates above critical pressures, typically ranging from 225 bar to 350 bar.

Many modern ships are turbine powered – and therefore still use water-tube boilers. But some of the biggest ships under construction have turned to giant low-speed two-stroke diesel (Heavy Fuel Oil) engines. Some other colossal ships are the 'floaty hotels' or "balcony-laden floating condominiums" otherwise known as **cruise ships**. Generally these are **not** seaworthy vessels and they usually avoid deep blue seawater. **Ocean liners** are not the same thing. The real ocean liners (like the Queen Mary 2) can safely cross an ocean, but are rare these days. The Queen Mary 2 uses diesel generators & and supplemental gas turbines to create what is called **integrated electric propulsion.** Three-phase electricity powers waterjet like impellers. The QM 2 uses some bow thrusters and 4 shrouded "propulsors", but has no rudder. The front two units are fixed but the back two podded "azimuth thrusters" can rotate 360 °s.

Water-tube boilers or hybrid (both fire & water tubed) boilers found their way into some automobiles, British trucks and Hungarian locomotives. US railroads tinkered with water-tube locomotive boilers, but quit. They decided that any efficiency gained, wasn't worth the extra maintenance. The next image shows a typical locomotive's **fire-tube boiler**.

https://en.wikipedia.org/wiki/Fire-tube_boiler#/media/File:Aufgeschnittener_Kessel.jpg
Thanks to: Rabensteiner

The Turbine

Above the caption says: *"The Great Fleet assembled at Spithead, for the King's Review, July 18th, 1914"*. This aerial image taken from a blimp, was photographed only three weeks before Britain entered WWI. Shown are some 400 English ships in the **Solent** (a strait between the Isle of Wight and mainland Great Britain) which happened to be in home waters at the time. The "capital ships" lined up a little right of center, the lesser cruisers to the left of center and so on. This image of the 1914 review is shown to lend perspective to an earlier "Review".

(Another vantage point of the 1914 fleet Review).

Fleet reviews were large but only occasional events. It was the Navy Review for the Diamond Jubilee of Queen Victoria at Spithead, on 26 June 1897 that is of special interest. Before 1897, the state of the art steam engine was the triple-expansion, reciprocating steam engine. That was the most powerful and popular steam engine technology that could be put in a ship at the time. Everything changed one day, when C.A, Parson's steam turbine-powered "**Turbinia**", crashed and disrupted the Admiralty's party (part of the Queen Victoria's Diamond Jubilee). (Few photos exist).

A painting of the 1897 Navy Review during the festivities showing the riotous boat in the center. (Shadowing Roman tradition, a Diamond Jubilee is a very rare celebration for a monarch, that survives long enough to sit upon a throne for 60 years). There were 165 naval ships, including 21 battleships at this event. At the center of the painting, the artist (Eduardo De Martino) depicts the revolutionary new steam turbine powered "Turbinia" streaking across the front of the flotilla like a naked person. As the uninvited Turbinia ducked and dodged between the big ships, a dozen or so slower naval picket boats chased her intently. The turbine powered boat was far too fast to catch; and could literally run circles around any other type of vessel there.

C.A. Parson's publicity stunt was audacious, and it won him the attention he sought from the Lords of the Admiralty and foreign dignitaries who were forced to watch. Soon he was selling steam turbines to the navy and commercial ship builders alike. Initially his propellers were insufficient to transform the power of his new engine into the propulsion he expected. Parson and an associate named Barnaby discovered in 1893, that the formation of vapor bubbles (cavitation) on the propeller blades was responsible for an initial sea-trial failure (unimpressive speed).

Parson set up a water tunnel (another 1st) in a laboratory and photographed propeller cavitation. His studies influenced changes in propeller design, their number and placement.

The Turbinia used Parson's 3-stage, axial flow, direct acting steam turbine. The successful version of the boat had 3 long propeller shafts with 3 propellers on each. Pressure was regulated somewhere between 170 and 200 psi. Speed was 34.5 knots (40mph / 64 km/h).

It's not much of a picture, but it is one of the few images publicly available that reveal what an early Parson turbine looked like.

A couple of more modern steam turbines below. The one on the left is a low pressure turbine, probably being constructed for the 1936 version of RMS Queen Mary.

Below is a picture dated Nov, 1970, found on Wikipedia and credited to a user named Bobtail. The blades are being installed and the turbine is being fabricated in an Allis-Chalmers factory.

Centered in an area between Wisconsin and Chicago, the Allis-Chalmers Company was formed in 1901 by an

amalgamation of four smaller companies. The company then was very likely the largest maker of steam engines in the world. The company grew huge and by the 1950s it had over 30,000 employees. Allis-Chalmers was a "major competitor in farm implements, construction machinery, and engines of every type". They built orange tractors and some of the turbines and generators used in the Hoover Dam. They helped build one of the world's first particle beam accelerators in the 1940s. They built many big steam turbines, as shown in the photograph. One this size, by itself, would be capable of serving several million people with electrical power. By the 1980s it was realized that the company was a little too diversified for its own good, and possessed no domination of any one market. By 1999 Allis-Chalmers had sold off its major business components, and closed completely.

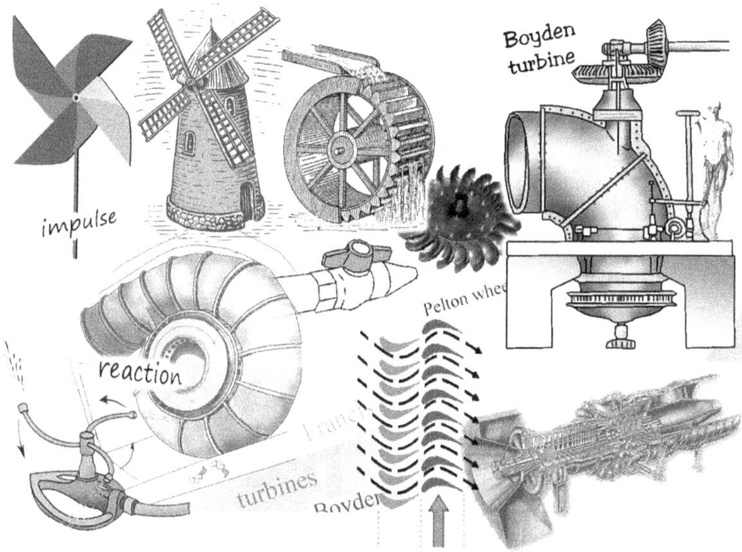

Two centuries ago the French were the first to use the word "turbine", which is taken from the Latin "turbo"- meaning that which spins or whirls around. A turbine is a wheel, turned by the force of moving liquid (water, steam, gas, wind).

Steam turbines are some of man's most powerful machines; ever. More than a century later, engineers today are still working on improvements to Parson's first example. The concept of any turbine is very simple, and examples have been with us for eons, in the form of water wheels and windmills. In essence there are just two types of turbine. All turbines work either by reaction or impact (or both combined). Turbines themselves to not create power, but change the force of motive fluids into a circular or rotary motion.

An example of an impact turbine would be a kid blowing upon a pinwheel toy, or Zephyr *(Gk. god of the West Wind)* blowing on a windmill. The impulse of air striking the blades makes the wheel spin. But a reaction turbine like a Francis turbine or a jet engine, generate torque by reacting to the fluid's mass or pressure. The differences between impulse and reaction forces in a turbine are often irrelevant.

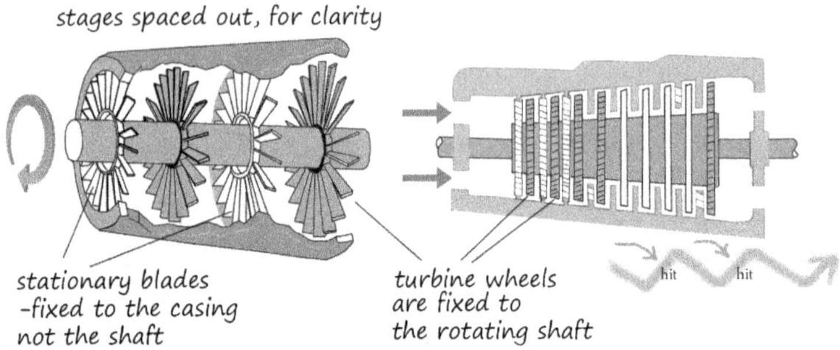

The configuration on the right side of the image above, resembles an average impact steam turbine, which would have about nine stages. What is not shown is a condenser, which swiftly cools the steam down back into water – creating in the meantime, a vacuum which helps pull more steam through an engine. Steam expands as much as 1,000 times its original volume during its short trip through a turbine. That's why each stage (set of stationary nozzles or vanes and moving blades or buckets) grows progressively larger in diameter. Older triple-expansion engines used successfully larger pistons in the same manner. To capture the last bit of power from expanding steam.

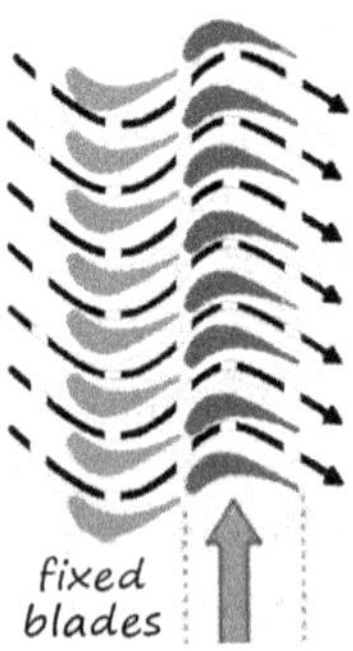

Nozzles and or sets of stationary blades fastened to the turbine casing, aim steam at the turbine wheels. The steam is guided so that it strikes the blades of the wheels at the correct angle. Steam turbines must be constructed especially well, because they are expected to run almost red-hot, nonstop, for months.

To say that the principle of a turbine "is simple and easy to understand" is one thing; to build one that works – is far from simple. The physics and mathematics involved when designing a modern gas turbine defies imagination. But then gas turbines require very specialized materials too. A common engine from an airliner, like the Boeing 747 for example, can have over 40,000 individual parts. To keep from being melted, many of those parts are made of rare alloys. Research for better and better alloys is a continual quest.

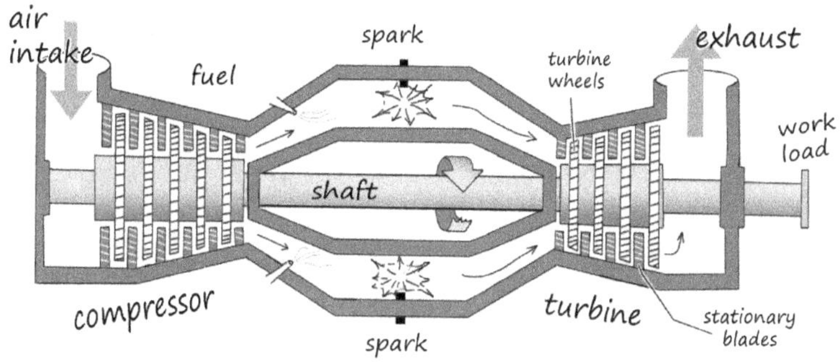

A average jet engine running near maximum thrust levels experiences approximately 4,100° F ...(2,300° C) in the hot section (combustion chamber). The melting point of iron is only 2,800° F, by comparison.

What engineers and metallurgist continue to do is to develop better nickel based superalloys (incorporating chromium, cobalt, and rhenium probably) to make their newest turbine blades. The story doesn't stop there because they continue to develop special thermal coatings, might cast parts using vacuum induction melting or use lasers to drill tiny holes in places to emit cooling air or thermal barriers. Even new directional solidification (DS) and single crystal (SC) technologies have been created to battle against fatigue and creep (by aligning grain boundaries in one direction).

Blade fatigue is a stress caused by vibration and resonance. Some other buzzwords in the industry are "fracture, yielding and creep". **Creep** is *"the tendency of a solid material to undergo slow deformation, while subject to persistent mechanical stresses".* The **yield point** is when the material in a turbine blade begins to deform elastically.

The turbofan (or fanjet) was invented to improve the fuel consumption of the turbojet. It incorporates both a turbojet (gas turbine) and a ducted fan (like a propeller).

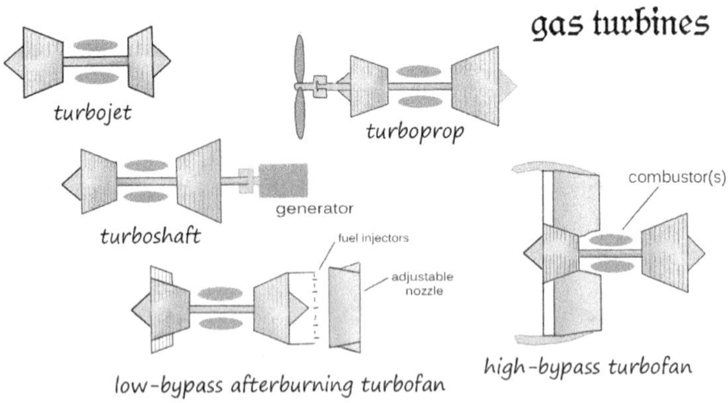

A low-bypass turbofan is the type of engine most likely to be found in a military fighter aircraft. Not all of the fresh air taken in is compressed and combusted; some passes by the engine (but still within the engine cowling) until it reaches the tail end of the engine. At that point (if afterburner power is called for) this air can be mixed with additional fuel and the regular exhaust gasses - to create additional thrust.

Turbo-shaft type engines like the big ones used in electrical power plants, come in at least two separate forms (conventional or "dry low emission" types). Conventional combustion or 'diffusion flame combustion' turbines operate at very high temperatures and can result in high thermal NOx formation (air pollutants - nitric oxide and nitrogen dioxide). These particular exhaust emissions can be reduced by injecting water or steam into the fray, but then a new set of problems emerge. DLE (Dry Low Emissions) or DLN (Dry Low Nox) turbines on the other hand are so named because they do not rely upon injected diluents to reduce emissions. Instead fuels are pre-mixed and combusted in a lean mixture (with an excess of air) to create lower NOx.

Finally (before returning to the main subject – steam) a wide spectrum of fuels can be used in gas turbines, provided a little adaptation. Almost all hydrogen containing petroleum fuels can be used – but diesel and kerosene* might be the most easily adapted fractions. LNG, LP, coal bed methane, Biogas fuel, Land Fill Gas, syngas**, crude oil, hydrogen and ammonia can all easily be used to power gas turbines.

*__Kerosene__ is the most commonly used jet fuel today. Kerosene was very valuable in previous times – as lamp oil. Many stationary engines and early farm tractors ran on kerosene however. For a long time another petroleum fraction '**gasoline**' was ignored as a byproduct with little or no use. But around 1900, the invention of spray carburetor accelerated the development of the gasoline engine.

** ***Syngas*** (synthetic gas) *"is a mixture of hydrogen and carbon monoxide, often used for producing ammonia or methanol. It can be made from natural gas, liquid hydrocarbons, coal or biomass"...* Solid coal can be pulverized and atomized and then burned in a gas turbine. Videos on the Internet have even shown pickup trucks and lawnmowers running from the smoke of a wood fire (gasification or pyrolysis of plant material).

<Returning to Parson's early turbines>

` From 1897 till 1908, Charles Parsons sold steam turbine engines to the British Admiralty and prominent British Shipping Lines alike.

But these were all **direct drive** applications to ships that wanted speed. *"Direct-drive turbine ships used more coal and water than steamships with reciprocating engines, and they were suitable only for services in which they could sustain speeds of 16 knots (30 km/h) or more"*.

One of Parsons' problems was, that he needed a way for turbines to <u>propel slower ships efficiently</u>. His solution was to create **reduction gearing**, as seen in the nearby image. Pictured above is the 1st such gearing installed experimentally (along with a turbine) in an old, steel hulled cargo ship. The reduction to the propeller shaft was 20:1. Rebuilt and re-launched by Parson's company in 1908, the 'SS Vespasian' worked as a collier for a few years.

Before 1908 then, turbine driven ships like the HMS Dreadnought (1906) and RMS Lusitania (1906) had propellers that <u>ran at turbine speed</u>.

Chronologically ordered

- Random steam machines that are neither ships or locomotives.

1771

Cugnot's steam wagon
(1771 'ish, Paris)

Nicolas-Joseph Cugnot (1725 – 1804) was a French Army Engineer, who built the first vehicle that could move about under its own power. It is rumored that in 1771 he apparently lost control of his second **steam wagon** (which is preserved in the Musée des Arts et Métiers, Paris) knocked over someone's brick fence and was soon arrested for dangerous driving.

1801

Richard Trevithick's creations

1801 - Puffing Devil

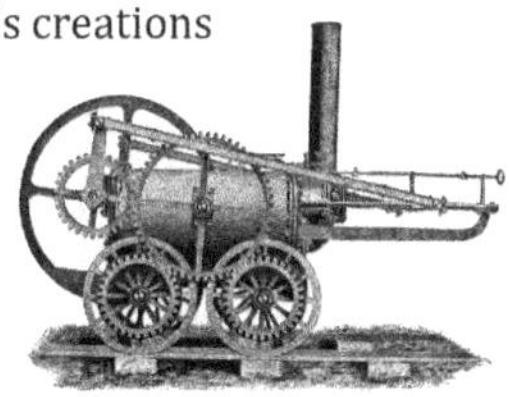

1802 - rail locomotive

1803 - London Steam Carriage

Richard Trevithick (1771-1833) was a gifted Cornish engineer and early steam pioneer. His development of higher steam working pressures allowed smaller engines to preform practical work. Some of his machines exploded though, which enabled his competitors (like the Boulton & Watt Co.) to criticize his work. His response was to fit his engines with at least two safety valves, one of which the operator could not reach.

1829

Between 1824 and 1836 an enterprising character named Walter Hancock (1799 – 1852) created at least 10 successful **steam carriage passenger transports**. He also invented a safer steam boiler that would break in half, before in could explode. His first "omnibus" as they were called, was a 10- seater which he named the 'Infant'. His later 'Enterprise' is shown above, the 'Automaton' on the following page. The 14 -seat Enterprise was used actively by the *London Paddington Steam Carriage Company* for two years (1832 – 1834).

"Bus" is a shortened form of the original word "**omnibus**". In Latin, omnibus simply means "for all." The French were the first to use the word as a noun (around 1820's), when describing long horse drawn carriages with many seats, that transported people about Paris. When used as an adjective, omnibus can relate to many things taken at once: as in a legislative 'omnibus bill' that contains many items.

Hancock's 1836 'Automaton' had 22 seats and required three people to operate it (a driver, a boiler-man and a fireman /brakeman). In the year 1836 the Automaton carried over 12,000 passengers during some 700 trips between destinations. It had a top speed of over 20 mph, but regularly traveled about 12 to 15 miles per hour. Heavy road toll fees (many roads in the 1830's Britain were turnpikes) and competition from new railways caused a decline in the steam omnibus business by the 1840's.

1831

The previous image is a cartoon; a characterization of mayhem in a possible future that the British cartoonist imagined in 1831. Well before the American Civil War. Some of the machines really existed, others were fanciful. Much of the cartoon's satire rested in the signs and text painted on the vehicles; text that unfortunately is not clear at this printing's compressed resolution. What can be seen is Hancock's productive influence, in the image.

1838

William Smith Otis (1813 – 1839) was a gifted, young American civil engineer who invented the first **steam shovel**. His machine could move almost 500 cubic yards of earth in one day. (One cubic yard is a lot of earth to move, by pick and shovel). Unfortunately Otis died at an early age (26 /from typhus) in the same year that he received his patent. His more famous cousin Elisha Graves Otis (1811 – 1861) invented the "safety elevator"; using mechanisms that prevented elevators from free falling.

W. S. Otis' Patent Excavator, 1844

1855

Way ahead of his time, a New Yorker named Richard Dudgeon (inventor of the portable hydraulic jack) also created one of the first self-propelled road vehicles built in America. At first his steam wagon (dubbed the 'Red Devil' in post-Civil War years) created so much pandemonium on the streets, that NYC officials confined his work commutes, to only one street. His first steam wagon was on exhibit in the "Crystal Palace" (NYC) when that burned down in 1858. Dudgeon (wealthy from making hydraulic jacks) moved to Long Island N.Y. for more privacy. He built another steam wagon (stored today in the Smithsonian) and used this to tractor around in his yard or to cart himself and neighbors back and fourth to the nearest village(s).

Richard Dudgeon himself, described the specifications of his second steam vehicle: in these words.

> *"It has a plain horizontal boiler. The furnace, four feet long and 17 inches wide; tubes, 16 inches long; shell, or diameter, 20 inches; cylinders, four inches in diameter, with a 16-inch stroke. They are hung on the smoke box at an angle and operate the cranked back axle. The wheels are three feet in diameter; link motion works the slide valves. Without any patents about it, it will go all day on any good wagon road, carrying ten people at 14 miles an hour, with 70 pounds of steam, the pump and fire door open, if desired. One barrel of anthracite coal is required to run at this speed for four hours. It weighs 3,700 pounds with water and fire to run an hour. It will go 20 miles in an hour on any good road. It is perfectly manageable in the most crowded streets."*[25]

25https://notorc.blogspot.com/2010/03/richard-dudgeon-forgotten-american_22.html

1855

PORTABLE STEAM ENGINE AND BOILER WORKS,

Fifteenth & Hamilton Sts., Philadelphia.

A. L. ARCHAMBAULT, Proprietor,

MANUFACTURER OF **Portable Steam Hoisting, Pumping & Saw Mill Engines,** with and without Wheels, from 4 to 30 horse power.

These Engines are arranged for driving Portable Saw Mills, Corn Mills, Threshing Machines, &c. FIRST PREMIUM AWARDED BY THE FRANKLIN INSTITUTE, in 1851 and 1853; also, by the PENN'A STATE AGRICULTURAL SOCIETY, held at Philadelphia in 1854; PHILADELPHIA COUNTY FAIR, at Philadelphia in 1855. Also, the NEW JERSEY STATE FAIR at Camden.

Also, manufactures to order Stationary Engines, large or small, Line Shafting, Mill Gearing, large and small Screw Jacks for Railroad Companies, and Machinery in general.

☞ REPAIRING PROMPTLY ATTENDED TO.

Even 1 ¾ centuries later, a portable boiler and engine combo like this would be highly desirable as a ***power station*** to a discerning, off-GRID homesteader.

The advertisement here mentions that it would be useful for pumping water, powering a sawmill or used as a hoisting winch. Connected to a belt and pulley system, such an engine could power a workshop full of 19th century tools. But connected to a good generator, controller and battery sink, every modern electric device and gadget found on today's store shelves – could also be used. <Inverters are not mentioned here for space considerations. It is difficult to adequately **invert** DC current into AC current. However a large number of AC hand tools and devices actually **convert** AC back into DC, before being used>.

If need be, a small engine could be mounted to a bicycle frame and then be pushed far into the backwoods along a deer trail; where roads don't exist. A little oil can, an ax to cut firewood and a source of water would be its only supply demands. In cold months the steam machine could also simultaneously help warm a domicile, cook food or preheat water for a bath.

No one manufactures small portable steam engines and boilers anymore because there appears to be no demand. That does not prevent a do-it-yourselfer from constructing their own.

1868

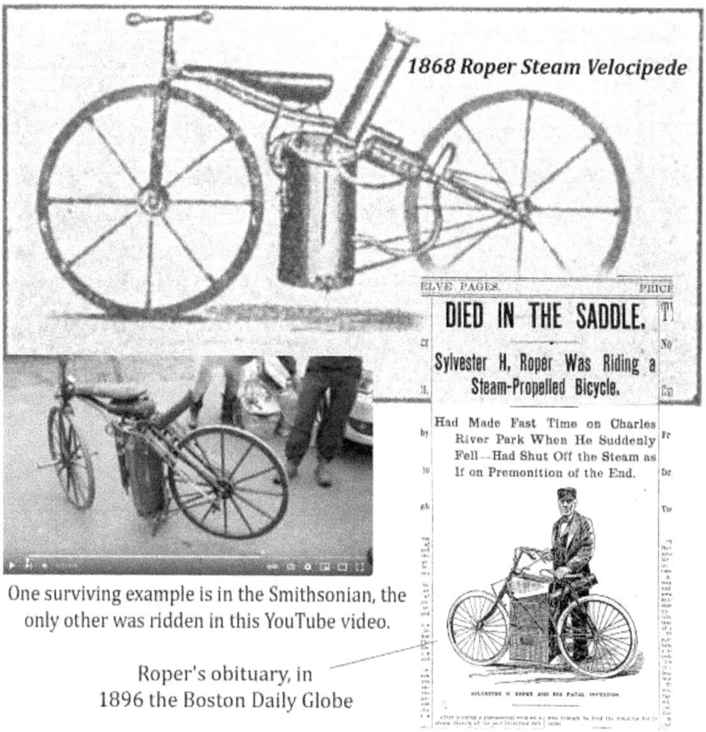

One surviving example is in the Smithsonian, the only other was ridden in this YouTube video.

Roper's obituary, in 1896 the Boston Daily Globe

Long ago a character named Sylvester Roper (1823 – 1896) invented the first steam cycle. It probably should be called a "steam velocipede" rather than a motorcycle however, because most dictionaries describe motorcycles as machines powered by internal combustion engines. Technically, steam motors are external combustion engines.

Roper proved to be a mechanical prodigy at the early age of 12, when he built his first of many steam engines. He designed and built an early steam carriage (4-wheeled) and drove it around Boston by 1863 (the Henry Ford Museum has one). He became a machinist at the Springfield Armory during the Civil War, Roper also invented hot air engines, sewing machines and shotgun chokes. Roper was racing bicyclist and riding one of his own velocipedes, when his heart failed and he crashed for the last time; at the age of 72.

1869

Robert William Thomson (1822 -1873) was a Scottish civil engineer who (among other achievements) developed and patented the first pneumatic "tyres" in 1845. These were more like rubber impregnated fabric fire-hoses that sat inside a stitched leather sleeve, which was then bolted to a wheel. His "ariel wheels" (wheels with tyres) were expensive to make and far ahead of their time. However later in life, Thompson developed what the British would soon call a "**road-steamer**' or a "**traction engine**". His land train locomotives could pull wagons along hard packed roads, and speedily if desired. One did for a time provide an omnibus service between Edinburgh and Leith. His engines were exported to Java, India, Australia and Canada. But from the 1870's it was an American entrepreneur named Douw D. Williamson that would be mainly responsible for developing the modern farm **tractor (steam plows** as they were called back then). A steam machine for tilling the soil; one that could deal with mud.

Two images below may look like tractors, but are merely stationary engines that were pulled to location before preforming their work. (Below in the upper right corner, the advantage of a longer exhaust flue – is that it should create more draft, and therefore a more vigorous combustion).

Two "Agricultural Engines"

(Scotsman, Andrew Meikle invented the first power-driven threshing machine in 1786, but the motive force was horses). A **threshing machine** or a thresher is a piece of farm equipment that separates grain seed from the stalks and husks. It beats the plant about to make the seeds fall out. Before such machines were developed, threshing was very laborious and time-consuming. *"Mechanization of this process removed a substantial amount of drudgery from farm labor".*

(A **reaper** cuts grass, and usually gathers it up or leaves it in a windrow. A modern **combine harvester** does everything at once – reaps, threshes and spits the grain into a truck or wagon, while the chaff is conducted in another direction).

1878

Mentioned before, Douw D. Williamson went to Scotland in 1869, to investigate the Thomson road engine[26]. He bought one to bring back to experiment with and also purchased the patent rights to make his own machines back in the U.S.. In

26https://www.farmcollector.com/steam-traction/williamson-road-steamer-steam-plow/

(Locomotive Works of Paterson) New Jersey, Williamson soon had a dozen of his own improved version of Thomson's steam engines produced. He sold or sent two to California, one to Minnesota and another to Pennsylvania. Then he sent two units back to Great Briton to establish and advertise his improvements.

1882

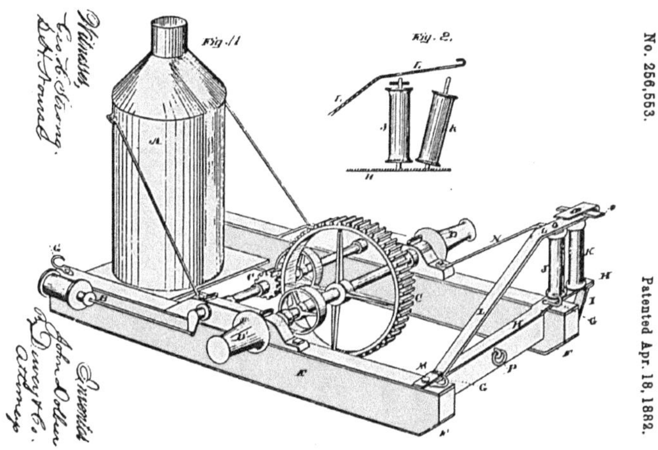

Why John Dolbeer received a patent for his "**steam donkey**" in 1882 is not clear. Similar small engines had been used aboard ships for years – as auxiliary lifting engines. These engines had been called steam donkeys, on ships too. Probably it was the fact that the vertical boiler, engine and winch were all mounted to a frame (which could then be attached to log skids). This allowed the the whole unit to be mobile. With a head of steam a donkey could drag itself anywhere. Once in position a steam donkey (the big ones were called "yarders") could be rigged with cables to spar trees, to drag or carry aloft huge logs.

John Dolbeer, had been a naval engineer before turning to logging. The arrival of his steam donkey revolutionized the lumber industry; especially on the West Coast. Dragging logs over "skid roads", with mules or oxen was out the window.

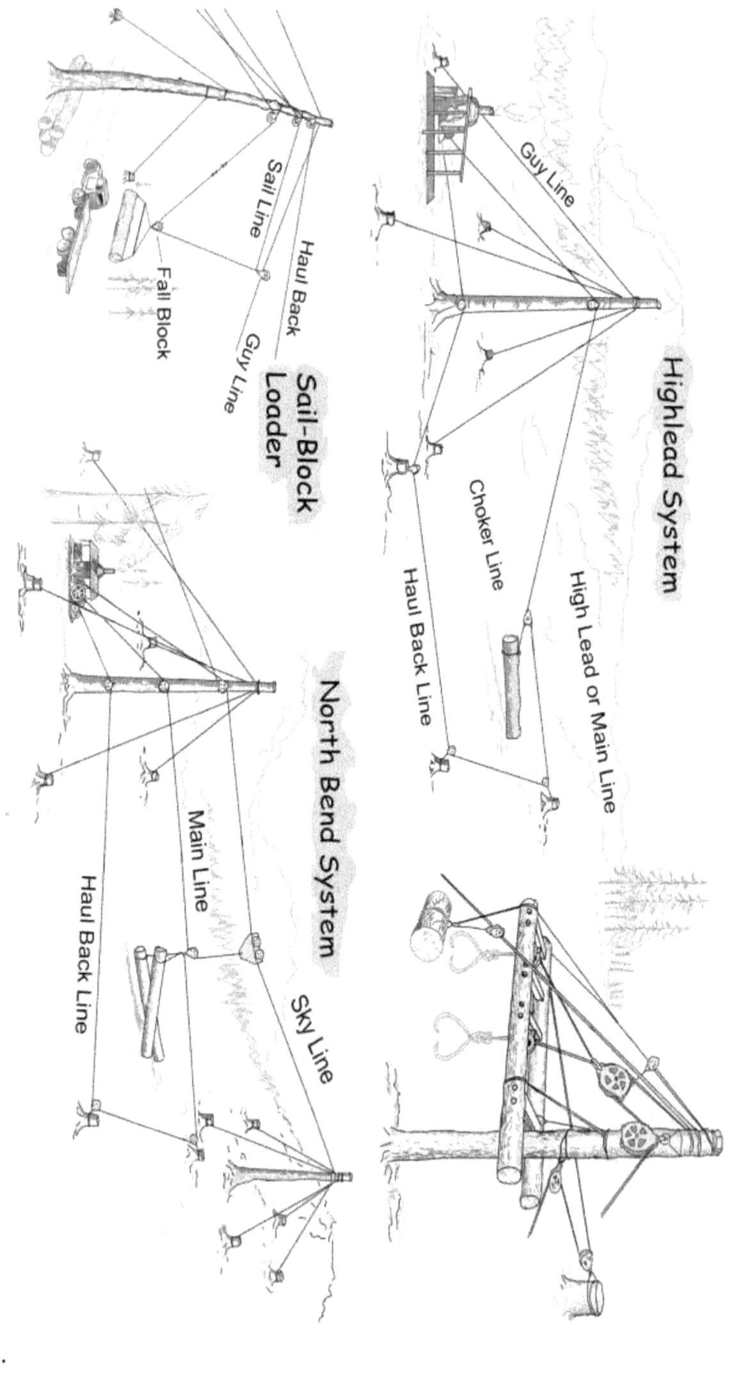
Highlead System
Guy Line
High Lead or Main Line
Choker Line
Haul Back Line
Sail-Block Loader
Sail Line
Haul Back
Fall Block
Guy Line
North Bend System
Sky Line
Main Line
Haul Back Line

simple Swing Boom

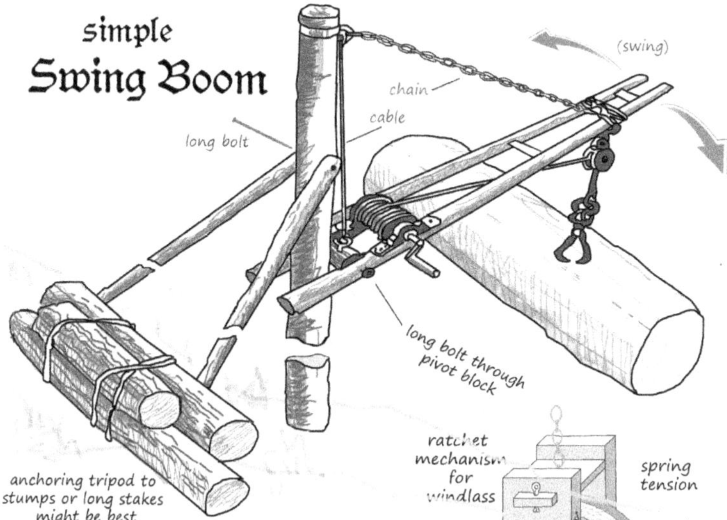

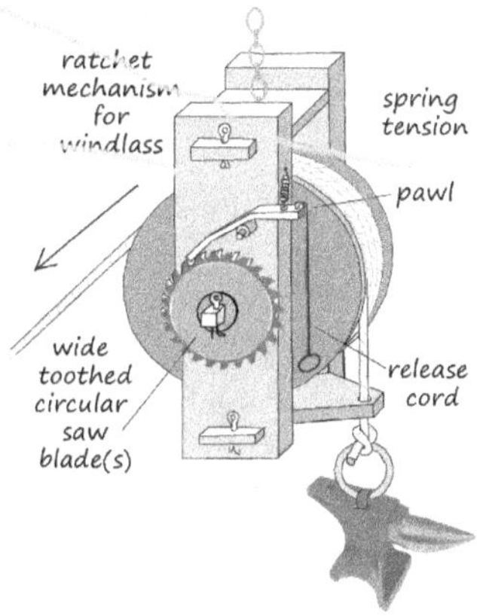

'skip page' forces new chapter to begin on a right hand (odd # page)

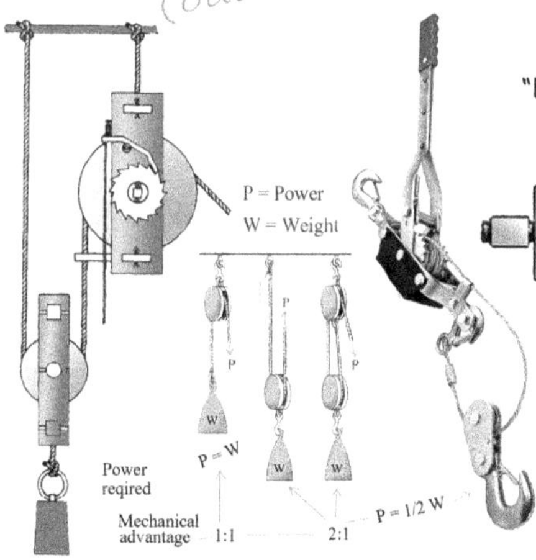

"Differential windlass"

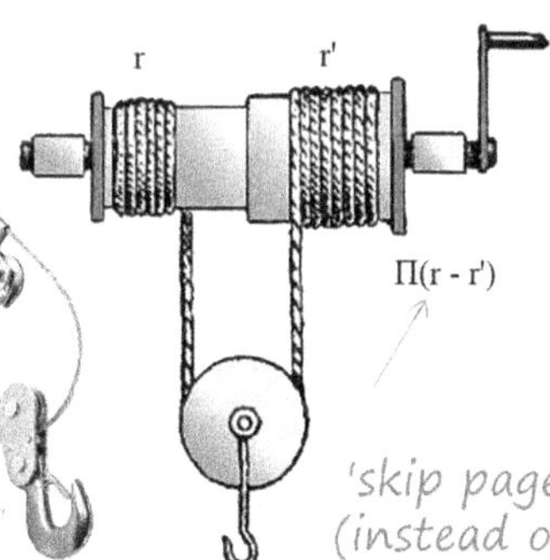

'skip page' (instead of blank page)

(This incongruous page is inserted - to force Chapter 4 (p.305) to begin on the right hand side of this book).

- all public domain images
this one by Sproat Lake,
near Port Alberni, Canada
1947

Even the most diminutive of these machines could drag big logs out of the woods with a chain or cable. Soon, 'skyline logging' or 'cable logging' could “fly” huge loads through the air.

Huge logs of redwood, cedar or Douglass fur, weighing up to 5 tons, could be transported.

* The distinction between "capstan" and "winch" is not always clear. Sometimes a machine is called both. A capstan on a ship's deck usually rotated in the vertical axis and sailors walked around it. But such a capstan may or may not have accumulated wire or rope on a drum. The definition of a winch sometimes – is that it does spool or accumulate wire or rope on its drum.

1885

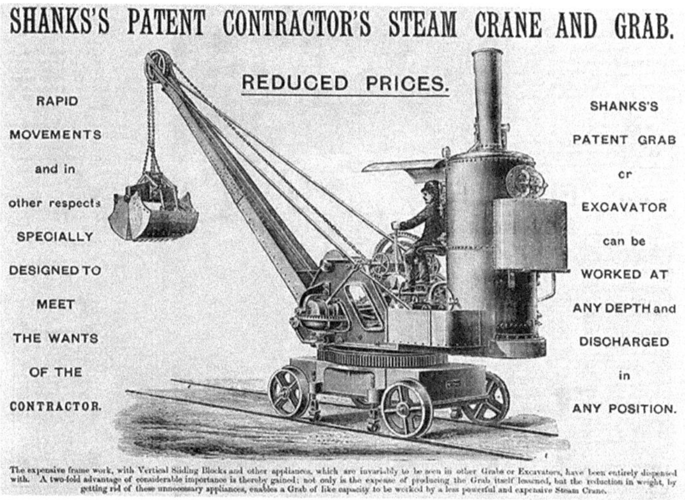

A few pagers back a drawing of William Smith Otis's 1838 steam shovel was shown. There were distinctions between 'cable shovels', 'draglines' and 'backhoes'. The image above is sort of a 'clamshell crane' or 'grab'. A dragline shovel pulled the bucket along the ground. Neither grab or dragline designs could apply downward pressure very well.

The advertisement above is like many others published by Alexander Shanks and Son[27]. *"The company Alexander Shanks founded was known throughout the world as significant manufactures of large steam machines used from the middle of the 19th century to the first quarter of the 20th century."*

1889

27 https://www.gracesguide.co.uk/Alexander_Shanks_and_Son

100-ton
derrick crane
steam donkey
engine

1890

Burrell Universal type Ploughing
Engine from around 1890
PATENT UNIVERSAL ENGINE
CHARLES BURRELL & SONS Lᵀᴰ

1896

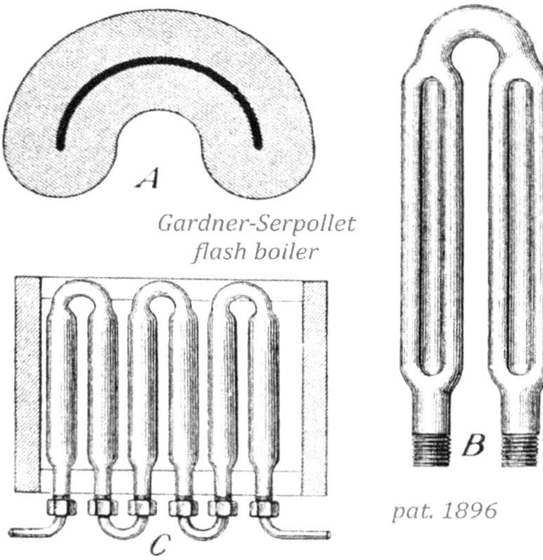

Gardner-Serpollet flash boiler

pat. 1896

In 1896 Léon Serpollet patented the **flash boiler.** An advantage of a flash steam boiler, is that it holds very little water at any given moment; which translates to a very safe or low explosion risk and a faster startup. However there is little to no reserve capacity of steam to draw from, in these systems.

Flash boilers can provide high thermal efficiency by working at higher temperatures and pressures. But more complicated controls are also needed for a machine like an automobile, to happily marry the necessary liquid injections (both water and fuel) – to the desired power output demand.

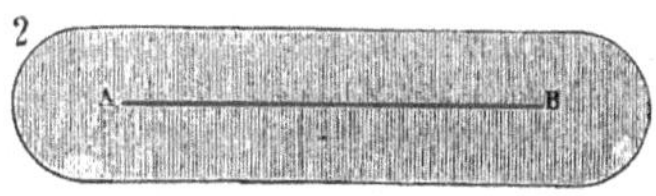

A steel tube was heated and then rolled flat. Then rolled into a spiral. 2) The resultant passageway was about 0.1 to 0.3 mm wide.

Water is pumped into one end, steam comes out the other.

1897

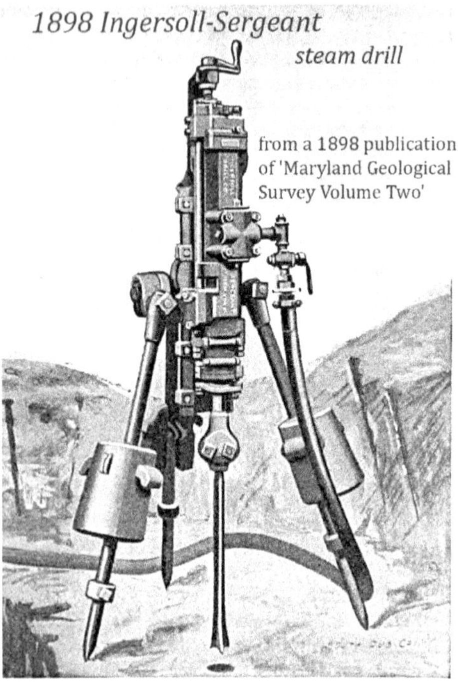

Simon Ingersoll founded Ingersoll Rock Drill Company in 1871 (in New York). He later joined with Sergeant Drill to form Ingersoll Sergeant Drill Company (1888). Brothers Addison and Jasper Rand, established Rand Drill Company in 1871. The company known as **Ingersoll-Rand** was created during a 1905 merger.

1898

Bright red double-decker **steam busses** were a common everyday sight on busy London streets.

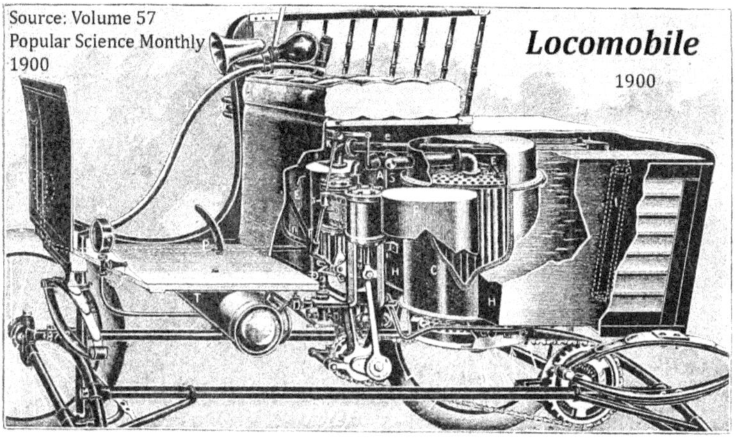

1900

1900 Locomobile steamer

At the turn of the century and for several more years, steam-powered and electric cars outsold gasoline-powered cars in the United States. The electrical cars of the period had a very limited range.

For a long while, gasoline was just an unmarketable byproduct from distilling crude oil - to get the much more desirable kerosene. Gasoline cars were noxious and noisy; they backfired, they had leaky mufflers and/or had exhaust cutouts which were manually opened for acceleration. Drivers had trouble mastering the new contraptions called clutches and gears. Not only that, gasoline cars relied solely upon hand cranks to start the engine. If the driver forgot to do something simple like retard the spark, the crank could kick back and hurt something. Broken hands, thumbs,wrists, shoulders, even a jaw or two were broken - back in the days before electric start.

Possibly the auto industry's single greatest advance came when the **electric starter** was patented in 1911. The first "self starting" gasoline powered car was a 1912 Cadillac.

1902

LORD HILLINGDON'S P.T.L. SERPOLLET DOUBLE PHAETON

Once working pressure was attained, early steam cars could be driven off, with high acceleration. Typically though it took half an hour to warm up a steam car to operating temperature. Some manufacturers like Serpollet (French) and later Doble (American) would developed "**flash boilers**" however, which greatly shortened the startup time. By 1902 there were many makers of steam cars: **Olds, Roper, Peugeot, Baldwin, Locomobile, Stearns, Jaxon, Century, Foster, Toledo** and **Pearson & Cox**; to name some.

1903 Steam Powered Lawn Mower
U.S. Capitol, DC

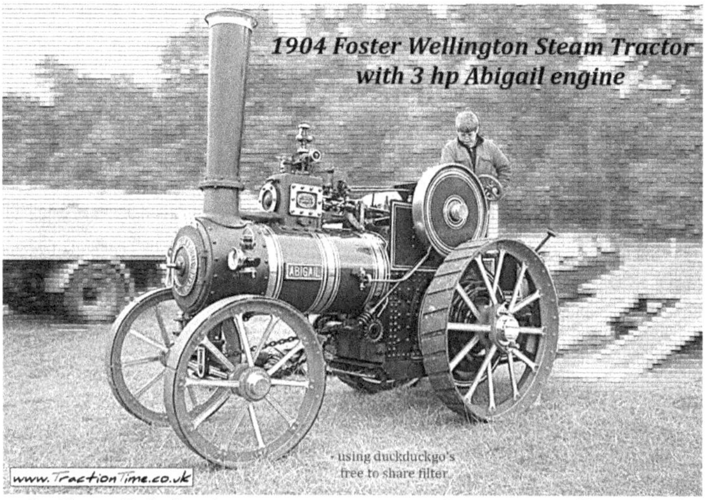

In Great Britain, the term ***steam tractor*** is more usually applied to the smallest models of ***traction engine*** – typically those weighing seven tons or less – used for hauling small loads on public roads. Although known as light steam tractors, these engines were generally just smaller versions of the ***road locomotive***; not meant to plow through the mud.

1906

An expert can determine the relative age of a steam locomotive design by observing the steam box. The center circle shows a slide valve that was common in the **1830s**, but the circle on the lower left inset shows technology that was prevalent in 1906. That could be either a piston and D-valve combination or a "low and high" pressure combo. The D-valve was single expansion (against the piston), but the more efficient, low-high (compound) expanded steam in two stageS.

1908

The landmark year in which **Henry Ford** introduced the most influential car of the 20th century. The **Model-T** itself was not gifted in any particular way. Almost every other car on the road (many were steam powered) out-shown Ford's car, in one way or another. Except price. Throughout 19 yrs of production, all "**Tin Lizzies**" (a nickname acquired by the Model T during a 1922 Pikes Peak auto race) shared the same internal magneto coil and ignition system. Before 1919 Ford didn't even offer electric start; even then only as an option.

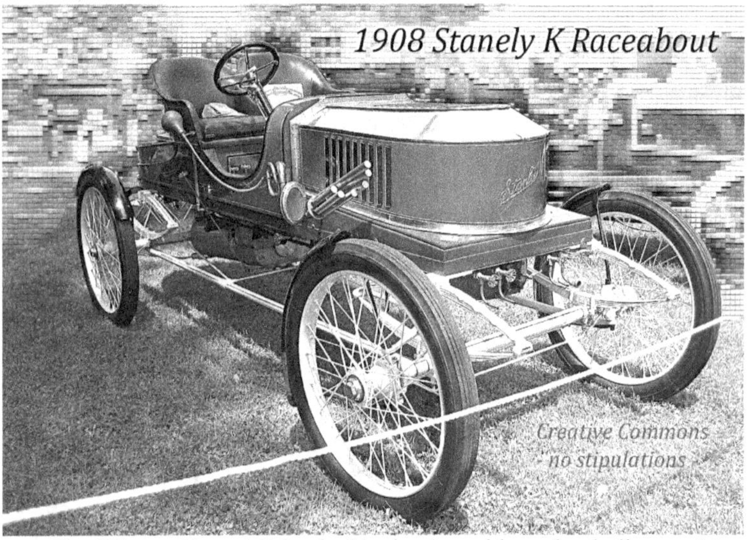

The "Flying Teapots" (early Stanleys) could reach a ludicrous 35 mph. Soon, towns and cities established 5 or 10 mph speed limits and policemen became infuriated with some speeders, which they could not catch. Soon there were several makes of steam cars to choose from; several were higher quality than Stanley. But for a short while Stanley actually was the top selling automobile in America. The name "Stanley Steamer" is notable or remembered because of old racing promotions, but also because of incorrect associations (after 1947) with a carpet cleaning company that adopted the name.

Because of their torque, steam cars outperformed gasoline

cars on mountainous roads. A gentleman's 30-hp roadster as shown above was capable of 60 miles per hour and could run more than 50 miles on a single tank of water. Such a car cost a very pricey $2,500, in 1908 though.

In 1906 a Stanley Steamer set the world's land speed record, at 126.66 mph. The Stanley brothers and a mechanic went back for more speed the next year; and (unofficially) reached 150 mph - before the car hit a hole and became airborne. The canoe shaped "**Rocket**" was demolished after it touched back down (on Ormond Beach just north of Daytona Beach) and the boiler kept rolling a long way into the surf. Apparently no one was killed, but the Stanley twins gave up racing after that.

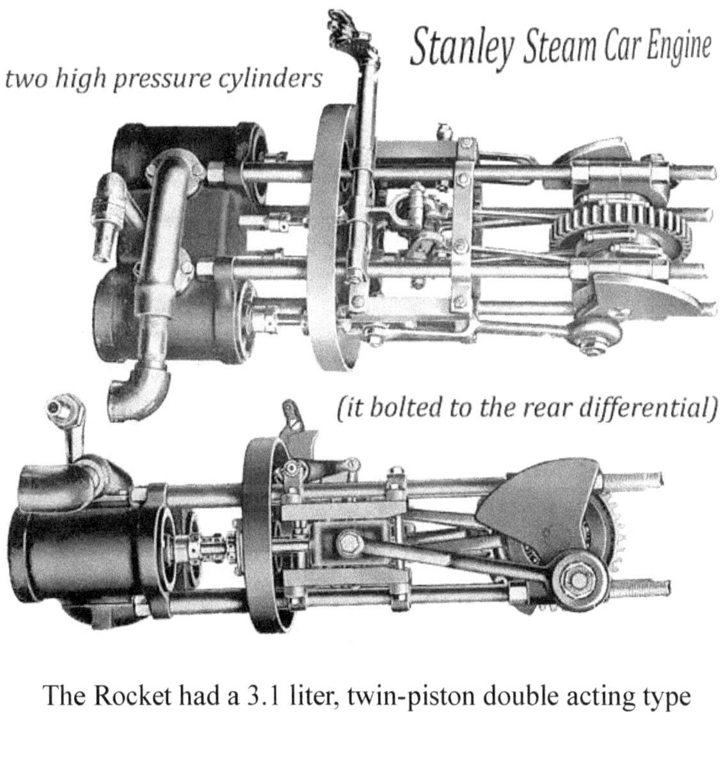

The Rocket had a 3.1 liter, twin-piston double acting type

engine that was capable of producing 150 hp @ 800 - 900 psi. The entire vehicle weighed less than 1,680 pounds.

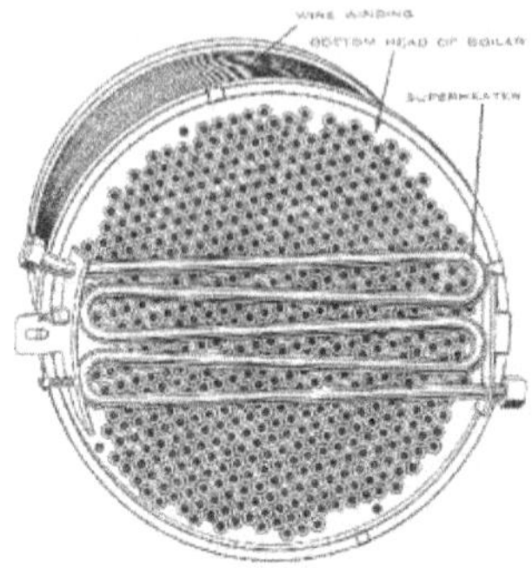

While the Stanley engine was laid out in a manner similar to that of contemporaries like White's and Dobel's, its boiler was unique. It was a vertical fire-tube boiler with a vaporizing burner underneath. The boiler was reinforced by several layers of piano wire wound around it, which made it strong and lightweight at the same time. Initially the vertical fire-tubes were all made of copper, but that changed when condensers (similar to car radiators) were introduced.

Where oil had been added to the water for lubrication before, this was all expelled to the atmosphere along with the steam. However a condenser cooled the steam back into water to reuse, which extend the car's range. When lubricant oil is re-circulated, it can cause oil-fouling that clogs expansion joints (around copper tube fittings) and creates leaks. The solution after that was to use welded steel fire-tubes.

Early Stanleys developed 4 ½ to 8 horsepower - at operating pressures of 400 PSI. Later the company offered three different engine /boiler combinations that routinely operated at 600 psi and were rated at 10 HP, 20 HP or 30 HP.

While more sophisticated steam cars like those made by White or Dobel would come along or were already existent, a fire-tube water-level boiler like the Stanley used, possessed a reservoir of power that the others (using water tube boilers) did not have. With steam up, you could shut the fire/fuel off and still have enough power to drive the car on another 2 or 3 miles.

Despite being a high pressure fire tube design, the Stanley boilers had a good safety record. They employed safety valves, their internal arrangement posed very little risk to the passenger and as far as anyone knows – one never exploded. (Some boilers would scorch occasionally; damage caused when allowed to run short of water. Or pin-hole leaks could develop down around the flue tubes where they interfaced with the flame. This could spray water down upon the hot burner grate and crack it; otherwise possibly creating havoc in the form of big clouds of steam.

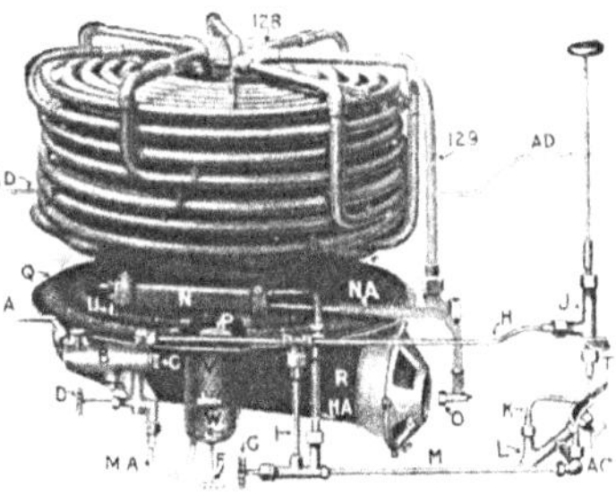

Fig. 42. Generator, Burner, and Fuel Connections Formerly Used on White Steam Cars

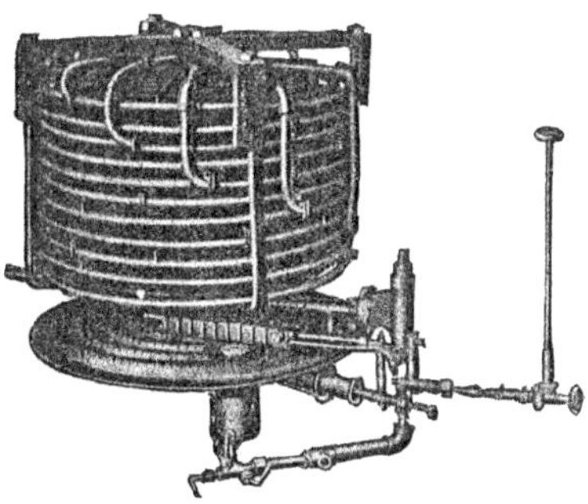

Thomas H. White's son Rollin, developed a form of water tube steam generator which consisted of a series of stacked coils that were connected at the top and (above, c. 1900).

From 1900 the **White Motor Company** manufactured sewing machines, bicycles, lathes and roller skates. For 10 years onward the White Motor Co. manufactured somewhere over 10,000 steam cars (mostly high quality, luxury and expensive). Their last steam car was built in January 1911 as the company made a transition to gasoline-powered vehicles. The company ceased all car production after WWI and concentrated on trucks. Then after WWII it concentrated only large trucks. By the 1950's White Motor Co. absorbed Oliver Tractor Co. and several well respected semi-truck brands like Diamond T, Autocar, REO and Sterling.

1909

President Taft's 1909, 40-horsepower White Model M steam-powered touring car

For transportation, President Theodore Roosevelt preferred the stately horse-drawn carriage to the loud, noxious or cloud forming automobiles of the day. His successor William Howard Taft however, embraced the new automobiles. In 1909 Taft had the White House's horse stables coveted into a garage and filled it with four automobiles (one Baker Electric, two Pierce Arrows and a White touring car). The White steam car might have been his favorite possibly, because he liked to use *"bursts of steam against pesky press photographers"* occasionally.[28]

> *"Difficult to operate in good conditions, the Steamer was often rendered useless in windy or cold conditions. Water lines froze, gas did not vaporize equally, and a good wind could snuff out the burner completely. Dedicated drivers hired mechanics to travel with them and most found they had to fashion parts and modifications to keep their vehicles on the road. Getting a good head of steam up in a cold engine was an arduous task. 'After a couple of hours I finally succeeded in getting the required amount of steam," wrote one novice driver, "but not until I had scorched my eyebrows and hands somewhat. . ."*[29]

28 Freeman, David W. (January 1993). "Cars of the Presidents". Popular Mechanics. **170** (1): 36–37. Retrieved 15 May 2013.

29 https://toledosattic.org/exhibit-collections/commercial-industrial-history/birth-auto-industry?showall=1

1910

This image is of a steam powered air compressor, being used in the Homestead Mine of Lead City, South Dakota. Image information sources say that the image was published anywhere between 1900 and 1910. However closer examination of the source image shows only Ingersoll as the brand name ("Rand" does not show). Indicating only that the compressor is older than 1905 (the year Ingersoll and Rand merged).

Most steam automobiles came with a large number of control gauges, nobs and valves. Not that these were difficult to understand and use, but tending to his teakettle often distracted a driver from keeping both eyes on the road.

1911

1912

The previous image is one slide from archival motion picture film footage taken during the construction of the Panama Canal. The original nitrocellulose film has deteriorated badly since 1912 and that is why this image is dark, dingy and out of focus. Several big Marion **steam shovels** like this worked in teams and rode on railroad tracks. Later once the canal was flooded, these same steam machines were fastened upon barges and used to dredge out the high spots, below the water's surface.

"They used to call steam "the hand of God" because it generates such tremendous torque, the force that turns a wheel". - Jay Leno

1912

The Stanley twins sold the rights to their 1st car to a new company named Locomobile. Then they helped manage Locomobile for about five years. During that period Locomobile built and sold over 400 steam autos. After that five years the twins bought back the right to manufacture steam cars under the Stanley name, while Locomobile switched to making only internal combustion automobiles. The climax of steam car sales peaked between 1910 and 1912. In 1918[30] the brothers quit the

30 https://racingnellliebly.com/weirdscience/daredevil-driver-sets-land-speed-record1906/

steam car business, but the Stanley Motor Carriage Company kept on going. Between 1902 and 1924 when the company finally capitulated to the superior convenience of electrical start and gasoline fuel, over 10,000 Stanley steam cars had been produced. Few exist today because most were loved and were literally, driven into the ground.

1914

Clam shell excavators, dragline excavators and steam shovels are not the same things. From 1884 the Marion Power Shovel Company (in Marion, OH) built large earth moving equipment. The huge steam shovels that dug much of the Panama Canal (shown earlier) were built by Marion. Some of the largest industrial machinery found in the world today, like the two crawler-transporters used to move around NASA's Saturn V and Space Shuttles, was built by this company. (That is until 1997 – when Bucyrus International, Inc. purchased Marion).

The 150 year old Case Corporation began making farm machinery in 1842. The company was officially named J.I. Case Threshing Machine Company. By the late 1890's and start of the new century, Case was the largest maker of stationary steam engines and steam tractors - on either side of the Atlantic.

(It was already mentioned that: in 1869 the American Douw D. Williamson brought back a “road engine” or “traction engine" from Scotland. Then for the next 9 years Williamson focused upon constructing a "steam plow"; specifically an agricultural machine for tilling the dirt).

1915

The Case company's line *"spanned dozens of models of stationary and steam tractor engines ranging in size from 6 to 150 horsepower, which employed both simple and compound operating cylinders"....* *"Case* (steam) *engines came in different sizes, including 30 hp, 40 hp, 50 hp, 60 hp, 65 hp, 75 hp, 80 hp, 110 hp and 150 hp".*

1919

Creative Commons Attribution-Share Alike 4.0 license: Thanks to Palauenco5 (left), Colin Creitz (right)

Between 1906 and 1909 four brothers while still attending high school in California, built their first steam car in their parents' basement. The fact that their father owned a metalworking factory, helped. The brothers assembled this car mostly with parts taken from a wrecked White steamer, but lacking as it did, the engine was of their own design.

Later, despite the fact that by the 1920's - steam car appeal had seen its heyday and had been surpassed by gasoline powered cars, the Doble brand would enter the market and come to represent the epitome of steam car development.

Doble engine sideview

Doble-Detroit Steam Engine: sectional view. [Photographic Prints] Retrieved from https://libwww.freelibrary.org/digital/item/44800

By 1923, a Doble steam car could be started from cold with the turn of a key. And then the heavy, stately, luxury car could driven away within 40 seconds! Model B could accelerate from 0 to 60 mph in just 15 seconds whereas a Ford Model T of the period took 40 seconds to reach its top speed of 38 mph (no 60mph).

The Doble Steam Car achieved its excellent start-up time by incorporating a highly efficient **monotube** steam generator to "heat a much smaller quantity of water along with effective **automation of burner and water feed control**". The cars featured several original patents. Because the cars had so much torque, they were "direct drive"; needing no gearing to transmit the engine's power. The Model E Doble used 3,000° F super-heated steam to create roughly 1,000 lb-ft of torque. Five to eight times what an average car today makes. (A disadvantage to this type of power-train is that there is **no neutral**. The engine is always engaged. Good brakes are important).

1924

Quoting next J. Leno (former talk show host, discussing a Doble, within his 150+ car collection).

(It) "*feels most similar to the Tesla P90D, which interestingly, with its batteries, weighs about 5,000 lb. As you're driving, you're watching the gauges. They show steam pressure and temperature, which both should be in the 750-850 range—that is, 750 to 850 pounds per square inch of steam pressure and 750° to 850° Fahrenheit. You control your speed with the throttle, a small wheel set inside the steering wheel. The Doble throttle is like the spigot on the side of your house, which is a screw valve. You turn one way to open, the other way to close. That little wheel controls the steam and the acceleration. There's no shifting. In many ways, my Doble feels like a modern car—*

quiet, solid, and I can go 70 mph with no problem, whereas with any other car from 1925 - 40 or 45 mph is the end of the world.
—Jay Leno,

A Doble Model E cost between $20,000 and $25,000 in 1925. That was 77 to 96 times, what a Ford Model T cost in that year. ("*The Model T's initial cost of $825 in 1909 fell to just $260 by 1925*").

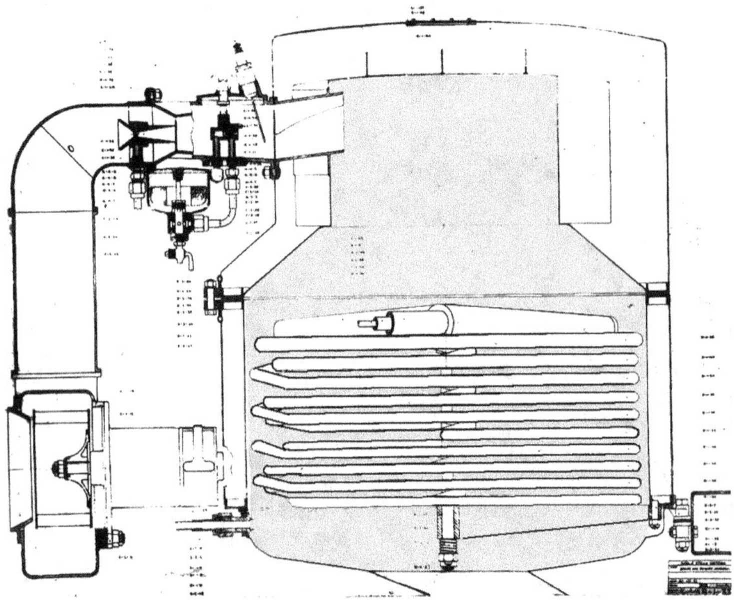

Doble-Boiler and Burner Assembly, 1922. / the AACA free digital Library

It's not much of a picture above but the general layout can be appreciated. The Doble flash type generator used a continuous steel tube 476 feet long. The coil grew wider – from bottom to to top. All the joints were acetylene welded. The fire was on top of the coil instead of underneath. Water was pumped into the bottom of the tube and steam was drawn off the top. Intense circulation supposedly prevented sediment or scale buildup inside the tube. (Water entered @ 29 ft./sec and steam was withdrawn @ apx. 600 ft./sec). Less than two quarts of water were in the tube at any one time. Venturi tubes and a splitter caused the flame to divide and create a whirlpool motion. The flame never came into direct contact with the tube.

1929

1930

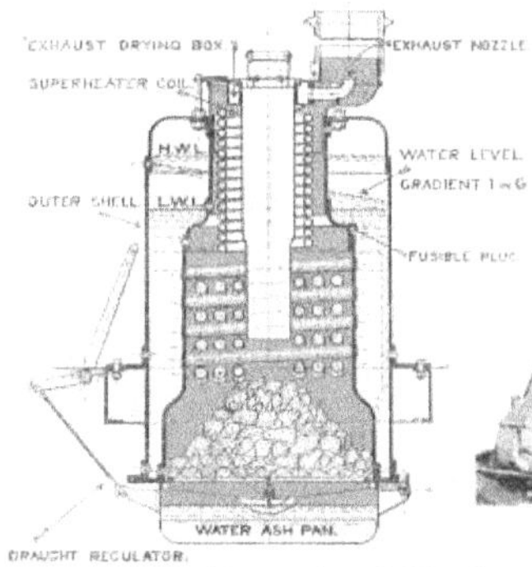

(modified images from Wikipedia material)

Sentinel steam boiler

Shown above is a typical upright or vertical, water tube boiler- a format that was used in most British steam lorries or steam wagons. The boiler was in the cab, positioned next to and a little behind the driver. A second person was oftentimes present to attend to the boiler.

Great Britain had no vast oil reserves like the U.S. The UK had to (still must) purchase and import its oil, and its large and thirsty navy took precedence in the supply line. Steam wagons like these were almost unique to the UK. They were very common before WWII, but were inevitably replaced by diesel engined trucks afterwards. Still, it was increased taxation played a large role in the influencing that changeover.

origional images published under Creative Commons Attribution-Share Alike 4.0 International licenses

On the left above is a Sentinel DG8, on the right is a modernistic looking 1934 Sentinel S4.

*"Initially, **imported oil** was taxed much more than British-produced coal, but in 1934 Oliver Stanley, the Minister for Transport, reduced taxes on fuel oils while raising the Road Fund charge on road locomotives".*

These 1930-ish steam trucks were quite capable and practical; for short hauls. An average one could push down a road at an easy 50mph clip. One could drive all day on a full hopper of coal, but might have had to stop here and there at a stream, pond of water trough, to refill on water. Water tube boilers like the ones most steam wagons employed, did not have condensers; so the steam was simply discharged to the atmosphere. There may have been no oil mixed with the steam for lubrication (a potential source of fowling), but the water sometimes used was dirty or questionable; so boiler design took that into account. Coal was loaded into the hopper from the top of the cab but the boiler feed was not automatic. The feeding of the boiler was done manually.

These steam lorries did require a bit of extra maintenance. For example the ash pan (which contained water to extinguish embers) needed to be dumped out daily, after use. The ash pan was inconveniently accessed from underneath, behind the front axle. And like the engineer of an old steam train, that circled his locomotive every morning with a big oil-can in hand, these lorries needed to be lubricated periodically in several places too.

1933

Another pair of brothers built a steam powered airplane that flew! Apparently, for the intellectual challenge of proving that it could be done. When Doble Steam Motors went out of business in 1931 (during the Great Depression), George and William Besler were there to pick up the factory's remains.

The brothers (who had engineering backgrounds) took an existing air-frame (from a Travel Air 2000 bi-plane) and replaced its motor with their steam motor. With 1,130 psi of steam the engine produced 90 hp, could best 100 mph and "could change altitude at will". The story and these pictures were published in July 1933, by Popular Science magazine.

(Apparently, in 1852 an airship (lighter than air, unlike an aeroplane) did fly over the skies of Paris - propelling itself with a 3-hp steam engine).

1954

The first nuclear power plant to generate electricity for a power grid, was Russian – in 1954. The first nuclear power station to sell commercially over a national power grid was located in England, in 1956. It was 1957 before a nuclear power station began to service the **U.S. electrical "Grid** (an interconnected network for electricity delivery).

modern steam turbine and generator

Nuclear disasters in Chernobyl, Ukraine (1986) and Fukushima, Japan (2011) have forced some countries to question the wisdom of building any new nuclear power plants.

> *"As of May 2024, there were 440 nuclear reactors in operation in 32 countries around the world. The United States had the largest number of nuclear power reactors in operation at the time, at 94 units".*

DIY Steam

After glossing over the history of steam power somewhat, it is now time to turn this chapter's attention to ideas that individuals themselves might be able to use. The purpose of discussing steam history in the first place is to better understand the applied science. This is a technology that has undergone centuries of study, trial and development and which once had emphatically practical usages. But today the potential of steam power is dismissed, or forgotten about and passed over. The complex society that we enjoy or abuse today was made possible by steam power, during the Industrial Revolution. There were no electrical or petroleum motors to drive the drills, saws or lathes back then. Working machines were driven by steam.

In the unpredictable future, there may again come a time when there is no commercial electrical power to be had. If that happens, then the ability to harness steam energy – on a personal level, might prove to be extremely valuable; if one knows how.

Considering what is feasible on an individual basis, one might notice that the means to build a small, capable steam machine are laying about, in every direction a creative mind might look. More machined metal parts, more pressure hoses, valves and couplings exist now than ever before. What is lacking is a decent guide on the subject, to share ideas and give modern examples. Not that this claims to be such a source.

A transportable steam power unit or station might be acutely practical to an "off-gridder" or survivalist. Such a machine could be used to pump water, run a sawmill, or drive a winch to lift heavy rocks or logs. Such an engine could mechanically power a workshop full of 19th century belt driven tools, or turn around and generate the power required to power 20th century ones. In the cold of winter a small, self contained steam power center might be wheeled inside to cook food, warm a cabin or heat some bathwater. Practical. Someone with a little engineering acumen could probably become wealthy by designing and marketing today, prefabricated units to paranoid "doomsday preppers".

Many used, machined metal air-compressors, motorcycle and automotive parts are available, that could be used for steam engine construction. Several youTube videos already exist that display how to successfully convert small weed-eaters or chainsaw motors into steam engines. Even an automotive shock absorber or motorcycle fork can be converted into a practical steam engine that can preform work.

Aside from an engine, the other mandatory part of a steam setup is the boiler. Stainless steel is the best material for boiler fabrication because it resist deterioration by flame and oxidization better than other alternatives. However SS is difficult to work with and might be unavailable. But many carbon steel pots and utensils like LP gas canister bottles can be easily converted to pressure boilers. Or someone might try to imitate the hi-pressure flash-tube or water-tube boilers like once used in Serpollet, Dobel and White automobiles.

Let it be known that a makeshift but capable ***arc welder*** *can be made with a few car batteries in an emergency. Such batteries are low in voltage but high in amperage. It is the amperage that allows an arc welder to function.*

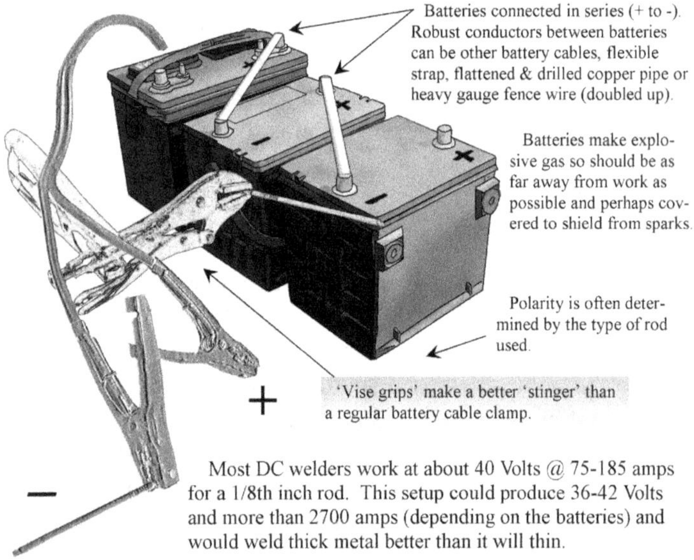

For someone not familiar with welding, a "6011" rod is a good all around, general purpose choice to keep handy in an emergency toolbox. This rod splatters a lot and does not make the prettiest weld, but it sparks and burns through dirt and rust when other rods won't. The 6011 also works with any type of current too, reducing confusion. An 1/8" rod is intended for thin metals while a 3/16" rod would require more current and would join thicker metals.

The polarity of DC current can decide if a particular rod will work at all; some are intended for (-) current while others only work with (+) current. {Polarity is easy to switch- if you realize you *should}. In general, **DC straight** (-) is supposed to melt and deposit metal smoother and faster, but **DC reverse** (+) produces a deeper penetrating, stronger weld. AC current flips its polarity 60 times a second (in the U.S.). The following chart is very superficial - meaning that it is missing much information.*

	DC-	DC+	AC	Penetration	Notes
6010		X		deep	The most penetrating rod. Burns through rust, dirt , paint and oil. Reverse DC only.
6011	X	X	X	deep	Utility rod. High splatter. Like 6010 but for AC as well
_ _12	X		X		AC or straight DC only
6013	X	X	X	low	For clean new metal. Soft arc, easy to clean slag
7014	X	X	X	med	Fast and smooth, easy
_ _15		X			Reverse DC only
_ _16		X	X		
7018		X	X	low	Low Hydrogen, quality welds, flexible weld—minimal cracking
7024	X	X	X	low	For flat or horizontal crack filling, common for 1/4" or 1/2" plate

This image above was also displayed at the introduction of this chapter. Even though it has a primitive cross-head bearing, it is still one of the plainest designs that one can come up with. This is a "simple", "horizontal", "reciprocating", “single-acting” - steam engine. What makes a single-acting engine, is the fact that steam pressure is applied to only one side of the piston.

Below though, is a double-acting design where the steam is alternately supplied to both sides of the piston.

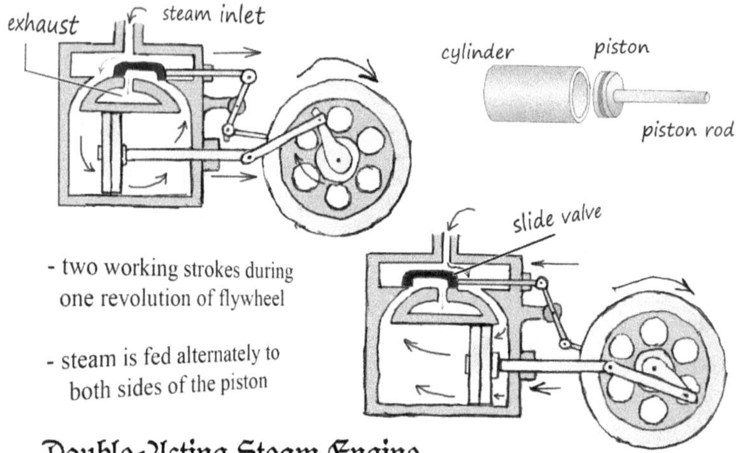

Double-Acting Steam Engine

More power can be created by a double-acting engine because each stroke is a power stroke. However single-acting designs can have advantages too. For example these can rotate faster because there are less linkages and fewer friction points in the way. Many high speed train locomotives used single-acting engines. Some single acting engines don't even need valves.

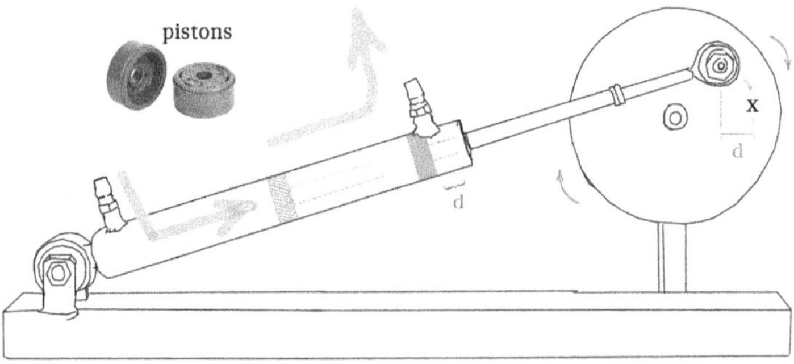

Instead the exhaust can vent through a hole in the cylinder, once the piston reaches the end of a power stroke. An adequate mass in a flywheel can return the piston to the bottom of a cylinder.

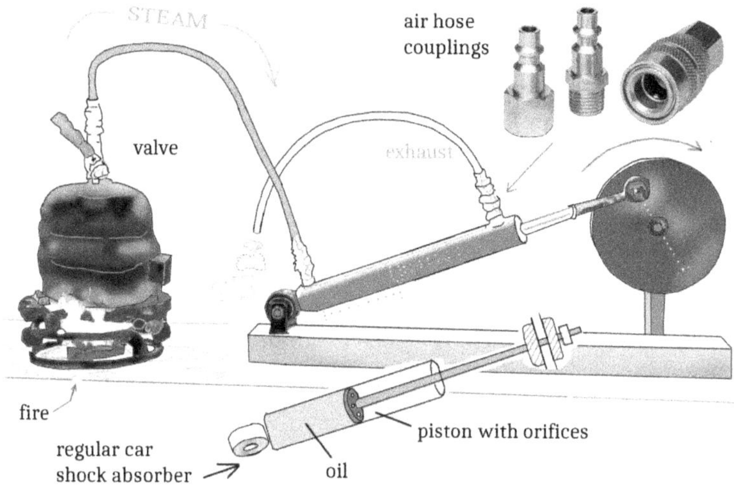

Below is a different type of shock absorber for comparison. These gas-over-oil shocks might not work like the simpler oil shock above. Sometimes there might be little way to determine what type of shock you hold – unless you cut one open to see.

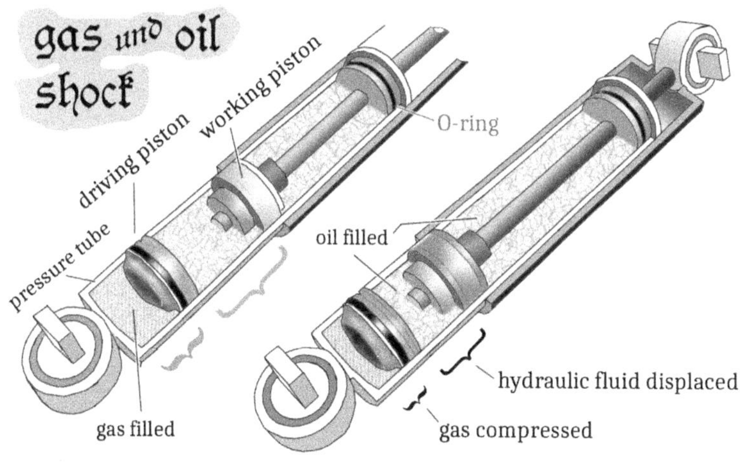

In regular old fashioned hydraulic shock absorbers (or dampners), the fluid is forced to squeeze through small holes or orifices in the piston. This dampens the impact of a bump, during both compression and return. A coil-spring, a leaf-spring or some other component of the vehicle's suspension is needed for the rebound stroke; to return the dampener to its ready state. The efficiency might be reduced by steam (instead of oil) leaking through the orifices in the piston, but that also might enable this very simple idea to work as it does, without complaint.

Converting Gasoline Engines

Sometimes old or almost worn out small gasoline engines have all the components needed to make a useful steam engine. And without requiring complicated metalworking. In the event of society's demise due to the collapse of the electrical power GRID nationwide, old fashioned manual metalworking tools will still work. A metal file may not be as handy as an electric grinder, but it will get the job done. Random car batteries might be joined to preform a bit of arc welding (as described earlier). If not, blacksmiths have been welding iron and mild steel for centuries.

Whether 2-stroke or 4-stroke, horizontal shaft or vertical shaft - small gasoline engines might (with the know-how) be easily converted into usable, practical steam driven counterparts. Already self contained, these metal engines can be anchored or tied down to foundations, bench tops or wheeled platforms. The power of steam can then supply the motivation needed for a myriad of applications.

Above, the crankcase of a horizontal shaft, single cylinder, 5 HP Tecumseh motor from a rotor-tiller. Many engines have a similar layout. A heavy and squat, torque insuring, cast iron, finned flywheel sits on the left.

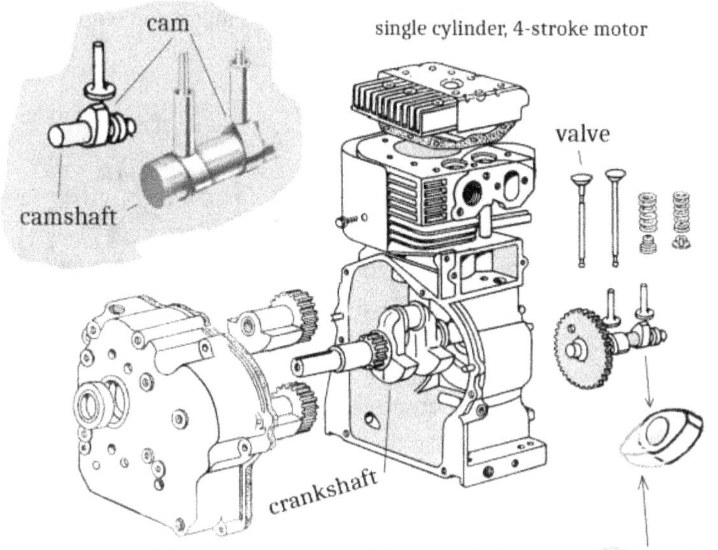

(To convert to steam: weld on and grind to shape new material to the backside of both cams; make the shapes equilateral).

Above: modifying <u>both camshaft lobes</u> will enable the valves to open and close at the right time for a 4-cycle steam engine conversion. What this does is to make every downward stroke of the piston - a <u>power stroke</u>; just as with a 2-cycle engine. But a real 2-stroke engine typically has no valves, so arranging for steam to enter the cylinder at the right time, would probably require creating a separate timing mechanism

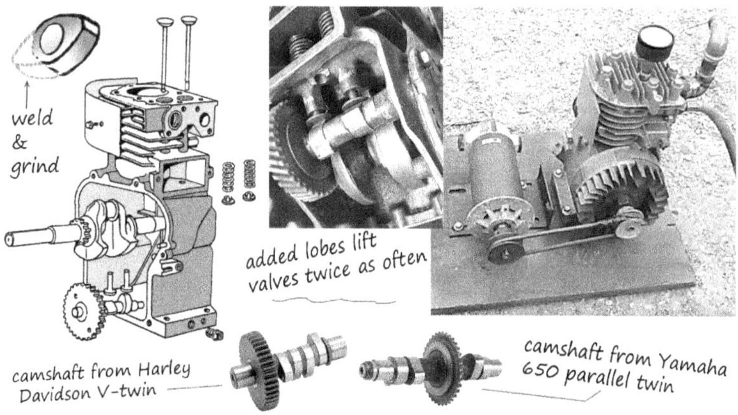

The two camshafts at the bottom of the picture above come from twin

cylinder, 4-stroke motorcycle engines. Engines that could be easily converted to steam in the same manner, and without being worried too much about timing afterwards.

The "head" (or engine?) of some **air-compressors** have quality metal cylinders and pistons too; just like simple gasoline engines. But these single stage air-compressor heads do not use camshafts, lifting rods or valves in the typical sense. In essence they simply rely upon a pair of check valves. Air enters on the piston's downstroke and gets pushed along on the upstroke. To make this work as a steam engine, some sort of external valve system would need to be improvised, and then adjusted for optimum timing. Many mechanics or engineers would toss the original top (the cylinder head) and replace it with a simple flat plate of metal. Then they'd drill and tap the flat plate for anchor bolts and threaded plumbing fittings.

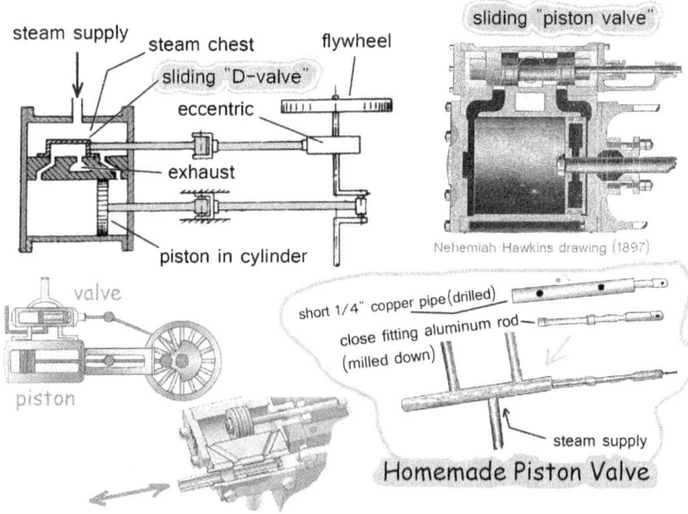

A "steam chest" or "box" is a reservoir, continually filled with ready steam. A sliding valve is a mechanism between the chest and the piston that passes steam once a valve moves over a port. Two simple types of sliding valves are shown above. These would be considered double acting; a single acting valve would have one less hole or port.

The term "PSI" is a general term but **"PSIG" (Pounds per Square Inch Gauge)** is a more precise term; it's relative to atmospheric pressure. PSIG indicates how much pressure is above or below the surrounding atmospheric pressure. For example, if a tire is inflated to 35 PSIG, it means the pressure inside the tire is 35 PSI greater than the atmospheric pressure.

Naturally 2-stroke engines can be converted to steam also. But these simpler engines have no timing valves of camshafts. In the next image, the intake port is plugged up and the steam gets ported into the spark plug hole. The work is not being harnessed because a lobe which is attached to the end of the shaft, is in the way. When the shaft rotates down, the point of the plywood lobe depresses the air chuck, and a shot of steam enters the cylinder. Here the timing can be regulated simply by altering the shape of the rotor or lobe.

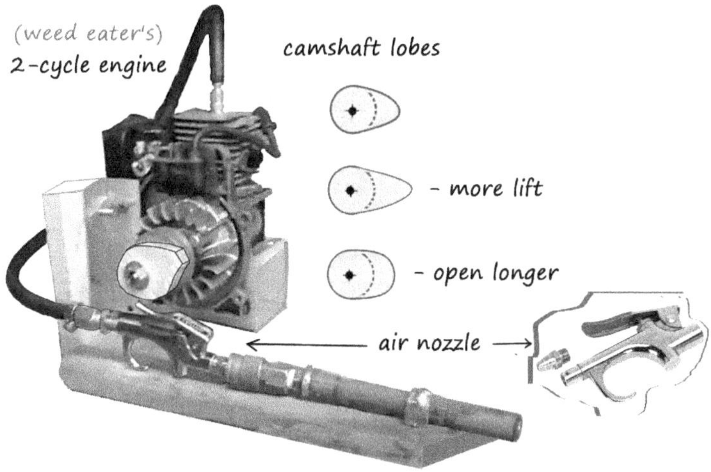

Both Harley Davidson V-twins and Volkswagen flat-four boxer engines are numerous; VWs were made by the millions. Both engines are air cooled, overhead valved, time tested, rugged, reliable, popular, bare boned, simple to understand and to repair. Both engines are over a century old in design. Their parts are easily separate-able (where most 'modern engines' are not) and therefore should lend themselves to adaptation in external combustion design; steam power.

Harley-Davidson's first motorcycle engine was a F-Head single-cylinder, in 1903. They introduced the "V-Twin in 1909. The engine has gone through many iterations. It has been surpassed technologically by numerous different, better performing engine designs. Yet this simple dinosaur of an engine is still a top choice in the motorcycle market.

Made by the millions, the Volkswagen beetle's engine was supposedly designed by Ferdinand Porsche before 1939. But he borrowed several ideas from companies like (the Czech) Tatra who pioneered the rear-engined, air-cooled flat-four with their 1936 Tatra T97. (The first flat-four or boxer four engine dates back to 1900).

A wide assortment of **high pressure gauges** and **safety valves** are currently available on the market. That does not mean they are cheap.

High Pressure Flexible Hose can handle up to 3,250 psi (224 bar) and 788 °F (420 °C). Such hoses come in several standard sizes and can be fitted with various couplings at specialty shops.

After a hypothetical crisis and collapse of society should it happen, the finer, high quality equipment won't be available by **mail order**, unfortunately. Which means that unless an individual has had the foresight to build a steam system or acquire the parts beforehand, only scrounged parts from pre-existing equipment can be used.

Almost all houses have water heaters so therefore a simple T&P (temperature and pressure) relief valve should be easy to locate. Although this will limit the pressure of steam system to 150 psi or less, that is still enough to do useful work. (Remember: in 1858 the SS Great Eastern was leisurely crossing the Atlantic Ocean with 25 psi (172 kPa) or less).

How much power from a converted engine?

The typical combustion pressure inside a gasoline engine cylinder ranges between (slow idle to racing) 160 to 1,750 psi. A more realistic average range is between 300 and 1,000 psi.

It seems logical that if you match gasoline's cylinder pressure with a steam pressure counterpart – that and equivalent horsepower should be exacted. Steam pressure in excess of 1,000 psi is certainly possible, but potentially very dangerous. Converting an internal combustion engine to make it run on steam, effectively changes it into an external combustion engine. Working fluid must be transferred to the cylinder. It is not the engine, but the boiler and pipes that become the biggest safety limitation.

A small gasoline engine with a modest displacement can be safely converted into a useful steam engine, abet with some loss of power. For a general idea; if one had an air-compressor head with about 11 cubic inches of displacement (apx. same volume as a 5 HP single cylinder rototiller type engine) then: using 100 PSIG steam @ 400 RPM should result in about 2 HP. (Two horsepower can be very useful, and is not to be scoffed at).

Boilers

The other important faction of steam power

Fire-tube boilers can be constructed of very simple components, but for safety's sake they are operated at lower pressures than are water-tube boilers. While usually less powerful than water-tube counterparts because of this, they also take longer to start up. But fire-tube boilers have the advantage of storing more energy; they can hold more water and store a larger, redundant supply of steam.

"Steam generators" are understood to differ from larger steam boilers. Boilers were originally large and were intended for continuous high-demand steam purposes, while steam generators were smaller, faster starting, more efficient and mainly more applicable to periodic usage. While steam automobiles like the Stanley, White and Dobel all had similar engines, their kettles for making steam differed widely. Stanley's used dense little fire tube boilers, made of many short vertical pipes copper pipes (initially). White and Dobel used water tubed steam generators.

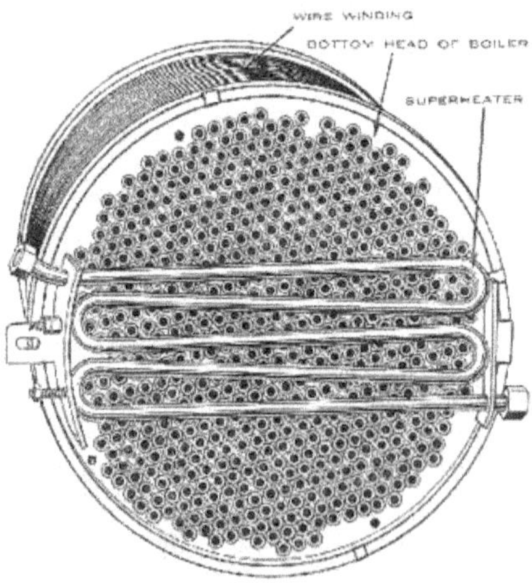

Although the typical melting point of tin-lead alloy solder ranges from 180 to 190 °C (360 to 370 °F) the copper pipes in these Stanley boilers were not soldiered. Instead they were swagged on both ends with iron ferrules. At first the boilers of both Stanleys and Locomobiles were operated at 200 psi. That figure was soon bumped up to 400 psi – and still provided a considerable safety margin.

The light skinned, water holding boiler drums were wrapped under tension, with three wraps of piano wire. *(The tensile strength of piano wire usually ranges from 230,000 to 399,000 psi, depending on the diameter)*. But to conserve water and to extend travel range, Stanley later was to add a condenser. Since lubrication oil was added to the steam, the attachment of a condenser caused oil-fouling in the expansion joints. So is event necessitated the use of welded steel fire-tubes. The move to steel fire tubes was accompanied by an increase in the Stanely boiler's standard working pressure (up to 600 psi).

> *"The boilers were reasonably safe since they were fitted with safety valves. Even if these failed, any dangerous over pressure would rupture one of the joints long before the boiler shell itself could burst. The resulting leakage would relieve the boiler pressure and douse the burner with very little risk to the passenger. There is <u>not a single documented incident of a Stanley boiler exploding</u>".*[31]

(There were numerous examples of "scorched boilers" where a low water condition has caused damage to the flue tubes. These boilers also on occasion developed pin-hole leaks at a flue, spray water into the flame, which enveloped the car in clouds of smoke and water vapor. The remedy for that situation was to use a boiler flue plug to block the leak off).

White's steam cars used multiple water-tubes, wound in parallel but connected at the top and fired from the bottom. Dobles used the most complicated automobile steam generator design, with a single 476 foot long continuous coil of steel tube, which was fired from the top down with a vortex or whirlpool of flames that never actually touched the coils. Doble's water tubes made super-heated steam and the cars worked happily in the 750 – 850 PSI range. These three American stream generators didn't consume wood or coal; they **burned kerosene** (mostly) because it was an efficient fuel that could be precisely controlled.

A half dozen different types of ½ inch copper pipe.

(wp = working pressure / bp = actual burst pressure, / A = annealed / D = drawn)

Type M	Type L	Type K
wpA - 505 psi	wpA - 613 psi	wpA - 758 psi
wpD – 1000 psi	wpD - 1206 psi	wpD - 1490 psi
bpA - 3000 psi	bpA - 3885 psi	bpA - 4535 psi
bpD - 6000psi	bpD - 7765 psi	bpD - 9840 psi

Annealed copper pipe is a softer and more flexible than hard drawn pipe. Annealing employs heat treatment to make metal easier to bend and shape. Hard drawn copper pipe is stronger and more rigid. The weakest annealed ½" pipe in the table has an actual burst pressure of 3,000 psi. But the approved working pressure for that pipe is only 505 psi (considerably less).

> > If one were considering constructing a coiled water tube type steam generator, consider that the (2026) spot price for 1/2" copper tubing at a

31 http://www.stanleymotorcarriage.com/Parts/Boiler.htm

lumber store is presently about $190 for 100 feet of coil (annealed). Also: 12 inches of 1/2" copper tubing holds 0.01 gallons of water, 50' would hold 0.51 gal and 100' would hold apx. 1.02 gallons of water. That may not be enough volume. Under working conditions probably less than half of a lengthy water tube will be filled with water, the rest with compressed steam.

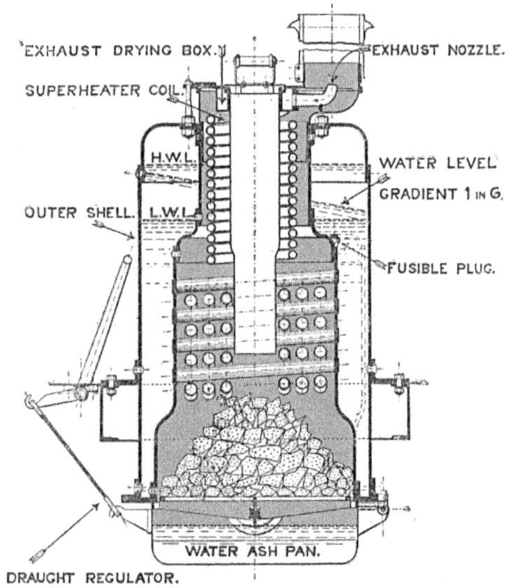

modified image from public domain / Wikipedia material

British steam lorries ran on just coal and water. This sketch is of a Sentinel boiler which worked at 250 psi. At one time there were some 140 different makes of British steam wagons or lorries. The majority used vertical water tubed boilers. The design is reminiscent of a Russian samovar. Typically such boilers had a firebox that was surrounded by a water jacket, then tubes which ran through the fire and finally a coil on top which made super-heated steam. A lorrie's boiler was usually fabricated from large and heavy boilerplate, unlike a steam generator.

Odds & Ends

Pressure cookers for food have exploded on several occasions throughout the years. It's usually a problem with dirty or clogged up safety valves. If one (the type that sits on a stove top) blows, the lid deforms and hits the ceiling, food flies all over room and the pot might deform the stove top itself. Pressure cookers generally operate at 2 atmospheres (14.7 outside + about 15 PSIG inside) which allows food to cook faster because the temperature (max 212 – 250°F) can climb above boiling. Usually pressure cookers will have two safety devices on the lid; a spring-loaded valve that opens automatically and a weighted valve. This weighted valve consist of a stem that protrudes from the lid and a calibrated weight (the bobber which sits upon it). This valve stem can serve as an overflow to release excess vapor pressure. At optimum pressure the bobber is intended to wobble back and forth over a little column of steam, thus regulated by releasing little puffs of steam with each swing. Removing the weighted bobber at this stage results in a forceful jet of steam that jumps up several feet high. In an emergency the spring-loaded valve might trip, but the bobber could blow off its stem and release pressure first.

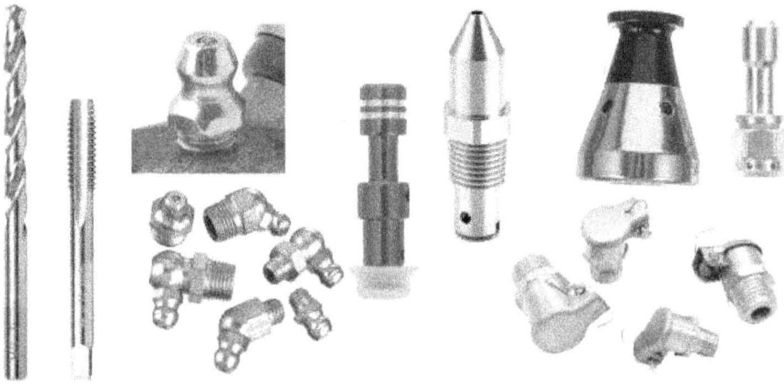

If constructing a steam system, some useful accessories can be added by drilling holes. Some holes might require tapping in screw threads and others may not. "Zirk fittings" or grease fittings might be installed near bearings in the engine or other points of friction. Old fashioned oil caps can also be found. Squirting oil from an oil can onto places of friction is part of a steam engine's daily maintenance.

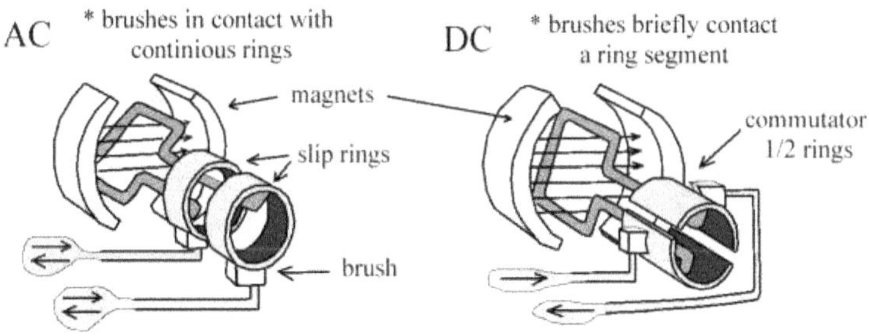

Back during the Depression it was not uncommon for remote settlers on the open range to rig up a windmill to generate a little power. They'd use a car's generator and a battery, to drive a light bulb or perhaps a radio. Unfortunately most automobile makers quit making generators by the early 1960's, and turned to **alternators**. The early car alternators were "**field wound** alternators", which made extra power. But now the new AC current needed to be rectified to DC current, before the energy could be stored in a lead-acid battery. Alternating current (switching back and forth between burst of positive and negative electron flow) however requires a **rectification** circuit to make it all DC voltage. In the previous image of a DC generator, current rectification was accomplished by using commutator rings.

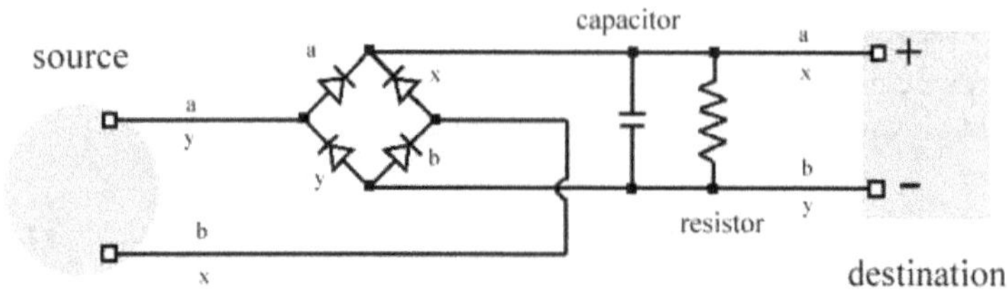

The simple circuit above is a "**full wave bridge**" employing a reservoir (smoothing) capacitor and a bleeder resistor. Diodes allow electron flow in only one direction. No matter which polarity or path the electrons take from the source, they'll end up rectified (DC) at the destination. The capacitor helps to smooth pulses of varying magnitude. Capacitors can hold dangerous, even deadly voltage, so the resistor discharges the capacitor safely. An older car might have an **externally regulated** alternator, which means that a separate, accessible voltage regulator is mounted somewhere like on the firewall (not built into the alternator).

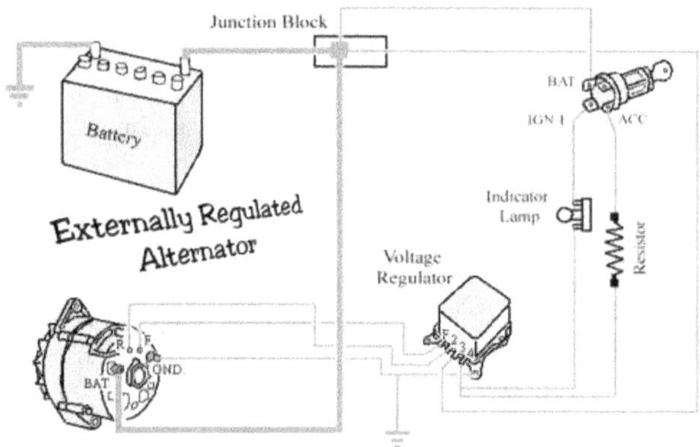

Alternators must waste some power, by creating a magnetic field current – that's used to induce voltage in the alternator's stator winding's. Originally and until the 1970's, up to 40 watts in a field-wound alternator was used to excite or energize the electromagnet field before it put any current out. Today modern car alternators typically waste only about 2 to 5 amps to excite or energize the electromagnet field. The alternators being put into autos these days are "**internally regulated**". Which means basically that the voltage regulating electronics have been reduced in size (to tiny IC (Integrated Circuit) dimensions) and placed inside the alternator itself. So not only are these new alternators harder to modify or experiment with, but they are also intentionally designed to be harder to dissemble and to repair.

Most contemporary car alternators have field wound stators and three sets of windings that generate three-phase alternating current.

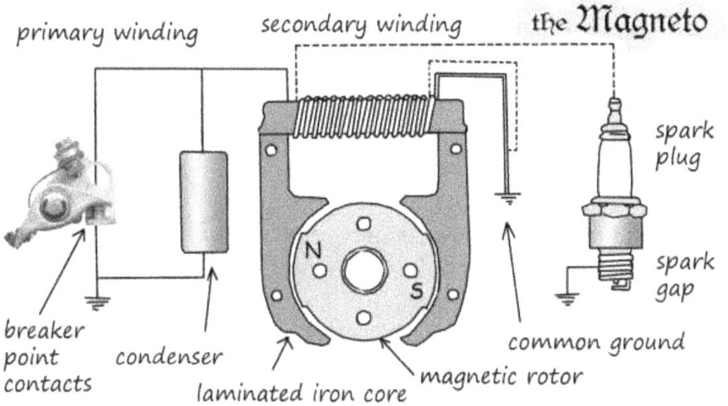

Above is a simple magneto as might be found in a common gasoline powered lawnmower. "**Permanent magnet alternators**" and "**magnetos**" are similar but have differences. Both produce AC current with stationary coils and rotating magnets. A permanent magnet alternator produces continuous AC. But a magneto generates (non continuous) periodic pulses of high-voltage AC. Magnetos are often chosen for ignition systems in simple engines because they're very reliable and don't need battery support. Stand alone magnetos power the spark plugs in chainsaws, lawnmowers, old Ford Model T's and almost all piston driven aircraft.

If converting a small 4-stroke engine to steam, it may or may not be worth the trouble to attempt salvage a little energy from the original magneto that came with the engine. A magneto common to a lawnmower or to a rototiller's engine as shown earlier, will be mounted behind or nearby the heavy flywheel. The flywheel will be mounted somehow with a magnet that when rotating, induces current within the magneto's core. This current won't be much (perhaps 5 to 50 milliamps, or 0.005 to 0.05 amperes). What the secondary winding on the magneto's transformer does – is to jump that tiny amperage up to a very high voltage, which can then fire the spark plug. Depending upon RPM or other factors – that voltage might be raised to 10,000 to 75,000 volts; a very high-voltage is required for a spark to jump an air gap. Since (volts x amps = watts) it seems evident that that little magneto could easily be made to power a couple of light bulbs or to top off a deep cycle battery sink. A single cylinder lawnmower's magneto might be commutated to use only the negative burst of voltage, to power a single spark plug.

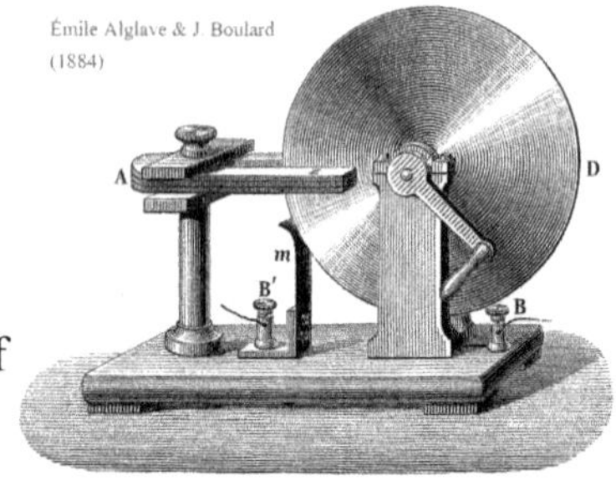

To the right is a 'Faraday disk' – the first practical electric generator. Actually a type of **dynamo** – that generates electricity by operating through a magnetic field. A copper disc was rotated between the poles of a horseshoe magnet, to produce a small DC voltage. But this contraption used no commutator. Today however, a dynamo is a generator that uses a commutator.

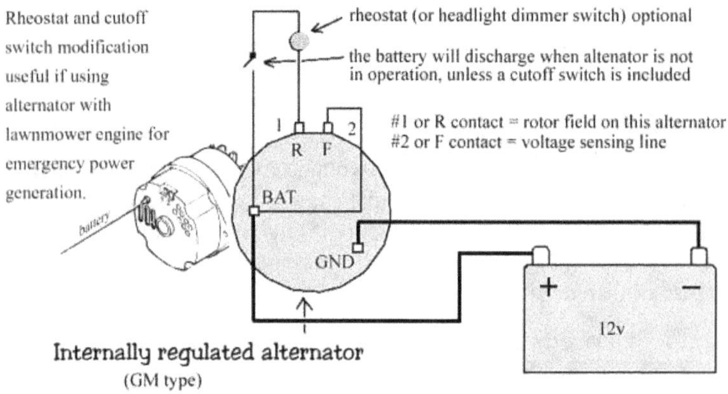

Driving a modern alternator with a windmill and belt today; won't work as easily as it did in the 1930's. Mainly because a modern car's alternator needs to spin about 600 rpm before its electromagnet field gets excited enough to put any current out at all. Windmills and watermills generally do not spin nearly fast enough. That means that if a car's alternator is used (not really a desirable choice of power generator to begin with) that some gearing will need to be rigged up.

One of many issues with wind-power- is insufficient wind persistence. Most often the wind flow is not consistent enough to justify the expense of erecting a wind turbine in the first place.

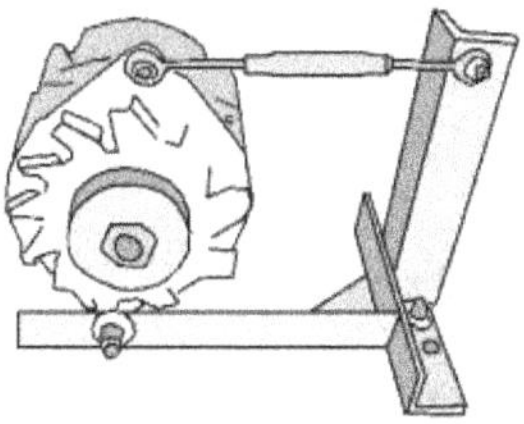

This image just shows how to weld or bolt some pieces of angle iron together, to make a simple framework for holding an alternator firmly. A turnbuckle is used to adjust a fan belt's tension.

Let's assume that gearing is needed to insure that an alternator's pulley rotates fast enough to insure good performance. Assume that the alternator's pulley is 3 inches in diameter. In the image below, let pulley B be the "driven pulley" at the alternator, and let pulley A be the "driving pulley" with an unknown diameter. *(Gearing ratios work the same way as pulley ratios, except that rather than* ***diameters****, the number of* ***gear teeth*** *are counted).*

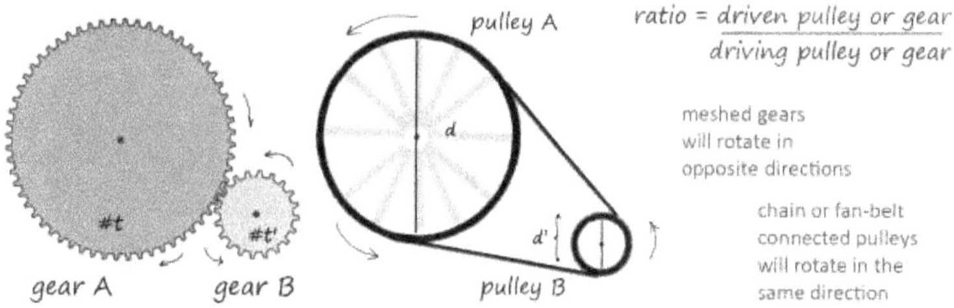

Pulley Ratio = (Diameter of Driving Pulley / Diameter of Driven Pulley) or in this instance: R = A/B.

Assuming that after much sampling of wind speed, that a particular windmill averages 58 rpm for long periods, how big must the driving pulley (A) be, to make the driven pulley (B), rotate at least 600 rpm?

This is a **continuously variable transmission** (CVT). An automatic transmission found in snowmobiles, go-carts and in some cars. How this might affect the output performance of a wind turbine, or of a small steam engine - is a subject for future speculation; because now the subject is being changed.

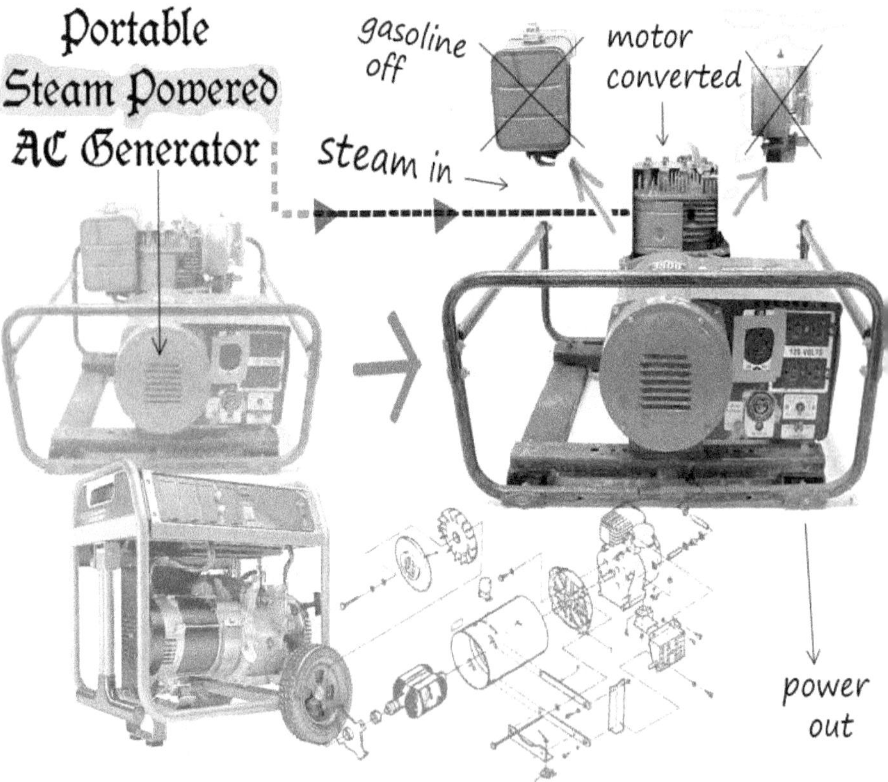
Portable
Steam Powered
AC Generator
gasoline
off
motor
converted
steam in
power
out

Chapter 4

Solar Energy

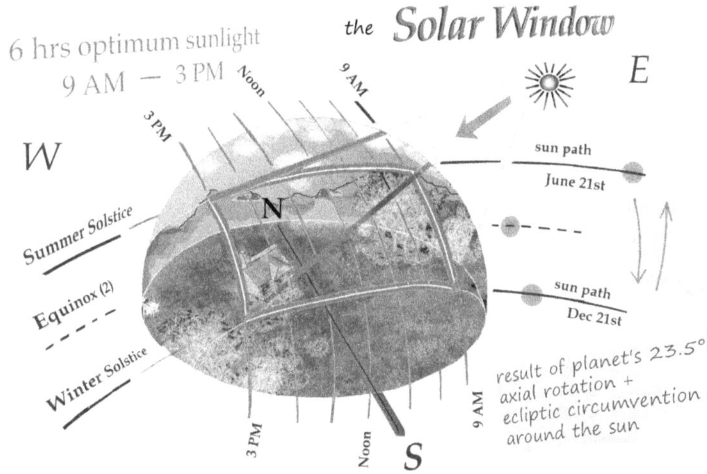

More than half of the solar emission aimed at us never strikes the surface of the earth. Instead, that energy is absorbed or reflected away; which is a good thing. Without this blockage by the atmosphere, humans would not survive exposure to the sun's full spectrum of radiation.

Even the best solar panels are inefficient. The glazing or protective substrate covering a solar panel will diminish the amount of energy obtained and converted. Then, for a typical PV panel, a big portion of the light spectrum that manages to reach one of its silicon cells cannot be used; but instead just manages to turn into heat, which further reduces a panel's efficiency.

But there are two separate categories of solar panel that can be fabricated: **photovoltaic** (PV) and **thermal collector**. Glass panels on a rooftop might all look the same from a distance, but inside the components can differ vastly. Solar research and technology has been gaining by leaps and bounds recently. New **hybrid panels** are entering the market – that smartly combine both PV and thermal collection at the same time!

This section (or chapter if it grows into one) is concerned with inspiring young builders to build smartly, by incorporating solar planning into their design from the beginning. Especially ways to make use of **passive solar** techniques, and how to fabricate their own space warming, air circulating thermal collector solar panels themselves if they wish. While most traditional suburban home owners and off-griders both will be interested in photovoltaic panels for electricity, these are currently far too complicated for a "do-it-yourself-er" to make at home. And expensive. But building homemade thermal collectors is just the opposite.

Since heating and cooling consume more power by far than anything else in a house, it makes good sense to at least mount some thermal panels on one's roof. This (thermal) endeavor begs to be discussed immediately, but the perfunctory task of explaining a little about photovoltaic technology should come first. Even an adequate cursory explanation though, is a daunting task.

A Reminder:

Without preaching a biology lesson here, the energy from the sun is directly responsible for all life on this planet. Most of the energy that mankind is able to harness today, was produced because of the sun. Oil, coal, natural gas and firewood are natural hydrocarbon fuels, produced by the prerequisite participation of organisms grown from sunlight. So called "**alternative energy**" like wind and water power - also come from the sun. Wind currents are caused by the sun warming the air; as thermals rise and the earth spins - these are moved and displaced by denser, colder air. Likewise the sun's energy is ultimately responsible for distributing moisture as snow and rainwater water to higher elevations. With the help of gravity, this water them moves downhill and supplies the kinetic energy used to power watermills and hydroelectric generators. Therefore wind power and hydroelectricity are really, just solar energy converted.

The sun appears to move. But it is we on earth that spin daily, while also orbiting around it yearly**.** The dome or half globe shaped introductory picture of this chapter attempts to explain the construct of a "**solar window**". (This same picture will be presented momentarily, in larger scale). This solar window is the optimum zone, from which a fixed solar panel can be expected to collect the best average energy year round. Outside of this box, solar collection will be diminished. For the East/West axis, a period of **six hours daily** is centered upon high noon. For the (North/ South) axis however, the box enfolds the optimal region of sky when considering the earth's cyclic **orbit of the sun**; from winter solstice to summer solstice and then back again.

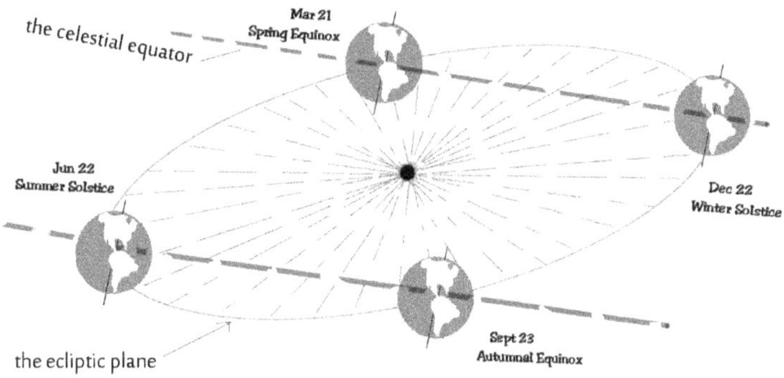

With a typical PV panel approximately 80% (or more) of any photons that actually manage to hit it, are wasted. That (arbitrary) 20% efficiency of photovoltaic panel diminish even more as it gets warmed up. And then a module (or panel), deteriorates or grows less efficient with age. Thermal collecting type panels are much more efficient than PV types, when considering energy conversion ratios. However, no solar panel is going to preform optimally unless it is **pointed correctly**.

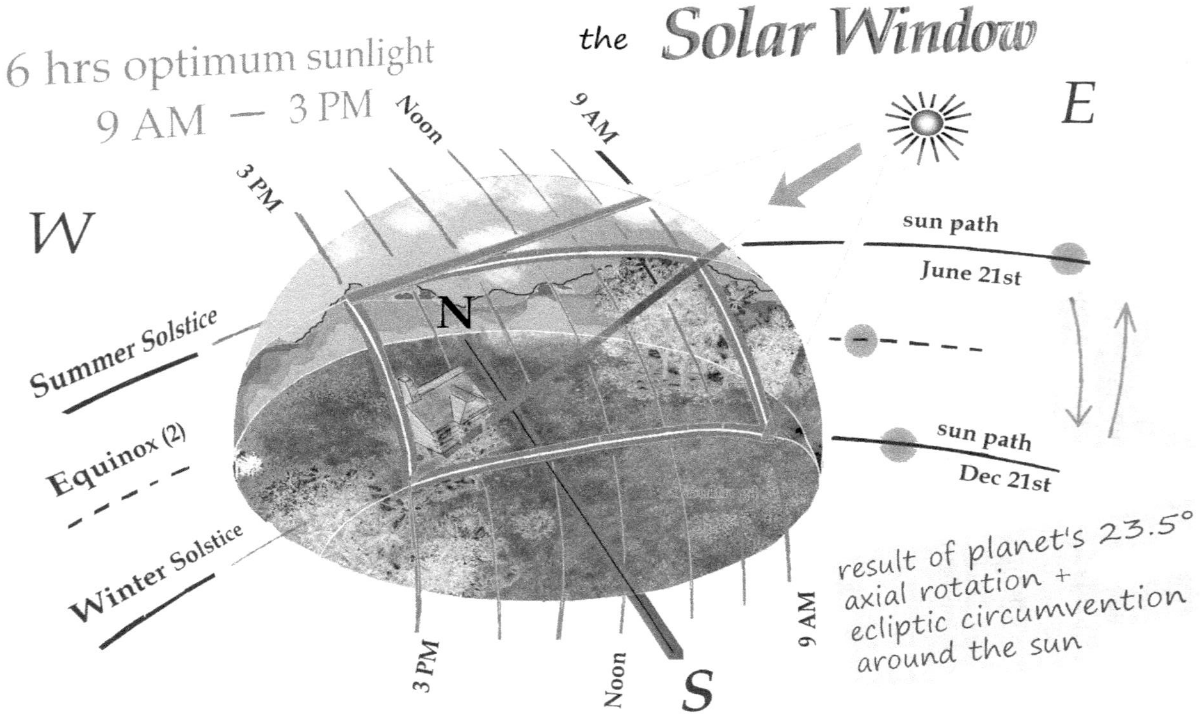
the Solar Window
6 hrs optimum sunlight
9 AM — 3 PM
W
E
N
S
3 PM
Noon
9 AM
Summer Solstice
Equinox (2)
Winter Solstice
sun path
June 21st
sun path
Dec 21st
3 PM
Noon
9 AM
result of planet's 23.5°
axial rotation +
ecliptic circumvention
around the sun

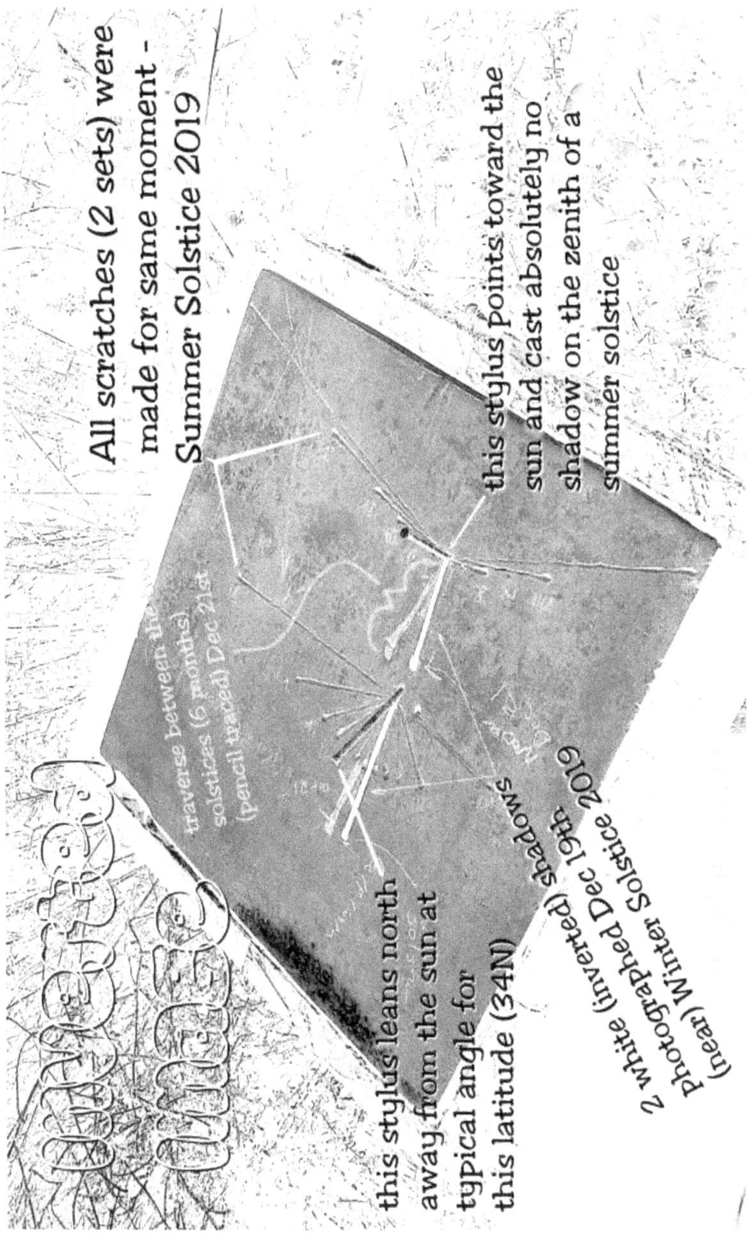

<Above>: Two nails inserted into a soft Mexican saltillo tile, that's been mounted on a tree stump. To study the sun's path...

A photovoltaic module or **panel** is made of a matrix of smaller individual PV **cells** - linked together. The number of cells might vary between 36 to 144 cells; but 60 – 72 cell modules are the most common configuration.

In an individual photovoltaic solar cell, an electrical charge is generated when photons excite the electrons in a semiconductor.

There are more than two dozen different types of photovoltaic solar cell technologies, and growing. But the market dominate technologies can be broadly separated into three general groups. Monocrystalline, pollycrystaline and thin-film.

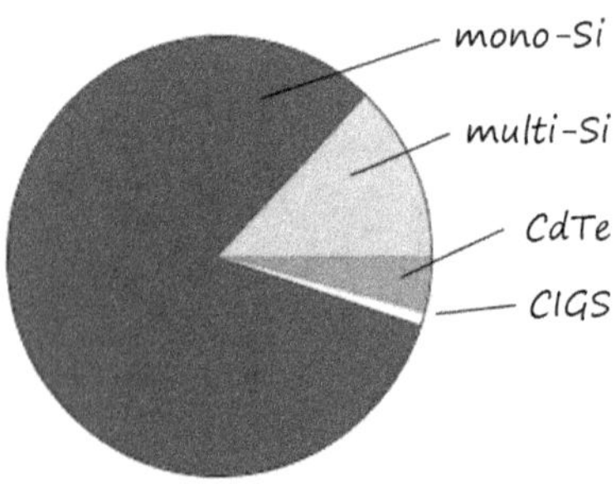

The original, the most expensive, the highest efficiency and the longest lasting type of solar cell is the **mono-Si** (**monocrystalline**) type. *[The "Si" in mono-Si stands for Silicon (atomic # 14), a prominent semi-conductor and the 8th most common element in the universe; (2nd most in the earth's crust)].* Silica is an oxide of silicon. But silicon (not the polymer silicone) is seldom ever found in a pure state. It has several allotropes, which help contribute to the confusing naming of emerging PV technologies. The "mono" in mono-Si means a thin singular slice or wafer of crystal cut from a silicon crystal. These crystals aren't found in nature – they have to be grown. In factories, and within laboratory like 'clean rooms'. Every effort is made to keep contaminants out of the molten silicon, as the crystals mature.

Molten silicon is "doped" with a trace element like boron or antimony, to alter the crystal's lattice (3-dimensional arrangement) and to thereby influence its electrical properties. Then in (the prominent Czochralski process) a process that is reminiscent of candle-making, a seed crystal is placed on the end of a rod and gently dipped into cauldron of molten silicon. The seed will influence the nucleation of the crystal lattice. Within an argon (an inert gas) filled chamber, this seed crystal at the tip of the rod slowly rotates, as it is pulled in and out of the hot 2,597°F / 1,425°C (past the melting points of gold, copper or silver) liquefied silicon, repeatedly. It grows slowly; taking from a week to a month. When expanded to desired specification such a monocrystal **ingot** (called a **boule**), might be 200 mm - 300 mm (8" - 12") in diameter, 2 metres long and several hundred kilograms in weight. The work is far from done.

In a a very delicate process but one that's reminiscent of slicing a salami sausage roll, thin wafers (0.2–0.75 mm thick) must be cut from from the main crystal with a diamond saw. The cutting is wasteful of the hyper-pure silicon and that expensive waste, now a dust – has been contaminated. Once cut the "wafer" might then be bombarded with radiation to implant ions and then acid etching and polishing might finally follow.

The first silicon photovoltaic cell was created more than 70 years ago, by Bell Laboratory (1954). Before that (during WWII) Bell Labs was using the **CZ method** (just explained / Czochralski's method was developed in a German lab in WWI) to make high-quality germanium crystals, used in radios.

There are several other methods used to grow crystals which have both merits and drawbacks, but the details won't be mentioned here. The CZ process alone accounts for 90% of all single crystal silicon used by electronics industry to make semiconductor devices.

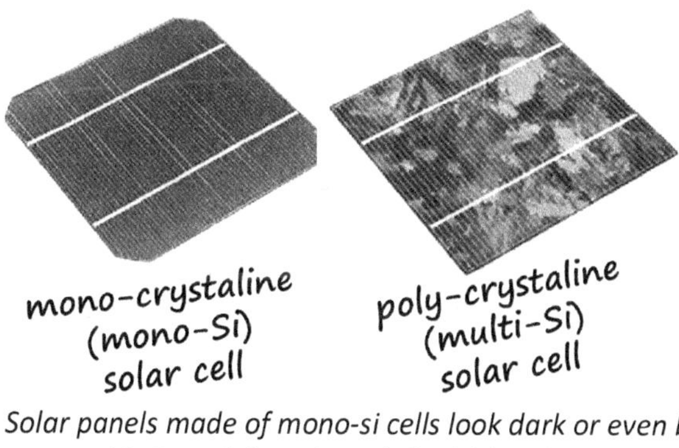

- *Solar panels made of mono-si cells look dark or even black whereas multi_Si modules primarily look blueish.*

- *By 1958 the Naval Research Laboratory used the first ever photovoltaic panel when it launched the Vanguard 1 satellite. The satellite used a tiny one-watt panel to power its radio. (The body of the satellite was only 6 inches in diameter, it had five antennas that poked out about 1.5 feet from the center of the body, and the hole thing weighed only 3.2 pounds).*

The next type PV panel (to be rated by popularity in the preceding pie chart image) is made with **multi-Si** type (or **polycrystalline** silicon) solar cells. Wafers cut from such crystals have a "metal flake" look rather than the homogeneous look of a monocrystalline wafer. Multi-Si panels are easier to make, cheaper to buy and are less efficient (15 -17%) than mono_Si panels (which are about 19% efficient). Multi-Si panels have been reducing in popularity in recent years, mainly because the more efficient and longer lasting mono-Si units have been coming down in price.

The other abbreviations in the previous pie-chart illustration are **CdTe** (Cadmium Telluride) and **CIGS** (Copper Indium Gallium Selenide) types. Both of these are termed "**Thin-film**" photovoltaic cells. A third type of thin-film PV cell is the **a-Si** (amorphous silicon) non-crystalline form of silicon used for solar cells, but presumably commands only about 0.1 % market share.

Thin-Film PV cells can be deposited in thin layers onto flexible substrates like glass, metal and plastic. Yes flexible glass (like Corning's 'Willow Glass') is a real thing; some smart phones use it.

So the 1st generation solar cells (mono-Si and multi_Si) can be 'PERKed'. **PERC** can stand for either - *Passivated Emitter and Rear Cell* or *Passivated Emitter and Rear Contact.* Basically it is a feature that adds a **back surface passivation** layer, which both reflects light back through the silicon to absorb a little more energy and also reduces heat a bit by reflecting wavelengths of light that can't be used anyway.

Most thin-film solar cells are classified as 2nd generation (CdTe, CIGS, a-Si, and GaAs {gallium arsenide}. Emerging 3rd generation cells include CZTS {copper zinc tin sulfide}, PSC {perovskite structured} and organic {small organic molecules or organic polymers that conduct}.

Above is a photo of **the failed 3,500 acre, Ivanpah solar plant** in California's Mojave desert. It has 300,000 mirrors mounted on 173,500 computer controlled heliostat turntables, which were pointed at three 459-foot towers. This boondoggle which cost some $2.2 billion was completed in 2014, and now is set to close in 2026, after failing to generate solar energy efficiently. Oh; and birds misfortunate enough to fly over it were fried or vaporized by the thousands.

> Solyndra was a manufacturer of cylindrical panels of CIGS (Copper Indium Gallium Selenide) thin film solar cells. Until it went bankrupt, that is. Based in Fremont, California the company was apparently not able to compete with conventional solar panel manufacturers of crystalline silicon. The company was accused of fabricating *"inaccurate information to mislead the Department of Energy"* in obtaining a government loan guarantee. Solyndra left taxpayers holding the bag for $535 million in loans after 2011.

Although new cell chemistries for solar cells are being developed every day, it appears that PV tech has hit a head-wall, or has more or less reached maturity. The practical efficiency maximums have sort of topped out somewhere around 22%. Thermal collectors and hybrid panels however remain more efficient at converting one energy into another.

> *"The silicon wafer in a solar cell can only absorb light in wavelengths up to 1,180 nanometers (nm), and higher-*

wavelength light waves pass through the silicon".

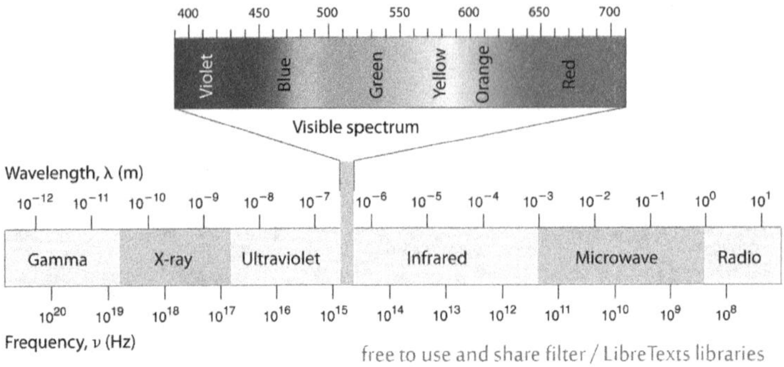

free to use and share filter / LibreTexts libraries

The visible portion of the electromagnetic spectrum is the narrow region with wavelengths between 400 and 700 nm. Glancing above, that means that almost every wavelength and frequency to the left of the visible spectrum is **useless** to a silicon cell, but still creates heat in a panel

There are several photovoltaic cell technologies and to explain all their unique differences in adequate detail would be tedious (so it won't be done here). But prominent types include Al-BSF (Aluminum Back Surface Field), IBC (Interdigitated Back Contact) , PERC (Passivated Emitter Rear Contact) and Bifacial. Then some of these can come in a "p-type" or "n-type" configuration.

The older, once common but now rarely manufactured photovoltaic cell is the **Al-BSF** (Aluminum Back Surface Field) type.

More recent **IBC** crystalline silicon (c-Si) cells are complicated to make, but introduce benefits including: reduced series resistance, reduced losses by shading and increased power output per square meter of surface area.

This may be an oversimplification but, a **PERC** cell is like sticking a mirror on the back side of a solar cell. So the cell generates many free electrons on the first pass of light, but then bounces the light back through the substrates to generate a few more.

Most PV solar cells and therefore panels, are monofacial; but **bifacial** cells capture sunlight from both sides. Panels made this way have clear glass coatings on both sides and are elevated and angled away from the ground or rooftop that they are mounted on, so that they can capture reflected and diffused light from the backside, also.

Most P-type and N-type solar cells are the same, with only slight manufacturing differences. **Boron** is used for doping P-type solar panels and these work fine in space; but deteriorate quicker than normal when exposed to oxygen in our atmosphere. N-type solar panels doped with **phosphorous** effectively become immune to the boron-oxygen defect. N-type solar panels also feature a higher conversion efficiency over their lifespan.

Thermal Solar

Solar Photovoltaic Thermal **Hybrid Panels** or **Photovoltaic thermal collectors**, typically abbreviated as PVT (or **PV/T**) can approach 85% efficiency because they capture both thermal and photovoltaic energy at the same time. Potentially a hybrid panel should be able to warm and circulate air, but most examples use liquid.

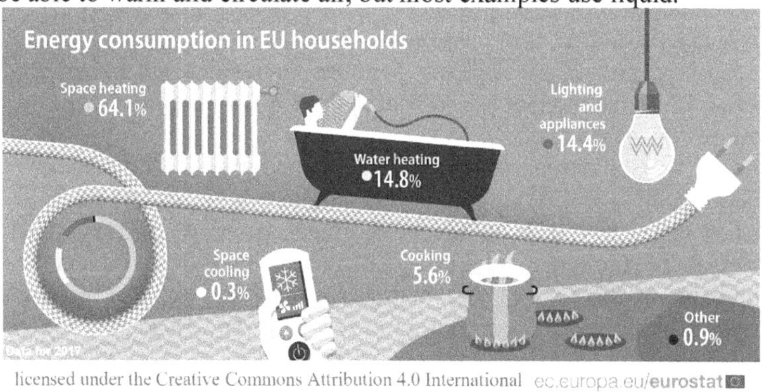

licensed under the Creative Commons Attribution 4.0 International ec.europa.eu/eurostat

The image above portends that the typical European household expends almost 70% of its energy consumption - by **heating**. The statistics have changed little if any since then (2017). However air conditioning was not mentioned. Which subtlety reflects that most European domiciles don't enjoy air-conditioning (due to prohibitively expensive electricity).

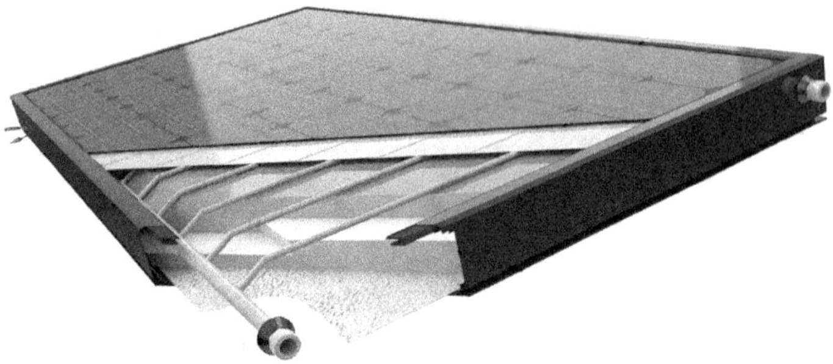

With a hybrid panel the high frequency light that passes through the silicon, strikes an absorber; a metal sheet beneath which is soldered to pipes that are usually copper. Unlike a typical PV panel, this type must be well insulated to preserve the heat energy and prevent its loss.

The fluid circulated in the piping is seldom ever water, which could freeze. Instead glycol or some other antifreeze agent is circulated in the panel, but is pumped into a heat ex-changer somewhere, where the heat (only) is then transferred to the house's hot water. The addition of the thermal feature likely adds only 20% to the cost to the manufacture of a regular PV panel. However the plumbing cost to connect this panel to a home may be an unwelcome expense. Hybrid (PV/T) solar panels can approach 85% efficiency because they capture and make use of both photovoltaic and thermal energy at the same time; processing the two energies in symbiosis. Since normal silicon PV cells operate most effectively in cooler temperatures, they can benefit from having that excess heat drained off or circulated away before they get too hot.

In big commercial solar farms where hundreds of PV panels are working, these might be cooled with fluid in many cases. If that fluid is hot water - then that can be cooled off and reused or simply dumped back onto the ground. For a private residence however, a problem might arise. PV/T panels might create too much hot water, especially in the summertime. Excess hot water might be very desirable however to a homeowner who also has a swimming pool.

Parabolic Trough

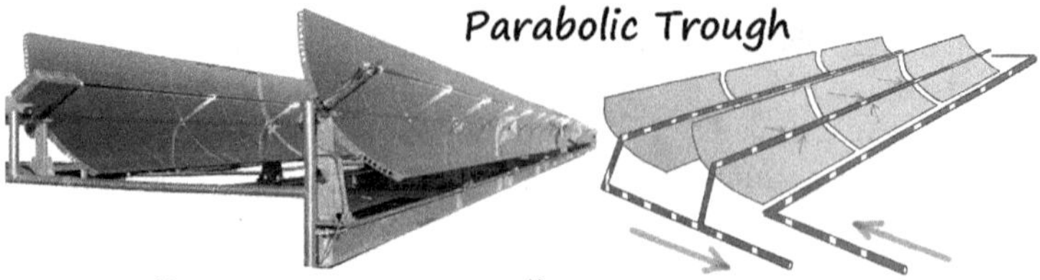

- a linear concentrator collector

Before dialing in on solar thermal plans for individual homes, a glance should be taken of the technology that big solar plants use. Not all big solar power plants use silicone photovoltaic panels to make electrical power. Some collect heat only and then make electricity from that. In general, three types of solar thermal power plants exist that do this: linear concentrators, solar power towers and solar dish/engines.

The name "**linear concentrator**" - because it is a long straight line of components. Looking like a long trough that cows or pigs in a feed lot might eat from, this type of solar concentrator pumps a liquid through a receiver or absorber tube that passes all the other reflectors down the line.

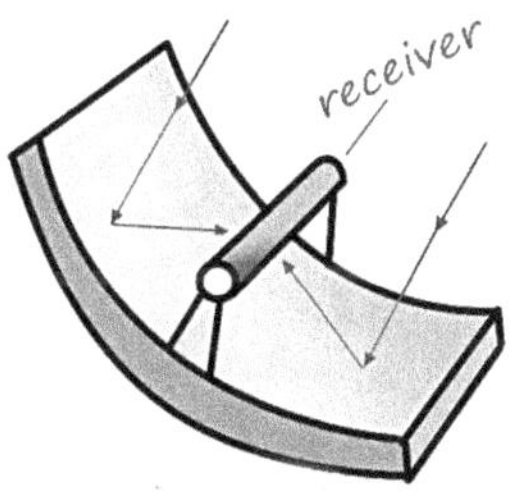

The hot fluid is then conducted to a steam generator, where steam is created, which spins a turbine; which makes electricity. The parabolic shape of the mirror pretty much ensures that all of the photons that hit it, will be re-directed towards the parabola's focal point; that's where the **absorber tube** will be. The long axis of each trough is aligned perpendicular to North-South, and minimal adjustment is needed East to West.

Twenty years ago the liquid in the tubes might have been a special oil, but now such systems are increasingly circulating liquids with better heat exchange properties, like molten salt. **Molten salt** is preferable for a couple of reasons, but mainly it facilitates the storage of thermal energy for later use. The molten salt used is not pure NaCl (table salt, sodium chloride) but a homogeneous mixture of several things, like potassium, sodium and calcium nitrates. This molten material is made less corrosive, its melting point is substantially reduced and it acquires physical and thermal properties that make it useful for energy storage. Molten salt works at a higher temperature but can still be pumped around and stored in tanks, just like water. And these salts are not allowed to drop in temperature, where they could solidify or freeze again.

Linear-Fresnel system

Another, less common type of linear concentrator is called a **Linear-Fresnel concentrator**. Here there is an absorber at the central focal point; just as with the parabolic trough. However the inclination of rows of moving planar (flat) mirrors is continually adjusted to keep the reflection focused on the absorber.

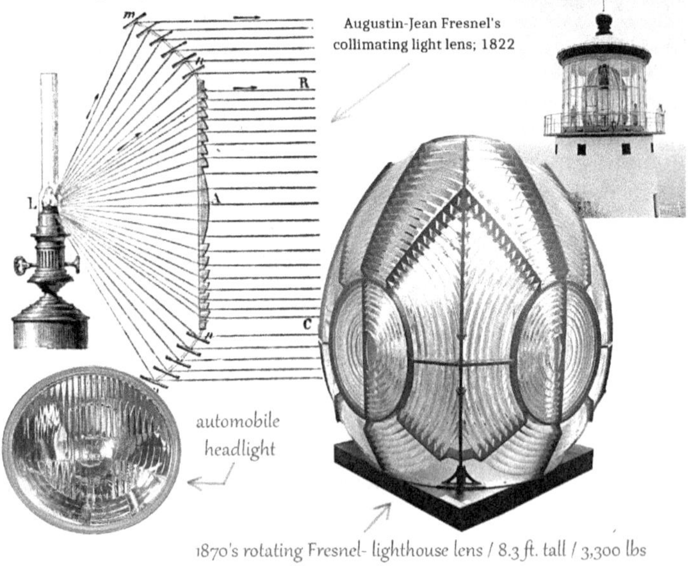
Augustin-Jean Fresnel's collimating light lens; 1822

automobile headlight

1870's rotating Fresnel- lighthouse lens / 8.3 ft. tall / 3,300 lbs
Musée national de la Marine, Paris /CeCILL image license (French)

At right is the first **"solar power tower"** named the PS10 Solar Power Plant. Near Seville, in Andalusia, Spain; construction on PS10 began in 2004.

Creative Commons Attribution 3.0 license

At the top of the 40 story tower reside a solar receiver and steam turbine that drives a generator. From the ground 624 big (1,300 sq.ft. or 120 sq.m) **heliostats** (moving mirrors) track the sun and reflect light towards the receiver.

PS10 receives very good sunlight year round, considering its 37°26' latitude. The twenty year old PS10 tower is rated at 10 megawatts electrical capacity and the facility can store solar energy for a short term (30 min) in the form of superheated and pressurized water. (Storage is within tanks, at 545° F or 285° C, and at 725 psi or 50 bar). The energy conversion efficiency of this power plant is approximately 17%. That does not sound like a lot, but remember that the optimal efficiency of a regular PV cell only ranges between 15% to 22%. When energy is converted from one form into another, about 30% of the original energy is lost. With a concentrated solar thermal system like this PS10 tower; turning sunlight into pure heat is one conversion, heat into steam pressure is another conversion and steam through a turbine to make electricity is a third conversion.

In recent years newer technology has come along in the form of working fluids like molten salt **or** molten metals, that can be used to store more potential energy away because they work at higher temperatures and don't require pressure vessels like saturated steam does. The melting point of table salt is about 1,474 °F (801 °C), but the molten salt we are talking about starts melting at a much lower 268 °F (131 °C). A solar trough or solar tower might heat this to 1,051 °F; whereupon it might be used to create electricity immediately, or stored away in a huge insulated tank to be used a week later.

Creative Commons Attribution 3.0 license. Author : Koza1983

Above, PS10 is in the background but the younger PS20 (commissioned in 2009) sits in the foreground. PS20 is a 20 megawatt (MW) tower and draws from a field of 1,255 mirrored heliostats.

Solar Dish/Engine System

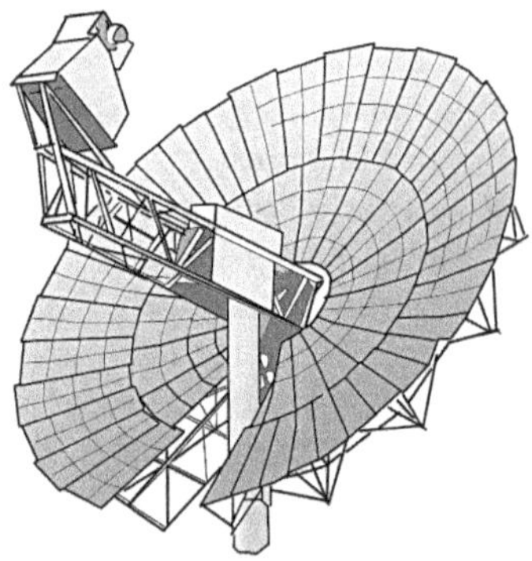

The **Dish/ Engine System** is usually deployed in arrays of multiple dishes that look combined, but each one is basically a stand alone unit in itself.

Capable of producing 3 to 25 kilowatts of electricity – if that is the intended work.

But sometimes mechanical work rather than power generation is the disk/engine's purpose. A dish/engine can convert more than 30% of the total energy in the sunlight reaching the mirrors, into electricity. The *solar concentrator* (the dish part) focuses light upon the *power conversion unit*, which is mounted at the focal point. This power conversion unit houses a *thermal receiver* and an *engine/generator*. The receiver passes thermal energy on to the engine/generator which converts that into mechanical energy first, and then probably into electricity.

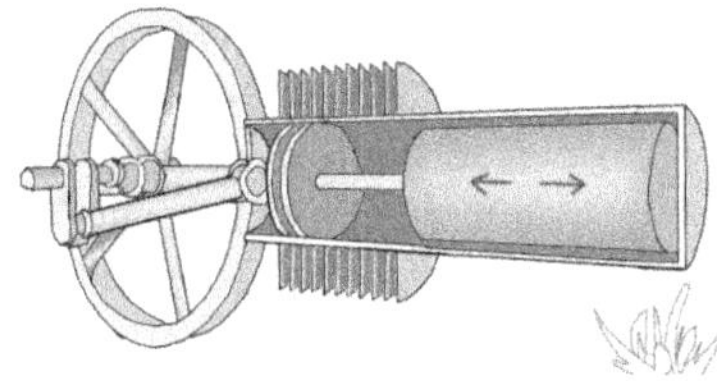

A **"Stirling engine"** uses heated air to move pistons and create mechanical power. It is essential for a dish/engine to track the sun throughout the day. Naturally this tracking (movement) adds complication and expense. Robert Sterling made the patent application in 1816, for an engine that worked by means of simply heating and cooling air. NASA invented a derivative solar-powered Stirling engine in 1976 for the purpose of pumping irrigation water. Today's Stirling engines might look much differently, but share the same principle.

Inventors have been experimenting with concentrated solar power for a very long time now; the idea is far from new. On occasion the ancients tried to blind their enemy with reflected sunlight. The inventor of the parabolic trough, a French academic named Augustin Mouchot was powering a steam engine with solar - in 1866. A decade later Mouchot won a gold metal at the world's fair (the 1878 Paris Exposition) for his work in producing ice, from concentrated solar heat.

A solar concentrator; Mouchot's largest sun machine at the 1878 Expo.

Next is a picture that is also over a century old. The dish powered a 10 hp steam engine by boiling water with reflected sunlight energy. The dish/engine was used to pump water for a farm, somewhere in or near Los Angeles, California, in 1901.

The caption included with the public domain image states: "*The mirror is mounted on an altazimuth mount, and after being pointed at the sun in the morning a drive system turns the mirror slowly to track sun across the sky all day. The engine requires about an hour to build up a 150 psi head of steam".*

Below is the world's largest **solar furnace**. It is located in the Pyrenees mountains, in France and next door to Spain. This furnace began operation in 1969 and has been used mainly as a research tool, although it can also generate 1 MW of electricity. It can obtain 4,530 °F (2,500 °C) in seconds. With a little more time it can reach and handle temperatures up to 6,330 °F (3,500 °C).

The first model for this style of furnace, originally grew during WW2, as an experiment to create an anti-aircraft weapon. Later, since it could heat up very quickly, this furnace technology was envisioned as being a practical method to use when melting ores.

More recently the furnace has proved to be very useful in aerospace research. This facility is unusual in that it bounces light between two sets of mirrors; some 9,600 of them. The concentrators in the parabola collect the reflected light from 63 sun tracking heliostats mounted on an opposing slope.

The "**GRid**" (Electrical Grid - a transmission system, an interconnected network of power lines and transformers for moving electric energy at high voltage) as yet, has no real facility to store electrical power. The GRid must constantly shift power here and there around the continent to supply demand. When demand exceeds the base load of an unvarying (constant output) power plant, **dispatchable generators** are hurriedly called online to makeup the discrepancy.

CSP (*Concentrated Solar Power / thermal solar*) technologies like the aforementioned 'parabolic trough', "solar power tower" and 'dish/engine system' are now being constructed with at least a modicum of **energy storage**. Excess or surplus hot oil or molten salt, can be pumped away into huge insulated tanks or underground reservoirs. When demanded, hot fluid is pumped toward steam turbines, which then produce electricity.

Spain and Portugal have 'set their caps' unsoundly at the present, upon the notion of going completely 100% solar. Murphy's Law (the adage that states: "anything that can go wrong will go wrong") was proven true once again. Spain (with 20,042 gigantic windmills) and Portugal (with another 3,000) combined with solar farms and hydro generated power projects, had an estimated 65.8 GW capacity online; hours before causing **Europe's largest blackout**. Spain's GRID had first gone completely solar-powered (for a few hours at any rate) on April 16th, 2025. But less than 2 weeks later, the embarrassing "**2025 Iberian Peninsula blackout**"[32] even struck parts of France and Belgium. All of Europe came close to being shut down due to Spain's cascading power problems. For 10 hours in most places (but 23 hours in others) subways, elevators and high speed electric trains - trapped their passengers. Cell phones, televisions and traffic lights went dark. Airport terminals became hubs of chaos and only hospitals equipped with backup diesel generators were capable of powering something so simple as a light-bulb.

No reason or excuse for this blackout has been agreed upon or forwarded by engineers that manage that GRID. But- *"The grid has some rather exacting operational requirements that the intermittent renewable generation technologies cannot fulfill".*[33]

Some engineers might simply blame the old GRID for its inadequacy. But it would also seem that politics and the Spanish GRID operator's over reliance upon renewable energy sources alone, was at fault. Missing were the traditional massive spinning turbines and generators with a constant torque; a

32 https://townhall.com/columnists/pauldriessen/2025/06/14/spains-impossible-dream-of-green-electricity-n2658547

33 https://www.manhattancontrarian.com/blog/2025-4-28-have-the-intermittent-energy-blackouts-begun

type of inertia that PV panels and windmills cannot provide. Their 'hell or high water' ambition to go completely solar, produced a grid that was more vulnerable to disruptions and subsequent frequency oscillations. Perhaps more molten salt storage, faster acting 'dispachable generators' and other improvements in technology can pick up the slack, in the future.

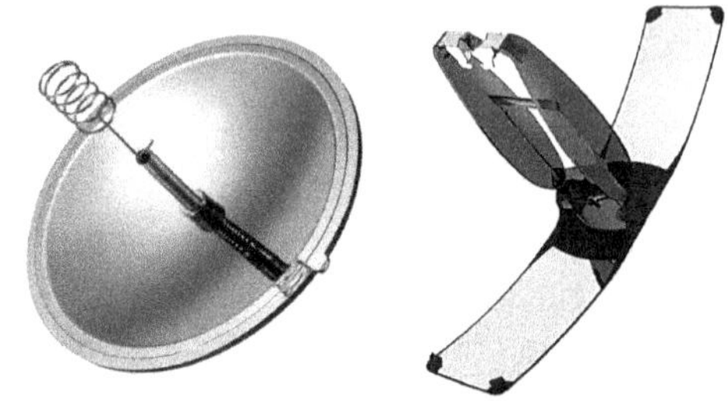

At left are two solar **cigarette lighters**. Both are tiny **CSP** (*Concentrated Solar Power* / thermal solar) devices. Both reflectors can be used to ignite cigarettes and with luck, perhaps campfires.

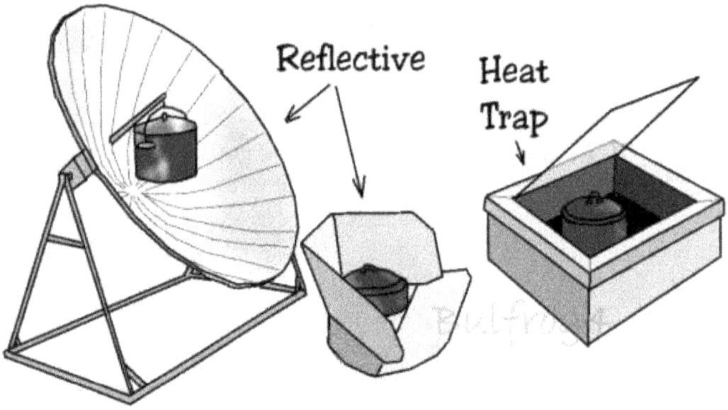

In the 1960's, "solar energy" suddenly became a fashionable new topic. People wishing to experiment were obliged to build their own solar ovens. A task that was easy enough to do then; and now. But today one can now order manufactured copies of those same ideas.

In the image above the parabolic disk shape on the left might be emulated by attaching aluminum foil to a parasol (umbrella). The shiner the surface, the better. The "**reflective oven**" in the middle can be made with flaps of cardboard covered in shiny foil or with newer reflective polyester films; **BoPET** (*biaxially oriented polyethylene terephthalate*) films like Mylar®.

The solar oven labeled "**heat trap**" can be made from a simple cardboard box and lidded with glass. But importantly this heat trap differs by requiring **insulation** and by being painted black on the inside. By trial and error it will be discovered that since temperatures inside such ovens can reach over 350 °F, materials like plastic and foam insulation might melt. Direct sunlight will cause cheap spray paint to fade. Incidentally all the pots holding the food in these examples should be metal, dark colored and lidded

Dutch National Archives, 1962 / Universal Public Domain Dedication

In order for CSP devices to survive any length of time outdoors, they need to be built well. The materials used will endure a pounding from the elements of mother nature. Repetitive high internal temperatures might do damage, if an oven is neglected and continuously exposed. But solar ovens or other cookers seem encouragingly practical for people living in impoverished conditions, who must cook but don't have electrical power, natural gas or adequate firewood.

Nowadays there are at least four basic types of CSP / solar cooker.

1) In a common **panel-cooker** at least three sides and a bottom are covered with reflective surfaces and the sun's rays are bounced toward the center. The whole thing can probably be folded up and stashed away. Panel cookers are slow and they don't preform well in low light conditions, but they are cheap and portable.
2) The box cooker or **box oven** differs by having a transparent lid. This traps the heat inside the insulated box. These can be high capacity (large volume) cookers that cook slowly and do not need to be attended to constantly. Some improved heat trap box ovens can reach 250 °F; even 350 °F (121 – 176 °C).
3) **Parabolic solar cookers** are disk shaped like satellite dishes or inverted umbrellas. The sunlight is reflected to a focal point where usually the bottom of a utensil, wok or pan is suspended. The heat can quickly become intense, easily reaching between 500°F (260°C) and 700°F (371°C). These cookers are fast and need to be attended to constantly. Generally they don't preform well in low light conditions and might be heavy and not very portable.

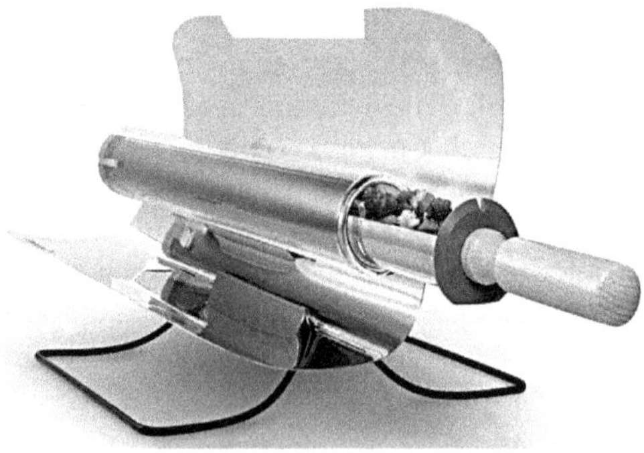

4) **Evacuated glass tube cookers** are something new. A double walled, tempered glass cylinder that has a vacuum and is dark tinted, sets inside some reflective panels. A tray of dry food can be slid in and out of the dark tube as can be a lidded canister (which must be vented to relive steam pressure). Perhaps lacking in capacity they still cook fast or in windy, cold or low light conditions. They are portable and light, but expensive.

A retired parabolic dish reflector which once captured satellite TV signals, can be converted into an efficient solar cooker. Ideally the dish should be mounted so that it tracks the sun. However fair results can be achieved by setting both the dish and the cooking pot on the ground and then focusing the reflected light. After some time passes, the sun will move and either the dish or the cooking pot will need to be moved also, to keep the heat concentrated.

The parabolic dish needs to be covered with some type of shiny, reflective, foil or film; by using contact cement or special spray adhesive. The 'feed support arm' (which normally holds out the 'feedhorn' and LNB) might need to be reconfigured. These types of antennas are "offset"; meaning the feedhorn is intentionally moved away from the center, in order to prevent blocking the signal. Satellite TV signals are very weak, because they are being sent from satellites in geostationary orbits, some 22,236 miles (35,786 km) away out in space. This all makes the aiming the dish very different from what might seem obvious.

Some people have repurposed satellite antennas, to get considerable gain (or longer reach), with terrestrial Wi-Fi equipment. These small dishes are very "directional", "high gain" antennas with a tight focus; which makes them difficult to aim. It can be expedient to mount the antenna dish upside down, when aiming (for Wi-Fi).

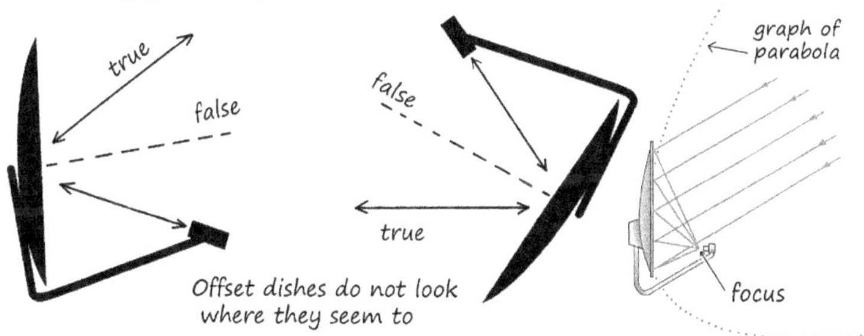

To find the hot-spot or **focus** for a dish that has been repurposed as a solar cooker however, a person just needs to employ a good deal of "fiddle around and find out".

For someone wanting to build their own parabolic reflector, determining its shape and establishing its focus is fairly straight forward (with a little math). The shape can be plotted out on graph paper, then transferred and cut out of cardboard. A cardboard model can be taped and glued and tested fairly expediently. Transferring a parabolic shape to flat metal is just as simple.

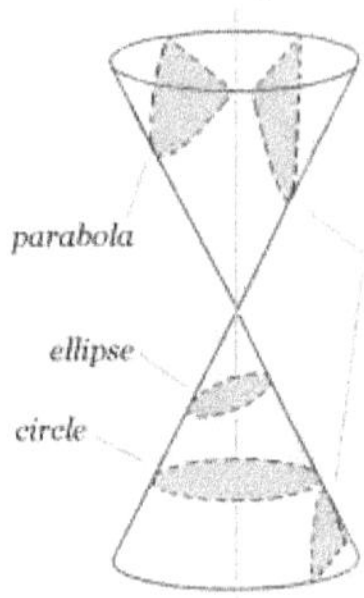

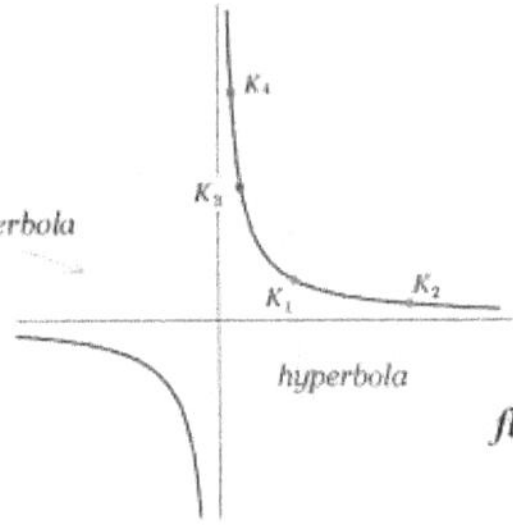

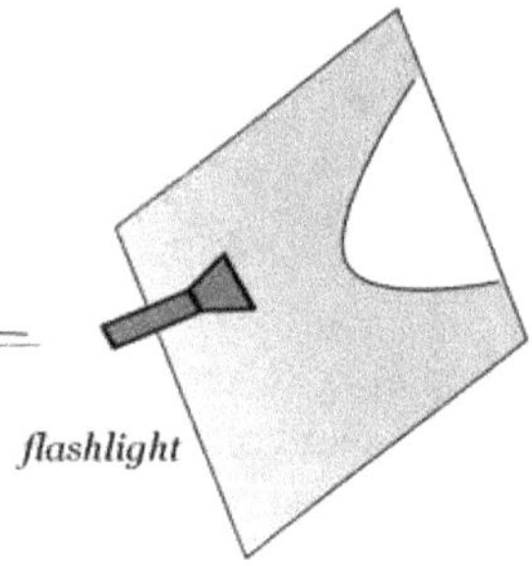

The Parabola

A parabola is a "conic section" or curve created when an imaginary plane intersects with a cone's surface. In geometry circles, eclipses and hyperbolas are defined as **conic sections**. A circle is simply a unique type of eclipse and a hyperbola has two components (or two parabolas as shown in the image).

Reflective parabolic surfaces have the ability to focus or **collimate** (to make parallel; line up, strengthen) waveforms; like sound waves or electromagnetic radiation, like light. A rooftop satellite dish antenna, or a dish-backed microphone that a detective or spy might aim, are parabolic surfaces at work. Internally, automobile headlights have parabolic shaped reflectors that can collimate the light emitted from a filament, into parallel light rays that then send a strong beam of light down the road at night.

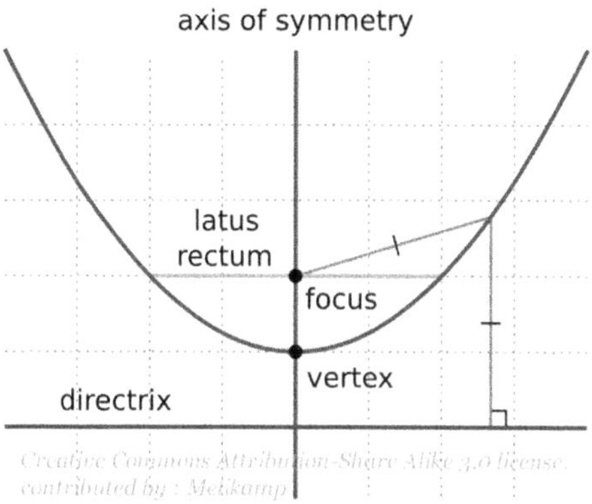

Algebraically the "U" shaped parabola corresponds to a **quadratic equation**. The general equation for a vertical parabola is $\mathbf{y=a(x-h)^2+k}$. For one that opens horizontally the equation is $\mathbf{x=a(y-h)^2+k}$; where (h,k) is the vertex of the parabola. The value of "**a**" determines both the width and direction of the parabola. If the vertex of a parabola is placed at the origin (0,0) of a graph - then the coordinates (h,k) can be dismissed; and the equation simplifies to $\mathbf{y=ax^2}$. (Or y = something x^2). In the following simple examples the coefficient "**a**" corresponds to the fractions ¼, 1/10 and 1/20 respectively.

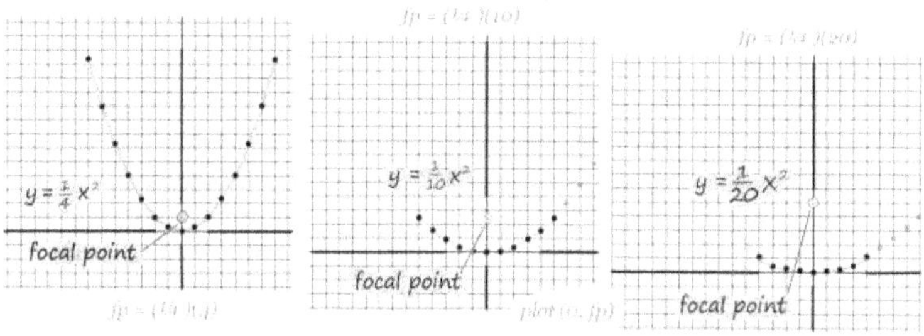

The **focal point** of photons from light bouncing from a parabola's mirrored surface, should collect at a focal length that is (¼) of the latius rectum's length, from the vertex. That means that in the above three examples, the focal points are { (¼ x 4) = **1**, (¼ x 10) = **2.5** and (¼ x 20) = **5** }. One can notice that as the parabola's curve flattens, the focal point will move out, or away.

----------------.----------------

If a coefficient of 1/40 is substituted in this simple equation, [y = (1/40)$(x)^2$] and the increments were measured in inches, then the hot-spot of a parabolic cooker should be 10 inches above the cooker's vertex. The feed horn of a typical satellite TV dish might be about 20" away from the vertex.

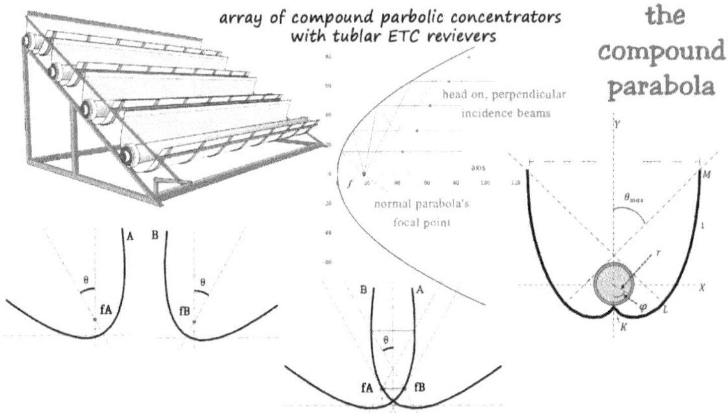

The normal parabola is ingenious and useful for collecting light and heat energy at the focal point. However only the light flying perpendicular into the parabola's "field" is collected. Indirect and ambient light is lost (potentially significant). The regular parabolic reflector must be continually adjusted to point directly at the sun to be effective. However the geometry of a **compound parabola** is more forgiving. Since much sunlight is scattered by the atmosphere, some of that otherwise lost indirect light can be captured by using a compound parabolic reflector. And a compound concentrator also requires less adjustment to track the sun. A tiny focal point is not ideal, when the target is a long and wide hot water pipe running above a parabolic trough type concentrator. Bur the reflector's curvature can be optimized to make light hit on all sides of a seemingly larger target. The solution is to use two parabolic curves as shown in the previous image.

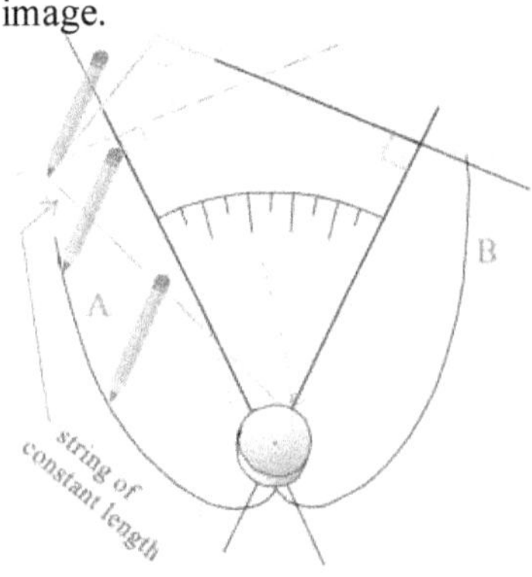

Without using math, two long lines can be drawn (as an X) on a sheet of paper, representing angles of incidence somewhere between 20° and 60° typically. Using a thumbtack and a fixed length of string (estimated), 2 parabolas can be traced; independently. One end of string is always kept perpendicular to the axis of symmetry for each parabola.

Solar Water Heaters & Thermosyphons

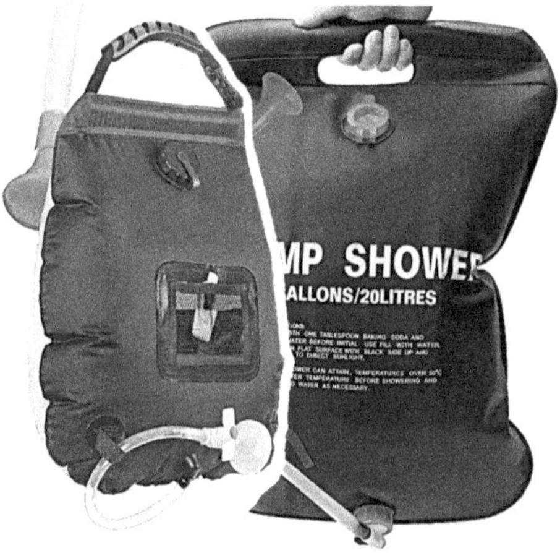

Many simple ideas work with success when exploiting only the thermal component of sunlight.

<<< Nearby are two rugged plastic water-bags. Both are dark colored and each has a drain tube, a valve and a little shower-head. The idea is to hang or set the bag in the bright sun at an elevated position so that once it heats up, one can then walk under it to take a warm shower.

Most people have noticed that a water-hose which has been setting exposed in the summer sun, can spray scalding hot water for a short time. When the faucet gets turned on, cool water pushes out and displaces the heated water.

The pictures above show coiled copper and either **vinyl** (polyvinyl chloride (**PVC**)) or polyethylene (**poly**) irrigation tubing. Irrigation hose is much cheaper than regular garden hose, but it's fairly stiff. Copper tube is expensive, polyethylene less so and vinyl tube is the cheapest practical solar tubing. Poly hose is less flexible than vinyl hose, but it is considered to be the more stable and longer lasting polymer.

Solar heaters for warming pools are somewhat like **closed loop** systems, where pool water is re-circulated through the solar heater. Pool water is not potable. Many swimming pools are not used in the winter, which reduces concerns over these heater's freezing, because they should be drained and not holding water.

If one lives in the tropics where the climate never freezes, then a hose can just lay out in the sun all year round. In the tropics a long continuously filled water hose laying in the sun might suffice as a home's only water heater.

In locations where winters can get cold, direct solar pre-heating of house water is rare. What is instead more common for advanced latitudes, is to **indirectly preheat** house water with heat exchangers or thermosyphons.

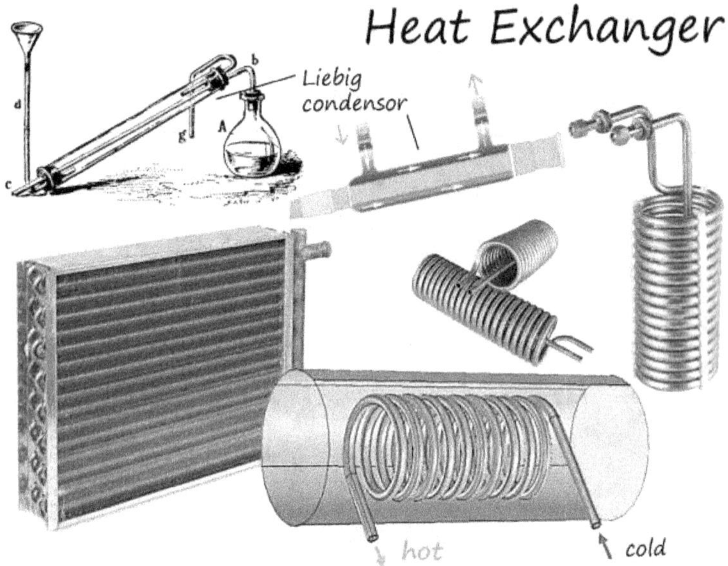

A "**heat exchanger**" and a "**condenser**" might look exactly the same at times, but preform separate functions. A heat exchanger swaps thermal energy from one 'working fluid' to another. The working fluids (either fluids or gases) don't mix. In a chemistry lab "Liebig condensor" or in a refrigeration or air conditioning condenser, the intention is to create a **phase change**. A condenser changes a gas into a liquid, while a heat exchanger just swaps temperature between separated fluids.

Short detour:

A quality **refrigerant** must have a much lower boiling point than water. The aluminum finned, copper coils at the back of a modern refrigerator hold a chemical refrigerant that is chosen for its thermodynamic properties. Some of the best preforming refrigerants have been banned though.

A century ago when the first manufactured refrigerators began to replace "iceboxes", ammonia and sometimes diethyl ether, were the first refrigerants chosen. But problems ensued when those early refrigerators, leaked gases into the kitchen. Later, safer **chlorofluorocarbons** (CFCs) were chosen for their superior heat absorption properties. But it was discovered that CFCs disturbed the high altitude atmospheric O_3 (ozone) which shields us all from overexposure to UV radiation. HCFCs (**Hydrochlorofluorocarbons**) then replaced the CFCs. Soon the HCFCs were themselves replaced by HFCs (**Hydrofluorocarbons)** – which are the dominant refrigerants today. But HFCs are still not ideal.

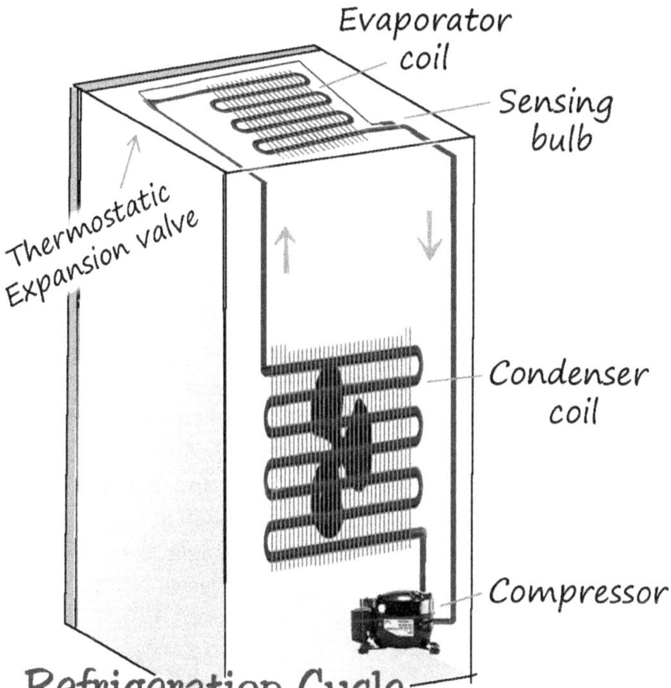

Refrigeration Cycle

The refrigeration cycle has four stages; evaporation, compression, condensation and expansion. To begin, a low pressure liquid (like an HSC) **evaporates** inside evaporator coils, absorbing heat from the interior of the refrigerator. [*When liquids evaporate - they cool things down; like sweat does for the body*]. The gaseous refrigerant is then pushed along and **compressed** (which causes it to heat up). Now the hot and pressurized gas is sent to the condenser coils at the back of the refrigerator. Here the refrigerant **condensates** back into a liquid, as its heat is released to the surrounding air. Finally the liquid refrigerant's pressure and temperature are further reduced as the refrigerant passes through a thermal **expansion** valve; whereupon the cycle repeats.

a TXV valve

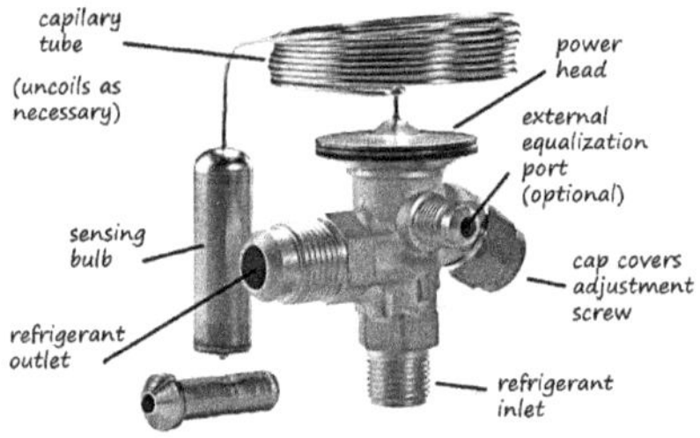

2 thermal snap switches

A **TXV valve** , or a thermostatic expansion valve, is a precise and expensive little valve that regulates the rate of flow of refrigerant to the evaporator of an air conditioning system. This device controls an air conditioner, refrigerator or heat pump. No complicated electronics, computer chips or IoT (Internet Of Things) technology is needed.

A refrigerator usually has two coils; an evaporator coil and a condenser coil. The evaporator coil sets between the TXV valve and its sensing bulb (which is hollow and filled with a refrigerant gas itself). Connected by a small but long hollow tube, the sensor is mounted at the evaporator coil's output and monitors the temperature. The expansion or contraction of gas in the sensor runs down the tube, presses upon a diaphragm to influence a little spring powered **poppet valve**, which restricts the flow of liquid refrigerant into the larger expansion coil.

An even simpler and much cheaper **snap disk thermostat** (or thermal snap switch) is all that is needed to control a solar heater. AC powered pumps or fans for a solar heaters do not need to work at night or on cloudy days. A snap switch either allows or disallows current to flow, based upon temperature. Even simple coffee makers use these type of thermostats.

A "**reversing valve**", that allows a HVAC "heat pump" system to reverse the flow of refrigerant. While a TXV valve adjust the refrigerant flow based on temperature and pressure, the reversing valve in the same system allows a change from heating to cooling mode; or vice versa.

Thermosyphon vs water heater

flat plate thermosyphon collector | ETC solar water heater | bank of ETC absorbers

A solar **thermosyphon** and a solar water heater are not quite the same thing. A typical solar water heater has forced circulation, but a thermosyphon works by **natural convection**. A thermosyphon uses passive heat exchange to circulate a fluid without the necessity of a mechanical pump. *"Natural circulation loops (thermosyphons) are flow systems heated from below and cooled from above, such that the heat sink is higher than the heat source".*

indirect solar water heater

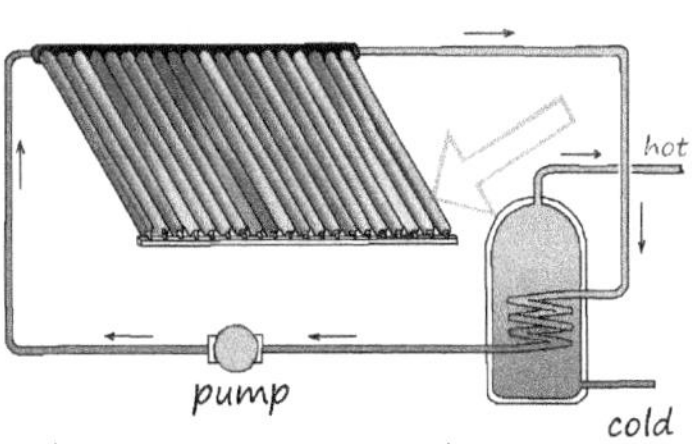

(has isolated refigerant loop)
(has a pump)

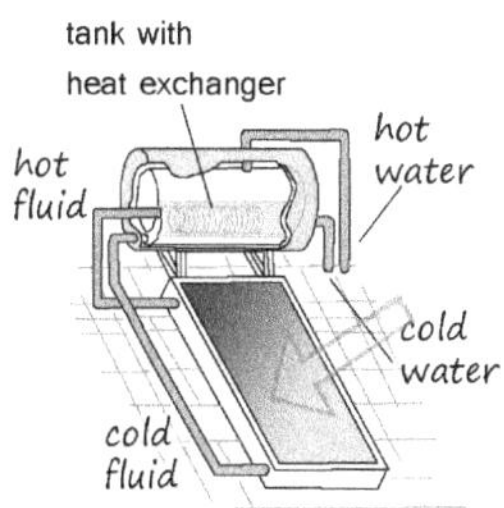

indirect thermosyphon
(has isolated refigerant loop)

A **direct** thermosyphon allows household potable water to flow through the heating elements of a solar panel. A water holding tank is usually elevated above the heat collecting panel, convection is caused by gravity as the warmed water rises, and must be replaced with denser, cold water. A small thermocline (a layer between hot and cold) might exist inside the tank. Pressurized cold water pushes the warmer water out, where it then enters the house's hot water line directly. This pre-heater becomes the feed for a hot water heater's tank. Potentially reducing a home's energy consumption in a useful way.

There are both FPT (**Flat Plate Thermosyphons**) and ETC **(Evacuated Tube Collector)** thermosyphons. FPT panels usually house black painted metal sheets, that are soldered to small copper pipes, and are behind an insulating glass panel. An ETC panel (each tube is already insulated by vacuum) may or may not be glazed.

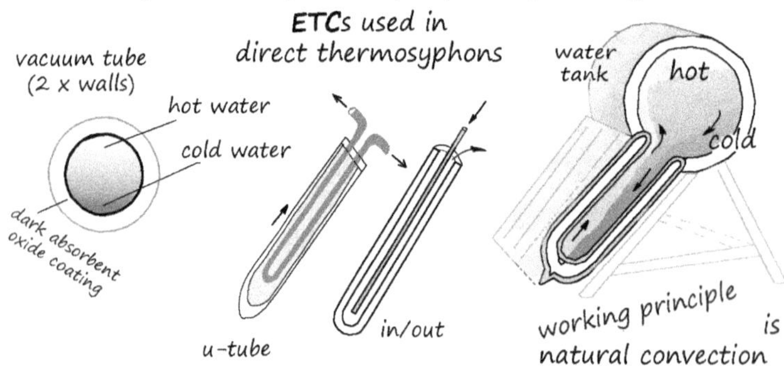

There are several different types of ETC elements. Some of these double walled glass tubes are for thermosyphons but most commonly are for water heaters (where circulation is forced).

In an evacuated tube condenser that uses a sealed metal "**heat pipe**", the head or bulb at the top end is its "condenser". It is where the vaporized refrigerant (often a form of alcohol)
pokes its hot head into bypassing water, but it does not mix with it. As the bulb exchanges temperature with the water passing through a **manifold** pipe, the refrigerant gas condenses back into a liquid and runs back down the tube.

There is a proclivity to name this type of ETC element a thermosyphon in itself – and therefore to a name the whole solar panel made from these elements an "indirect" thermosyphon panel. But this just adds to confusion because then, more than one fluid would be involved in the whole thermosyphon's 'natural convection loop'.

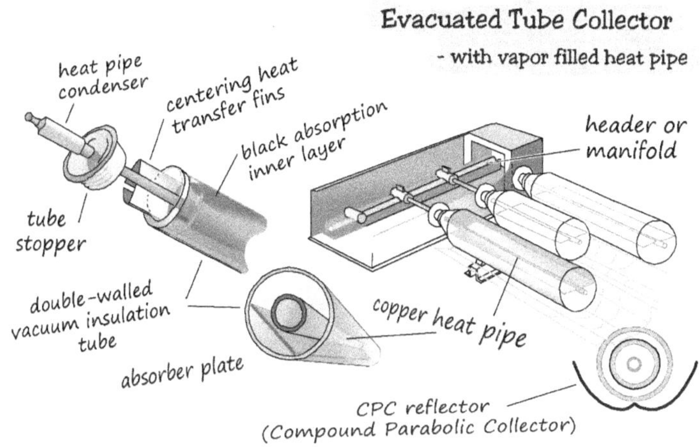

These tubes are difficult to make and therefore expensive; but recently mass production in China has brought the prices down. Some of these double walled glass tubes may have a good vacuum seal that holds nitrogen gas inside. Other ETC tubes might be painted (at the bottom tip end) with a little silver colored barium, which is there to remove any oxygen that seeps into what should be a vacuum.

Borosilicate glass is used in ETC tube construction. Borosilicate glasses are known for having very low coefficients of thermal expansion – which means they can withstand temperature differentials without fracturing. Pyrex ® cooking glass utensils became synonymous with borosilicate glass in America – since about 1915.

White or shiny surfaces reflect light but dark surfaces absorb light. Some ETC glass tubes are tinted darkly with an aluminum nitrate coat. Its called a high-emissivity coating (HEC); where both the absorption and emission of thermal radiation is improved. Other solar glass tubes might be entirely clear, but house absorber plates or fins internally, which are painted darkly instead.

----------_----------

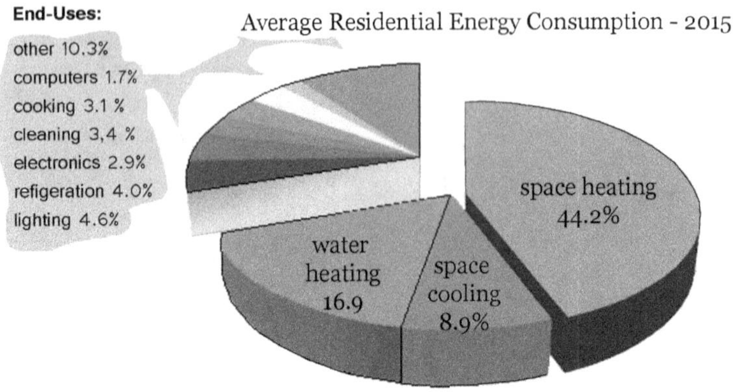

Looking at this graphic of typical household energy usage, shows that about 17% of the average energy budget is spent just on - heating water. Interior temperature control plus hot water easily equate to the lion's share of a home's energy expenditure.
For someone wishing to exploit solar energy, it should be noted that concentrating just upon PV solar panels alone, fails to address the main problem. **The largest expenditure is heat**.

Converting PV generated electrical power into heat, would be grossly inefficient. Besides, thermal collecting solar panels are already far more efficient in that area than are PV panels. A panel made with homemade photovoltaic cells, would be impossible for a hobbyist to accomplish alone. However constructing a quality, homemade thermal collector is easy to do.

With some 2"x 4"s, some plywood and some double pained glass windows, one can make entirely practical water or air circulating heaters. These can be added to or retrofitted to existing buildings. For the lucky individuals that have the opportunity and wherewithal to build a new building or home from scratch, they have a chance to smartly exploit solar energy from the beginning. Many good passive solar ideas exist, where at the time of construction - a wall, a floor or an entire room can intentionally be turned into a working thermosyphon of sorts, for the rest of the building. A choice to simply install more glass windows on the south facing side of a house (when appropriate) is an example of an opportunity that might go missed. But one that potentially could provide many needed heating calories in the future winters to come.

Heat Pump

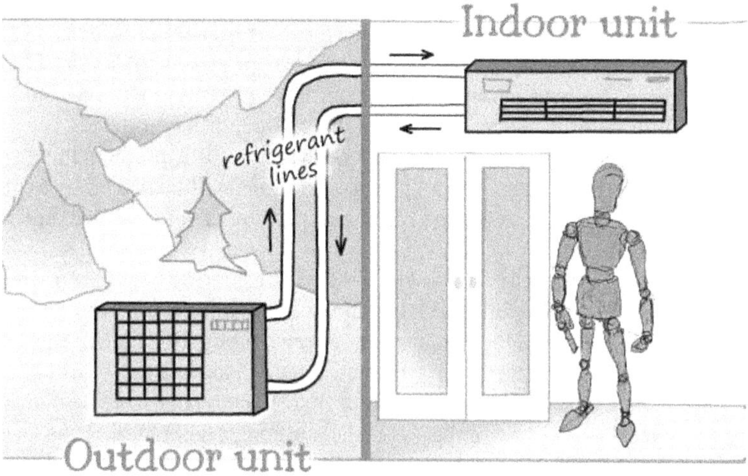

A "heat pump" is a **HVAC** system. Which isn't saying much, because HVAC (**Heating, Ventilation, and Air Conditioning**) is a broad term. Although it may seem to have a magical ability to extract heat from the outside air even on a cold day, a heat pump is in principle, nothing more than a glorified air conditioner. As with the 'refrigeration cycle' explained a few pages back, a heat pump has two coils; one to evaporate and another to condense. The main difference between an air conditioner and a heat pump, is that the heat pump has a special valve that can reverse the system, by reversing the flow of refrigerant. What was once the evaporator coil now becomes the condensing coil, and vice versa.

A heat pump works by using the refrigeration cycle to "*take heat from the outdoors to warm your home, or move excess heat from inside and dump it outside to cool your home*". A heat pump does not burn fuel or use power to make heat. It does use electrical power to move heat back and forth. But since it takes much less energy to just transfer heat than it does to generate heat in the first place, heat pumps are considered efficient HVAC systems for most **temperate climates** (*regions with moderate temperatures that are not extremely hot or cold, typically experiencing four distinct seasons).*

In Europe heat pumps are the default choice for heating and cooling installations in new building construction. It is predicted that by 2030, approximately 70% of European homes will be heated by heat pump. The U.S. market share in heat pump systems is far behind, but playing catch-up.

There are a few caveats however when considering heat pump systems. They're efficient and adequate until outside temperatures really begin to drop. Once outside temps reach about 35 °F, then the heat pump begins to need help. This might mean assistance from auxiliary heat strips in the unit, or from traditional heaters - like oil or gas furnaces, stoves or fireplaces. If the ambient temperature drops below say -14 °F (-10 °C) then the heat pump must work much harder, which means that it must draw more power, and becomes less efficient.

Ground-source heat pumps (or **geothermal heat pumps**) exchange temperatures with the earth. Some buildings have been successfully heated and cooled with this genre of heat pump for more than 50 years now. Although efficient, a geothermal installation is a fairly rare and expensive route to take; mainly because of all the digging required. The deeper the collector pipes can be placed the better, generally. The ground's temperature (deep below the surface) is very consistent, with little fluctuation.

There are "ductless" mini-split system, central "ducted" heat pump, and "packaged" heat pump systems.

First off, the **packaged** version represents everything in one box; one self contained unit. It would be the type normally installed on the exterior of a large commercial building, but rarely a home. The other two standard types of heat pumps are both "split" systems. Meaning that the components of the system are split into outdoor and indoor components (units).

A ductless **mini-split** system is the easiest to install. After refrigerant supply lines from the outside unit lead through a window of penetrate the exterior wall, they then only need to reach the "<u>head unit</u>" *(which includes a heat* *exchanger and fan, and which supplies conditioned air directly into a room)*. The head unit is usually mounted high on a wall. A mini-split system can only deliver heating and cooling to the one room with a head unit. So occasionally one might see mini-split installations where multiple head units are paired with the same outdoor unit.

The inside components of a **ducted** or **central heat pump** are meant to fit inside a traditional ducted, forced air system. Here the conditioned air is distributed into vents that range throughout a house. Such a unit can also be retrofitted into a preexisting forced air system. Afterwards the original existing furnace can just sit back in silence, unless it is needed for fallback heating.

It would seem that both PV solar panels and solar thermal panels – can power, be tied to or can otherwise compliment a heat pump system.

Solar Air Heaters

The notion of heating air with a "**solar furnace**" has been around since at least the 1970's. Apparently though, it's been a less popular notion than heating water with solar, for some reason. But when considering pie charts and energy consumption statistics, the heating of air would seem to meet the larger need. There are some commercial companies that will install solar furnaces or sell the parts and accessories to do so. But the concept is so simple and the parts so affordable that there is very little solar furnace competition in the commercial marketplace. Probably not enough profit margin involved.

The happy news is that almost anyone with some basic fabrication skills, can easily build their own. And under the right circumstances, solar air heaters can be exceptionally effective and reliable devices to help maintain the warmth of a building.

There are two versions of solar air heating thermosyphons, if you will. They differ only in that air is circulated behind the absorber in one, and made to pass through the absorber in the other. Here the absorber must be perforated to allow air to past. In the leftmost example of the next image to come (of a back-pass collector design), the absorber could be something as mundane as a solid sheet of roofing tin. (Painted black and enclosed behind double-pained glass preferably). Either way, no fan is needed because a natural convection will pull cool air up from the floor and push heated air out from the top. A thin, flexible flap of celluloid or plastic sheet can draped over the output vent to maintain flow on only one direction; no back-flow is allowed when the temperature drops at night.

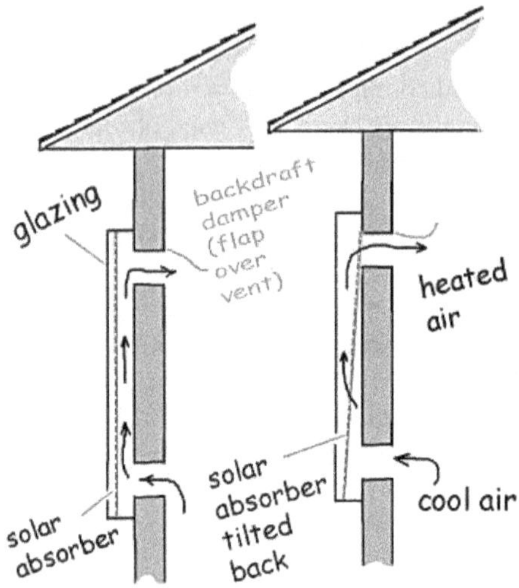

In northern latitudes (above 40° or more?) , the wintertime sun has a rather low orbit. So it might be sensible to mount these wintertime heaters on south-facing walls - rather than on rooftops.

Absorbers can come in all shapes and sizes. An absorber collects heat and transfers it to the air. Often some thin walled layer of aluminum is chosen, because aluminum transfers heat quickly. But many other materials will work adequately.

Some examples have included metallic soffit vent (shown above), rain gutter downspouts, stove or gas vent pipe, tin cans, beer and soda cans, wire mesh or even bug screen.

In essence a solar furnace / solar air heater is just an insulated box, with a transparent top. It needs to have a frame, be insulated, be painted matte black inside, have an absorber, have both an air inlet and outlet and have glazing on top.

To be worthwhile, a solar air heater needs to be fairly large. A wood **frame** of 1"x 6" or 2"x 6" lumber is commonly used in it's construction. Since the backing of most homemade panels will probably be plywood or OSB, the most frequent dimensions for its frame are 4' x 8' (the standard size for American wood sheathing / paneling).

Insulating the panel is very important, especially at the back. The highest "R-value" (Resistance value; resistance to the flow of heat) the better. But considering space limitations – 1 or 2 inches of **insulation** is all that can be crammed in. There are many types of insulation. Some of the cleanest and most applicable insulation to use would seem to be one of the new closed cell, rigid foam board products found at the lumber store. These come in a variety of thickness, have different R-values and are made in every color of the rainbow. But they are not all created equal.

The real test for insulation inside a solar panel, is whether or not it can stand up to the maximum heat possible inside a solar panel. It can approach **200 °F** in such panels, especially in the summertime when the device is not circulating air. Most rigid foam board **sheathing will melt** at such temperature. There are *polyethylene, expanded polystyrene, extruded polystyrene* and *polyisocyanurate* types of foam board insulation to choose from. Only polyisocyanurate (which is usually backed with foil on both sides) is expected to stand up to the task. "Polyiso" foam might not be available in local lumber stores.

The absorbers and exposed surfaces inside a solar furnace panel should be darkly painted, preferably in matte (non reflective / flat) black. And that **paint** should be heat tolerant (as in a quality automotive engine paint) because those surfaces will receive a brutal pounding from sunlight.

The choice of **absorber** for a homemade thermal panel is not as critical as someone might suspect.
The material used must take up thermal radiation from the sun and then transfer heat to the air that comes into contact with it.

^ *The emissivity of the surface of a material is its effectiveness in emitting energy as thermal radiation.*

The absorptivity and emissivity of different materials differ widely. In general, a good absorber will also be a good emitter, while a poor absorber will be a poor emitter. (Exceptions being intentionally altered "selective surfaces").

`A collector tube made of drilled out and glued together **aluminum cans** might have a slightly superior heat transfer coefficient than a similar tube made with steel or **bi-metal (tin cans)** containers. But the differences are slight and fairly inconsequential in the great scheme of homemade thermal-solar panels. A bigger benefit might come just from the turbulence created inside a column constructed of aluminum cans (typically joined with silicone adhesive); something that would be missing in a straight walled absorber column of tin cans or (rain gutter) downspouts.

An absorber composed of a couple layers of wire mesh or **bug screen** - has proven to be popular and efficient. Some comparisons have shown screen absorbers to be superior to cans, at heat exchange. And additionally, a screen or soffit vent type collector is much easier to construct than one made with cans.

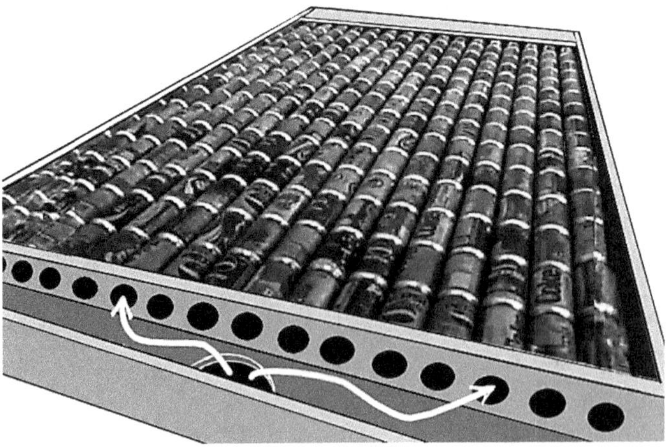

There is a good deal of labor involved when working with aluminum cans. They often need to be rinsed out; if not they will pass along odors to the circulating air. They need to be drilled or cut out at both ends, but in doing this often crushes some because these cans are so feeble and thin walled. They need to be glued together (with silicone caulk usually) into columns, but to keep these columns straight is a delicate task. Another problem is with the manifolds at either end of the columns. Even if a 2" x 6" board is used for the manifold, it might tend to split because the holes are drilled so close together. Even drilling the holes in the right place and then fitting the cans is not that straightforward, especially if the insulation behind the cans interferes by being flexible.

It seems much more expedient in the long run to just choose metal bug screen or metal soffit as the internal absorber. Importantly, to warm a room it is more effective and preferable to circulate a large volume of modestly heated air immediately, rather than to wait around for a smaller volume to be well heated first.

So panels with big surface areas, big diameter ductwork with short lengths of travel and strong blower fans; are the way to go. The choice of glazing is important too.

Warm surfaces are partially cooled when they emit thermal radiation to the passing air in a solar heat collector. (At times you can actually see **thermal radiation** being emitted. For example when a blacksmith pulls a glowing, red hot horseshoe out of a fire).

The **emissivity** of the surface of a material is its effectiveness in emitting energy as thermal radiation. Emissivity is graded from zero, to the maximum value of 1.0. Ordinary black paint has high solar **absorption** and high thermal emissivity. The emissivity of a surface depends on its chemical composition and geometrical structure. Normally glass has a very high emissivity; but this can be changed with a special coating. ***"Low-E windows"** with transparent low-emissivity coatings emit less thermal radiation than ordinary windows.* A Low-E window might be fine for summertime, but be less than desirable for wintertime – if you expect radiant heat to come through a window from the outside.

The concept of a thermosiphoning panel, bolted to an exterior wall, was illustrated a few pages back. There, a natural convection caused by warmed air rising, will exit an **air outlet** while simultaneously pulling cold air into an **air intake**.

The alternative to natural convection by thermosyphon, is to push the air by electric fan. A "duct fan" like the one shown here is merely a booster fan that fits inside or inline with ventilation ductwork. (A larger house for example might have a centralized HVAC unit with a big blower fan – but some rooms might be so far away that assistance from inline duct fans is needed to supply good circulation). Fans come in many different diameters, shapes, sizes and choice of current (AC or DC). HVAC type fans often come in assorted output performances. "CFM" (Cubic Feet per Minute) is the standard gauge or metric for rating fan output performance in the US.

A blower fan on a solar thermal panel needs a **thermostat** to <u>automatically</u> tell it when to turn on and turn off. When the sun goes down and temperatures drop, the blower does not need to work. The blower for the heater does not need to work during the warm summer months, either. While an expensive, remote, programmable or non-programmable or Wi-Fi enabled thermostat could be used for the job, a simple, very practical solution exist. Just use a "bi-metal disk thermostat" or "**snap switch**", which is a thermally actuated switch that cost only a few dollars.

When a bi-metal disc is exposed to its predetermined calibration temperature, it snaps and either opens or closes a set of contacts. Breaking or completing an electrical circuit. An automatic reset snap switch can either open or close its contacts as the temperature increases. As the temperature falls back and returns to a specific reset temperature, the contacts will return to their original state.

Examination of the simple ingenuity behind a common "drip" style coffee maker will show that this same type of thermostat is doing the work. It's fastened to a heat element that is soldered to a hollow tube where the water is boiled. Once the coffee is made the switch might turn off. But afterward it will flicker on and off again as necessary, to keep the coffee and the warming plate that the carafe sits on, at the correct standby temperature.

Bi-metal disk thermostats can be found in catalogs or supply houses, but might have labeling that can confuse. Take this example: *3/4" Snap Disk Fan Control, Cut-In -110 Degrees F, Cut-Out - 90 degrees F.* What does this mean? Why is the cut-in temperature higher than the cut out temperature?

Well simply explained, this particular fan switch is off at night but will cut in and allow current to flow to the fan's motor, once the temperature inside the solar panel reaches 110 degrees. It will allow current to flow all day until or unless the temperature inside the panel, drops down to 90 °F. There is a wide selection of preset temperature ranges to choose from. If you think that your solar panel should work harder, say until the temperature drops to 80 °, then your solution is as simple as purchasing a different switch.

A few more notes

*A snap switch or the sensor of a more complicated thermostat should be mounted inside the thermal panel, but out of direct sunlight.

*In the summertime when extra heat is not called for, the simple snap switch controlled panel needs to be disconnected from power manually, otherwise it will keep working.

*If they are not being used, thermal solar panels will take a beating from sunlight and built up heat. It would be a good idea to shield or cover them in the summer.

This panel was constructed on the top of a two story roof. Everything was cut and measured on the ground, but individual components had to be carried up a ladder. The panel sits in chairs that lean back (northward) at the optimum angle for the local latitude in winter. Nothing touches the masonry chimney at its back, but the chimney will prevent the panel from being blown forward or backward because of high winds.

Behind the panel are two unseen duct fans; one pushing from the input vent, the other pulling air from the output vent side.

One photo shows the "chase" that was needed to conduct and protect the insulated ductwork which comes up the back-wall and around the chimney.

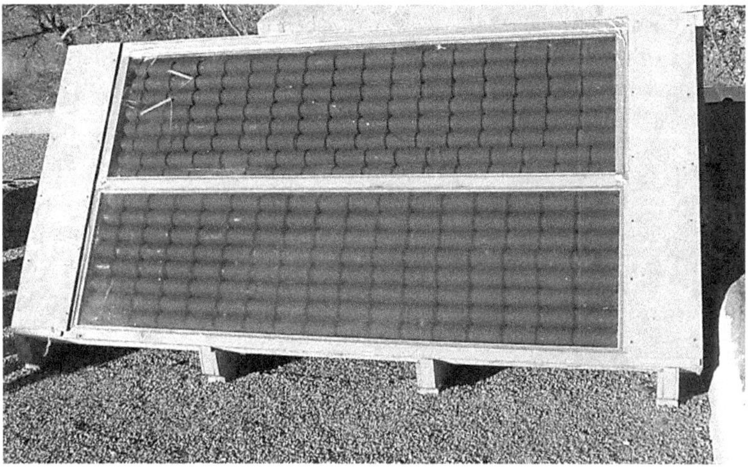

Several lessons were learned from this project. The roof location was selected because there was little decent choice – other structures blocked the most desirable southern exposures. The glazing used were two very tall double-pained windows (purchased very cheaply at auction). About two years later the glass cracked (probably due to hail) and the panel was reconfigured to fit a more common glass dimension. Although the panel works and provides heat to a remote bedroom, performance could be better. Little can be done about the remoteness of the downstairs room in this case, but stronger fans would force more heated air through the system. Larger ductwork would have been more expensive and would have complicated the original construction, but it would have also allowed an increase in the flow of air and better overall heating of the room. Again, it is better to force a large cubic footage of mildly heated air, than to wait around and insist upon well heated air – which will come in smaller volume.

Above is evidence of an experiment using a small, dirt cheap computer fan. It's from a PC's PSU (Power Supply unit) and it outputs 0.3 amps @ 12 volt DC. [If (**V**olts)x(**A**mps)=**W**atts then this fan burns about 3.6 **W**].

The diameter of this tin-can pictured, fits perfectly inside a 4" wide, dryer hose vent. It is a silent fan (now "ducted") that produces a wimpy airflow. But enough airflow perhaps to displace the air inside the volume of a hallway closet, within one hour.

> *Twenty of these dirt cheap fans might somehow produce a useful heated airflow. While safely expending less energy than would one old fashioned, incandescent light bulb.*

Glazing

To close in and help insulate the front side of a unit but still allow the sun to heat the solar absorber, one needs a translucent or transparent glazing. But there are new glasses or glazings out there now that are just wrong for the purpose of solar heating.

Types of glazing (an Old English word for glass) include: fiberglass, acrylic, polycarbonate and glass. All at one time or another, have been used to cover skylights, window openings, and greenhouses. Each form of glazing has its merits and detractions. Some types are inappropriate for thermal panels.

Fiberglass paneling in either corrugated or (rolled) flat form was a popular building commodity back in the 1960's & 70's. People used it for canopies over the house's front porch or over a car port. Others might have used the translucent sheathing as a skylight, when inter-dispersed with regular sheets of tin roofing on their hay barns. Some made greenhouses, covered entirely with clear fiberglass. Fiberglass paneling or sheathing is still available today; it is just rarely found now in most hardware or lumber stores. Because acrylic or polycarbonate sheets have become more popular. Not that these are superior.

This fiberglass product whether flat rolled or corrugated, is made from polyester resin and fiberglass. And the finished material is specified by the unusual format of "ounces". In the U.S., fiberglass cloth is specified by its **weight per square yard** (6 oz. fiberglass weighs 6 oz. per yard^2). In the world of boat hulls and surfboards, fiberglass fabric comes in an assortment of thicknesses and weights; (4 oz, 6 oz, 8 oz, 10 oz, 18 oz, and 40 oz are some common denominations). The thinnest fabric + resin would be appropriate for providing a clear and permanent finish over a wood tabletop for example. The heavier fiberglass fabrics would be used to repair or strengthen boat hulls.

The most **popular resins** to use when forming or laminating fiberglass cloth or mat, are *epoxy resin, polyester resin and vinyl ester.* Various ratios of **MEKP** (Methyl Ethyl Ketone Peroxide) are used to control the hardening of polyester resin or vinyl ester. Epoxy resin though can be cured by an assortment of **hardeners**, ranging from either aliphatic or aromatic amines, to acid anhydrides and phenolic compounds (not hydrocarbons).

As a glazing consideration for thermal solar panel, fiberglass should have a lifespan between 10 and 15 years; commensurate with the expected lifespans of both acrylic and polycarbonate. If advertising is to be believed though, FPR is more stable than either PVC or polycarbonate. FRP (***Fiber Reinforced Polymer***) **:** – *where once the fiber was historically drawn from molten glass – but now can include filaments from carbon fiber, basalt or a newfangled synthetic called 'aromatic polyamide' (or aramid, a new, strong, heat-resistant fiber).*

Corrugated fiberglass panels can be expected to survive a much broader temperature range that the competition, without concerns of warping, excessive expansion and contraction or cracking and brittleness. Flat rolled fiberglass supposedly provides excellent light and solar energy transmission. But there seems to be no double-wall versions of fiberglass, as there is with polycarbonate. No glazing is going to last forever – considering the steady pounding from nature's elements (hail, oxidization, destructive UV light). But in terms of longevity, fiberglass does not take a backseat to the alternatives.

* Glass fibers have been evident since antiquity but in 1932 a researcher discovered by accident that glass fibers could be made en-mass by blowing a jet of compressed air through a stream of molten glass. The company he worked for merged with another to form Owens Corning in 1936. Owens Corning patented this process, using it to make glass wool, sewing thread and yarns containing glass fibers. "**Fiberglas**" (spelled with one "s" – German style) is or has been a registered trademark for that world famous, pink glass wool blanket insulation, installed in the walls of newly constructed buildings.

There are several established and distinct forms of plastic; as in *polystyrene* (**PS**), *polyethylene* (**PE** / high and low density versions), *polypropylene* (**PP**), *polyethylene terephthalate* (**PET**), *polyvinyl chloride* (**PVC**), *polymethyl methacrylate* (**PMMA** or **acrylic**) and *polycarbonate* (**PC**). But only a few of these have been applied to window glazing or roofing panels.

Acrylic was invented in 1928. Shortly afterwards chemists and industrialist in both England and Germany developed a synthetic resin which polymerized into acrylic glass, in the 1930's. Today this thermoplastic is called *polymethyl methacrylate* **(PMMA)**. The English version was registered and marketed under the trademark "Perspex", the Germans trademarked their version as "**Plexiglass**" and later DuPont in the U.S. trademarked their version of the product as "**Lucite**". PMMA found many uses during WWII, but most noteworthy were in its application as shatterproof windshields and transparent bubble canopies for machine gunner turrets and aircraft pilots.

Ethylene dichloride (EDC or *dichloroethane*) was a liquid used from an eyedropper or syringe to almost instantly bond or "**cold weld**" the edges of Plexiglass® (PMMA) together. Since it was identified as a potential health hazard though, ethylene dichloride was removed from the market about 40 years ago. Many similar examples exist.

Cyanoacrylate was discovered by a chemist attempting to develop plastics for the military during WWII. Today we know cyanoacrylate as **"Super Glue"**. Variants of cyanoacrylate are many, but some are sold today for the purpose of replacing EDC in the bonding of PMMA. Acetone or nitromethane can be used to debond cyanoacrylate welds; but at the risk of damaging certain surfaces possibly.

The original "**Gorilla Glue**" was a polyurethane adhesive that worked indiscriminately upon foam, metal, ceramic, glass, stone and wood. Unbeknownst to most users possibly, this polyurethane adhesive can be undone or dissolved with acetone, alcohol or paint thinner.

Many of the chemicals used to make plastics have eclectic names and are rarefied; as are some of their **glues and solvents**. However the typical hobbyist might possess or have access to milder or safer chemicals at home – which can be used to experiment with - upon certain types of plastic. For example gasoline, naphtha, xylene, toluene, paint thinner, lacquer thinner and mineral spirits are generally considered to be consumer safe. Some of these might have profound effect upon a sample plastic, or none at all. (Experimentation will tell).

Plastic Recycling Symbols

1 PET (POLYETHYLENE TEREPHTHALATE) - (or PETE) / beverage, vegtable oil and salad dressing bottles / easy to recycle

2 HDPE (HIGH DENSITY POLYETHYLENE) - milk jugs / bleach, detergent, shampoo and motor oil bottles / grocery bags / certain grades are effected by toluene / easy to recycle

3 PVC (POLYVINYL CHLORIDE) - plumbing and sewer pipes / thermoformed blister packaging / wire or cable insulator or jackets / garden hose / window frames / may dissolve or deform from gasoline / rarely recycled

4 LDPE (LOW DENSITY POLYETHLYENE) - squeezable bottles / bread bags / Ziplock and frozen food bags / sometimes effected by toluene / seldom recycled

5 PP (POLYPROPYLENE) - straws / caps / syrup and medicine bottles / containers for hot liquids or foods because of high melting point / modified by toluene / sometimes recycled

6 PS (POLYSTYRENE) - egg cartons / disposable cups and plates / carryout containers / CD cases / potential carcinogen / disolved by xylene, toluene and gasoline / seldom recycled

7 OTHER - many plastics that don't fit previous 6 categories are lumped together here / polycarbonate plastics / PLA (polylactic acid) biodegradable thermoplastics from plant material / ABS (Acrylonitrile Butadiene Styrene) durable impact-resistant thermoplastics easily molded into shapes / LEGO bricks / sunglasses / nylon / large capicity water bottles / rarely recycled

The topic of plastics is fascinating, but too intricate a subject to discuss without involving a knowledge of chemistry . For almost a century now plastics have been under continuous development and the products thus produced are found daily in every facet of our "civilization". The significance of plastics is enormous. Yet plastics might have some unappreciated and threatening downsides. Centuries from now some historian might look back and name this period "The **Age of Plastic**" (as per - the Stone Age or the Bronze Age). That future historian might reflect upon how our current preposterous, present variant of civilization, allowed itself to be harmfully impaired by these synthetic materials. Imagining this planet being utterly cluttered with plastic trash that is decomposing at a glacial pace, it might become apparent in the future that the human genome itself was eternally diminished. Because of the hormones disrupted or the questionable nanoparticles of plastics now copiously ingested. We eat, drink and rub up against these synthetic polymers daily. That microscopic synthetic stuff does not rinse out of our bodies; it stays to become part of us...

Back to Thermal Solar Panels

*There are two main classes of organic hydrocarbon compounds: aliphatic and aromatic. **Aliphatic** compounds include: alkanes (saturated), alkenes (unsaturated with C=C double bond), and alkynes (unsaturated, have triple bonds). Some example aliphatic hydrocarbon compounds are paraffin, methane, ethylene, propene, hexene and acrtylene. **Aromatic** (polycyclic) hydrocarbons include substances like fluourene, pyrene, benzol, toluene and naphtalene.*

Polycarbonate was almost discovered by mistake, at the same time that PMMA came out. But not until 1953 did more study and the mixing of polyester resins with aliphatic hydrocarbons result in the first polycarbonates. "Lexan ®" (not Lucite ®) is a well recognized trade name for polycarbonate.

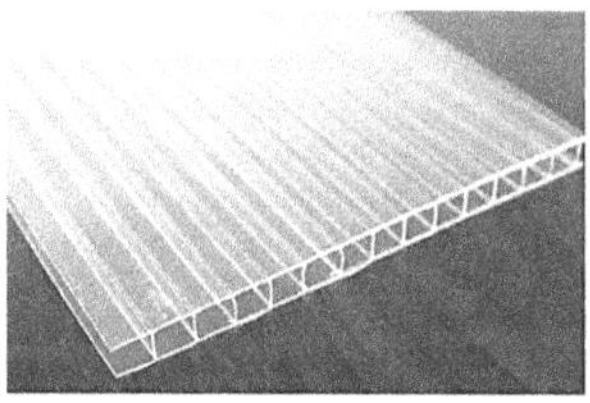

Acrylic and polycarbonate are both thermoplastic polymers and the two share similarities. However they differ in their chemical structures; acrylic is made from poly{methyl methacrylate} (PMMA / $C_5H_8O_2$), while polycarbonate $(C_{16}H_{14}O_3)n$ contains repetitive carbonate groups in its structure. This difference contributes to their unique properties, such as acrylic being more scratch-resistant and brittle while polycarbonate is more impact-resistant and flexible.

Comparing Acrylic and Polycarbonate

- *Of the two, acrylic (PMMA) is less expensive and offers better clarity, while polycarbonate (PC) is stronger and much more impact-resistant.*
- *Polycarbonate blocks almost all ultraviolet light and is vulnerable to damage by UV and ozone attack. It can yellow over time when exposed to sunlight, whereas acrylic can be polished to restore its clarity. Some polycarbonate roofing and window panels are **UV protected**, but this is only for the plastic itself - not the people underneath.*
- *Acrylic is less prone to scratching and more transparent. Both thermoplastics can be manufactured to provide excellent optical clarity, but these plastics cannot compete with glass as the best collector glazing.*
- *Polycarbonate can be very flexible when acrylic might tend to crack. But acrylic is generally stronger in terms of tensile strength.*
- *Polycarbonate supports wider heat tolerances. It is used in bullet proof glass and automobile windshields. It has been certified as food safe for many food and beverage containers; yet since it contains suspicious cancer related bisphenol A _ (BPA), its use in the food industry is declining.*
- *PMMA (acrylic) is commonly used in domed skylights, illuminated advertising signs , optical fibers, swimming pool enclosures, aircraft canopies, instrument panels, and luminous ceilings.*

PVC (Polyvinyl Chloride) is another extremely popular plastic polymer (alongside polypropylene and polyethylene). Rigid PVC products range from credit cards to plumbing pipes, siding, window frames, plastic bottles, cling wraps, flooring, medical gloves and clothing. There is some crossover between products made with ether PVC and acrylic or polycarbonate. But not enough to consider PVC a realistic alternative glazing material for a thermal solar collector.

Glass

The raw materials used in glass are generally cheap and plentiful. It is actually the fuel used to melt the glass, that becomes the biggest expenditure. Not all glass is alike. Within the last 36 centuries or so, many new variations of glass – each featuring unique physical properties, have been introduced. It would behoove a prospective home builder or solar panel maker to study the differing properties in glasses, if intending to purchase some. But remember that the advertising world is full of hype because hype sells.

Broadly, the three most common categories of glass are "fused silica", "soda lime" and "borosilicate". **Fused silica** (or silica glass) is basically quartz. **Quartz** is an oxide of silicon. Silicon dioxide (SiO_2) is either the first or second most plentiful mineral in the earth's crust. Pure silica glass is the most thermally stable glass made. It is used in the windows of spacecraft like the Space Shuttles, and in optics like binoculars and telescopes. However, because the high fabrication temperatures are difficult to work with, (@ or above 3,000° F) fused silica products are rare.

Soda lime glass on the other hand is the most commonplace type of glass. Soda lime is cheap to make and is found in most windows, bottles and jars. But soda-glass is not very stable thermally, and tends to shatter when heated.

Where does the name "soda lime" glass come from? Well, this glass is still about 75% silica (SiO_2), but to make it easier to work with, a **flux** (about 15% of volume) of sodium carbonate (Na_2CO_3 / **soda ash**, laundry washing soda) is used to lower its melting point; substantially (to about 1,300° F). By itself though, soda glass is water soluble, which is not good. So **lime** (calcium carbonate / limestone or $CaCO_3$) is therefore added (10%) to stabilize the final glass. (Soda and lime, silica glass).

Many other substances can be added to molten silica to change the properties of a glass. **Lead oxide** can be added to produce a glass that is resistant to X-rays.

Manganese dioxide can be added to remove the green coloration caused by iron in the sand.

Borax (the mineral *sodium borate*) is added to produce a glass that has low thermal movement and is resistant to thermal shocks. For example Pyrex ® is the trademark of such a **borosilicate glass** that was introduced to the American market in the 1920's. Pyrex and other borate glasses are used in cookware, ovenware and laboratory containers because they don't crack or shatter easily when heated.

The best glazing solution for a solar, air heating panel might probably be a double pained glass French door panel that is full pained (not divided) and without a "Low-E" coating.

It is often possible to purchase good glass like this second hand at auctions, thrift stores, garage sales or from classified advertisement websites like Craigslist.com. It makes little sense to be gouged by high prices for the cost of new double pained glass at lumber stores. Any windows will work but a large door sized, double pained glass is sort of optimal from a logistical and construction viewpoint. Insulated, not too big to move, not too big for one sheet of 4'x8' plywood to cover at the back, but still a large enough light permitting area to make a useful amount of heat, that can be conveyed into the home.

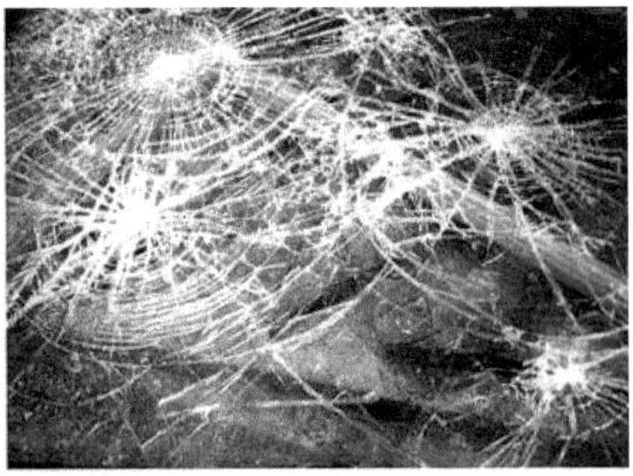

From an energy conversion standpoint, glass makes the best glazing.

But Murphy's pessimistic Law hints that objects like falling tree limbs, baseballs or large hailstones might fall from the sky at any moment. A fiberglass or polycarbonate glazing might ameliorate or eliminate breakage for a long while. But no solar panel will preform forever. Periodic maintenance will still be required. The damage and decay caused by oxidization, atmospheric precipitation and harmful elements from sunlight itself, are inevitable and do accumulate. Sooner or later repair or replacement of the toughest glazing will be required. Whatever the wavelength (UV, Visible, IR or X-Ray) sunlight over time, has a very deleterious effect on most materials.

Glass ages, and weakens like anything else. *"Strength reduces on aging, as surface imperfections increase, whether by chemical attack or simply mechanical abrasion; glass which has weathered for some years being much weaker than new glass".*

Warning : off topic detour ahead

A little study on glass. Glass we take for granted, because it is so ubiquitous. Effective solar panels can be built without expending too much consideration over the nature of the glass used; so long as it is big enough. But since this is a book with yet many empty pages to fill, the next few will be dedicated to briefly explaining some glass terms and technologies. Perhaps some reader might find the trivia useful. The focus of topic will return to thermal panel glazing, after a dozen or so pages of meager mental meandering.

Glass Properties

There are many expressions and specifications to know. Consider these very obvious differences between glasses.

-thickness & weight, density,
-thermal properties, fire performance,
-coloration, fluorescence,
-abrasion -resistance,
-refractive index, light-transmission properties and resistance.

Then one can contemplate terms like “absorptivity”, “reflectivity” and “emissivity”. A popular trend in new construction glass, is to brag about its emissivity; as in “**Low-E windows**”. Conversely, high-emissivity coatings (HECs) might be applied to the insulating ceramics on a spacecraft, to facilitate radiative cooling and protection of the underlying structure.

Flat glass is often evaluated for its heat and sound insulation capabilities. Here specifications like **U-Factor, SHGC** and **VT** are stipulated in advertisement literature.

Common Glass Types

Earlier it was stated that the three most common categories of glass are “fused silica”, “soda lime” and “borosilicate”. Historically that might have been true, but now it seems misleading insofar that, there are now new composite glasses that fit into none of those categories. Consider the thin glass substrates on smartphones for a moment. Are smartphones, i-Pads and portable laptops not common? Aluminosilicate glass is now one of the most-used glass types in the commercial and industrial sectors.

The following terms deserve at least some explanation.

Historic Glass

- blown glass
- molded glass
- crown glass
- broad sheet glass
- lead glass (which is not the same as) leaded glass
- flint glass
- glass wool and glass fiber.

Modern glass

- flat-drawn glass
- rolled glass
- float glass
- annealed glass
- toughened or tempered glass, heat-strengthened glass
- laminated glass, safety glass, bulletproof glass
- aluminosilicate glass

Early glass making

It is not “So Simple That A Cave Man Can Do It”, but toying with glass is something mankind has been doing for thousands of years now. Classic archaeological periods like the “Stone Age”, “Bronze Age” and the “Iron Age” for instance are differentiated by technologies that were permitted by or resultant of, prior improvements in the control of fire and heat. Step by step the progress of controlling heat in foundries or furnaces, allowed for the melting of harder or better metals and therefore the fabrication of better tools. One can imagine that as the years passed, every form of rock or mineral was eventually subjected to a heat test. It was obviously discovered at some point, that baking certain forms of mud made **ceramics,** and that cooking sand made **glass**.

In the fire pit a crude amorphous blob of molten glass could be rolled around with a stick, or be pulled out of the fire and be flatted a bit, before it cooled down and hardened. Many years probably past though, before someone thought to use a hollow pipe to blow air into a lump of molten glass. That blowpipe had to be long enough to keep the person away from the heat, yet rigid enough to support the weight of the molten glass as it was swung about and being worked on. What were the first blowpipes made from thousands of years ago? Clay?

Curiously enough, even flat glass windows had to be blown first. Not until the beginning of the 20th century could humankind find a way to mechanically fabricate flat glass without using a blowpipe.

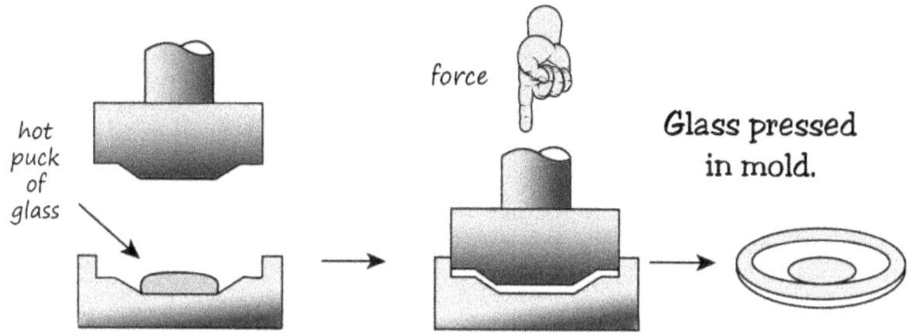

Although a blob of hot glass alone can be shaped by dropping and pressing it into a mold, it appears that the first shaped molded glass products were actually produced by blowing glass into clay molds; around 50 BC.

Above are examples of "pressed glass". Most (glass) drinking glasses, plates, cups, saucers, ashtrays and other assorted kitchenware knickknacks fall into the pressed glass category.

The next image to come shows how early slab glass (or sometimes "Norman slab glass") was made.

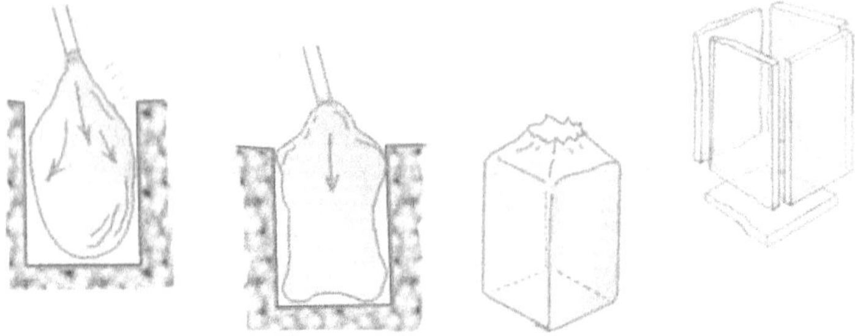

Glass was expensive and relatively unaffordable until the 17th century. Before glass became common, animal skins (cow and pig mostly) were split and then shaved down careful to make a translucent window covering that would permit light to enter the home.

Crown glass was another early type of glass used to make windows. The glass was apparently blown into a hollow globe, set down, cut off and then spun on something like a potters wheel. Centrifugal force help flatten the disk. Afterwards the disk could be cut, ground and polished; to be held together with **'lead came'** like any cathedral's contemporary stained windows would have been.

Lead **came** is drawn into only two shapes, “H” or “U”. While the H-came is used to connect two separate pieces of glass, the U-came is used only for the outside edges of a proper leaded or stained-glass window.

One artisan became famous for his complex stained glass windows. His unusual process amounted to pouring molten lead over the glass pieces as they laid flat; and then troweling and scraping the excess off.

Today lead-came can still be purchased and people still make attractive stained glass windows. A less expensive process to making small stained glass windows though, is to shape thin strips of copper foil around each individual piece of glass and then use a soldering iron and lead solder to bond the copper joints together. Then that can be bonded to or sandwiched between other sheets of clear glass – to make a weather proof exterior window.

Creative Commons Attribution 3.0 License (pictures)- Thanks to: vagabondblogger.blogspot.com

The demands for constructing a glass kiln are not high. In the above images a very simple glass kiln (near Cairo, Egypt) is shown. It is built with adobe bricks and mud mortar. A few fired bricks and concrete cinder blocks might be thrown in. The heat source is probably natural gas (both plentiful and cheap near Cairo), which is simply piped into the back of the furnace.

Again, a major expense for glass factories is fuel. The first British settlement in America was an industrial venture that resulted in the 1608 **colony of Jamestown, VA**. Commercial glass-making was a primary motive for the undertaking. The American coastline had vast forests of wood and long sandy beaches. Back in England, glass factories were stagnating because wood was becoming scarce and expensive. Serious coal mining was still in the future. Oh, glass-making knowledge was also high-tech and secretive back then. German and Polish glass-makers were even bribed and smuggled into the Jamestown adventure – because of their knowledge.

* In 1696 a **Window Tax** was introduced in England and Wales. It was basically a property tax, in lieu of an unpopular and unenforceable income tax. A levy was charged for every window, the more windows you had, the wealthier you obviously were and the more tax you paid. This action led to many property owners bricking up some of their window openings. France established a similar tax which would not be repealed until 1926.

In the above image on the left, is the Betz Building (Philadelphia, PA, c.1892) and on the right is the iconic Flatiron building (Manhattan, NY c.1902). Both buildings (and others worldwide) were built using glass windows that were produced by what is called the "**cylinder sheet"** glass method. A technique improved little, from what ancient Roman glassblowers were doing ages before.

It was difficult and took a long time to mechanize the production of plate glass. In 1904 Emile **Forcault** would patent a new method for drawing glass into larger sheets. However it would be 1915 before any company could put his vertical "**drawn sheet"** glass process to work. The method involved drawing molten glass directly from the furnace, by grabbing it from a "debiteuse" (a rectangular clay block, which floats upon the molten glass). A series steel rollers pull the hot (but cooling) continuous sheet up and squeezed it to the desired thickness at the same time. Later in 1952 the "**float glass**" process came along (where molten glass is floated over a bath of molten tin). This is the most common form of commercial plate glass manufacture today. But this is all "putting the cart before the horse". Up until 1915, buildings like the Betz and the Flatiron (named for it's triangular shape which was reminiscent of an antique laundry iron) had large windows that were all cut from hand "blown glass".

From: Weltall und Menscheit / by Hans Kraemer (ca. 1880)

This illustration above comes from the five volume 1880 encyclopedia <u>Weltall und Menscheit</u> (Universe and Humanity) and explicates the cylinder sheet method efficiently. In this 19th Century glassblowing workshop, compressed air (doubtlessly from a steam driven air compressor) is enlarging a globe of molten glass. (Glass blowing is hard work, and takes a good set of lungs). The worker on the left is elongating his glass, from a sphere into a cylinder – within a pit. Once the cylinder reaches the desired length, the work is taken to a table or a rack of an annealing kiln, where the ends are cut off and the cylinder split down the middle with shears. Gravity and manual assistance help the glass lie flat. "**Annealing**" the glass means: cooling it down slowly, in a managed way so that it does not crack.

Above are public domain images from a 1906 edition of Scientific American (magazine). On the right, a sliced tube is being flattened in an oven. Below, glass workers in the Chance Brothers' factory are elongating hot glass cylinders with centrifugal force. They are swinging the glass back and forth in a pit. A mechanical 'swing set' is being used to support the weight.

The picture above is dated 1910, but 59 years earlier this same location (the Chance Brothers company) was responsible for making an enormous amount of cylinder sheet glass for the Crystal Palace. Tasked with an order for almost a million square feet (996,786 ft^2) on short notice, the company had to import extra workers from the Saint-Gobain glassworks in France.

THE CRYSTAL PALACE IN HYDE PARK FOR GRAND INTERNATIONAL EXHIBITION OF 1851.
Dedicated to the Royal Commissioners

The **Crystal Palace** was a glass conservatorie of monumental achievement in almost every respect. It took only sixteen months from start to finish to design, manufacture and erect the modular Crystal Palace, on site. It would sit upon some of the most valuable ground in the world. The humongous and airy glass building was then filled with eight miles of tables, for fourteen thousand promoters which displayed more than one hundred thousand different exhibits.

The Great Exhibition (the first World's Fair) opened May 1, 1851. It was a grand success. And profitable. In the five months before the exhibition ended, more than six million paying customers would walk through the Palace's doors. Then this biggest building in the world at the time was itself sold off, removed from Hyde Park and reassembled in another location of London.

The largest plate glass (here - cylinder sheet glass) that could be efficiently made at the time measured 10 inches by 49 inches. The Crystal Palace's configuration and dimensions came from the fact that it used some 293,000 of these identical panes.

Leaded glass was mentioned before, where it was mentioned that crown glass and stained glass both, are held together by channels of lead came.

- The oldest stained glass window still in existence, supposedly resides high on the end wall of St Peter's monastery at Monkwearmouth (north central England). Benedict Biscop founded the abbey in 674 AD. The window is small and circular (about 12" in diameter). The colored glass was brought to the Kingdom of Northumbria (middle England) during one of Benedict Biscop's five trips – back and forth to Rome, Italy. (That seems like a lot of travel for the 7th century). In its day the little abbey has witnessed un-Christanized Celts, Frisians, Angles, marauding Saxons, 9th century Viking raiders and the 11th century Norman conquest.

public domain: Wikipedia

Lead Glass is commonly called crystal or **crystal glass**. It is a variety of glass in which lead replaces the calcium content of a typical potash (soda) glass. Original lead glass traditionally contains between 18% and 40% lead (II) oxide (PbO) by mass. A more desirable "**Flint glass**" (crystal glass- famed for its clarity and brilliance) contains a minimum of 24% lead oxide. The addition of lead oxide to glass raises its refractive index and lowers its working temperature and viscosity. Aside from attractive optical properties the addition of lead oxide to potash glass renders it more fluid and more workable than ordinary soda glass. Less air bubbles are trapped – leading to a more clear, almost flawless product.

- The use of authentic lead glassware to serve food or drink has become rare. The concern is that some lead might leach out. The small amount of lead that could possibly make its way into a drink (like wine) would depend largely on the amount of time that the liquid has spent inside the vessel (like a crystal decanter). While these fears are overblown in some cases, no amount of lead is safe to digest. In contemporary glasses now called crystal or flint glass, the lead oxide is likely replaced with titanium dioxide, zirconium dioxide or a combination of barium oxide and zinc oxide.

- Lead glass is or has been used for radiation shielding laboratories and hospitals.

Lead or flint glass is much heavier than standard soda lime glass. It can also be distinguished sometimes by the drawn-out chiming sound is makes when tapped by a piece of silverware. Lead crystal glass sometimes also reflects a rainbow like prism of colors when struck with light.

Mirrors

Between the 16th and 20th centuries mirrors were commonly "tin-mercury amalgam mirrors"; where tin foil and liquid mercury (liquid tin) created the reflective surface. Later aluminum and silver were used. For a reflective backing, aluminum foil can be glued to glass.

But silver foil is expensive and impractical to waste on a mirror. Since silver can not be directly bonded to glass, liquid tin is applied first. Then a thin layer of silver nitrate can be sprayed on – that will bond.

Between the 16th and 20th centuries mirrors were commonly "tin-mercury amalgam mirrors"; where tin foil and liquid mercury (**liquid tin**) created the reflective surface. Later aluminum and silver were used. For a reflective backing, aluminum foil can be glued to glass. But silver foil is expensive and impractical to waste on a mirror. Since silver can not be directly bonded to glass, liquid tin is applied first. Then a thin layer of silver nitrate can be sprayed on – that will bond.

A hobbyist at home can make a mirror by coating a clean pane of glass by using a "Tollens' reagent" and a reductant (a reducing agent – an electron donor). For example mixing and applying a solution of **silver nitrate** and ammonia. The silver ions are reduced to metallic silver, forming a reflective layer on the glass surface. This process is called a "**silver mirror reaction**" and it is used in other products like thermos jugs; (where *conduction* of heat is stopped by a vacuum in the container's glass, *convection* is minimized by a stopper and heat loss by *radiation* is stopped or reduced by a mirrored surface on the glass that reflects heat back into the liquid).

With a factory mirror fabrication process the glass is cleaned more methodically and polished with cerium dioxide (an oxide of a rare-earth metal - common in glass polish). Liquid tin and silver nitrate are sprayed on the glass. A reducing agent is then sprayed on that surface to convert the sliver into a metallic state. (A reducing agent is a chemical species that "donates" an electron to an electron recipient; reducers have extra electrons and oxidizers lack electrons). Finally a layer of copper solution is sprayed over the silver layer – to prevent tarnishing and then another paint (usually gray) is sprayed over that to provide weatherization.

Silver nitrate **film stock** was used by Hollywood movie studios, up until 1951. This same chemical that stains glass yellow, will **stain** human skin blue. Then silver nitrate can be used to form notoriously volatile silver acetylide, silver azide and silver fulminate. Very sensitive explosives, as are copper versions of these salts. In fact they were all very capable but now obsolete **primary explosives** used by militaries in previous centuries. (Primary explosives are the little ones; like a blasting cap that is needed to start the more powerful secondary charge).

Silver nitrate can be purchased online, but it is not cheap. It is an antimicrobial disinfectant that is also used to remove warts. This silver dissolved in nitric acid, has many interesting applications.

Silver nitrate based **indelible ink** is used to stamp the back of the hand, for recognition or identification that can't be washed off easily. Blue or purple colored stamp on the skin. But then on glass: "**silver stain**" (silver nitrate) or "**yellow stain**" was used vigorously by stained glass window makers since the 14th century. The stain in the phrase "stained glass", comes directly from this nitrated silver that actually penetrates the the glass surface a bit. Silver stained glass ranges from light yellows to dark oranges or browns, depending upon how it is applied. On blue glass, silver stain creates green colors.

Two-way mirrors have a very thin reflective layer on the backside. Normal mirrors by contrast, have thicker reflectors and opaque layers of paint, that won't let light pass through the glass. With a two-way mirror – light can pass through in both directions. The illusion of a two-way mirror is made possible by different lighting between two rooms. If the room behind a two-way mirror is dark, and the other room brightly lit, a viewer in the darker room can see through the glass. An observer in the bright room can not see through the mirror (even though a weak image is coming through) because bright light and reflection in his room is overwhelming the weaker light.

Then there is something called the "**true mirror**"; which essentially bounces the image from one mirror onto a second. The light bending and the image reversal done by the first mirror is undone by the second mirror. So, if holding up a sign or label to a true mirror, the text can be viewed as intended; not reversed. True mirrors can be purchased or be simply constructed by framing two mirrors together at approximately a right angle. Pundits praise how uplifting and enlightening it is to view oneself, one's true reflection in such a device. Such individuals, who've probably spent many hours studying themselves before in reversing mirrors, somehow think that reality has been improved.

The **Mirror Self-Recognition Test** *(MSR)* was first proposed by a psychologist in 1970. It is used to determine if an animal recognizes itself in a mirror (most can't). Sometimes just called "the red dot test" an animal is anesthetized and marked with some very obvious paint, but on the backside or somewhere not usually viewed by the animal. Then the animal's behavior is observed when it walks past a mirror. If the animal investigates or touches the mark when it sees its mirrored reflection, then it is considered to have self-recognition. The test results are mixed. For example most monkeys and baboons fail the test but some chimpanzees, orangutans and gorillas have passed. Bottlenose dolphins, some Asian elephants and even mice have passed to show self awareness. Most birds including the venerable African Grey parrot and the New Caledonian crow (both considered very intelligent otherwise) have failed the test. However individual species like the Indian greynecked house crow and Eurasian magpie have succeeded. Young humans, dogs, giant pandas, octopuses and most other animals have failed the test.

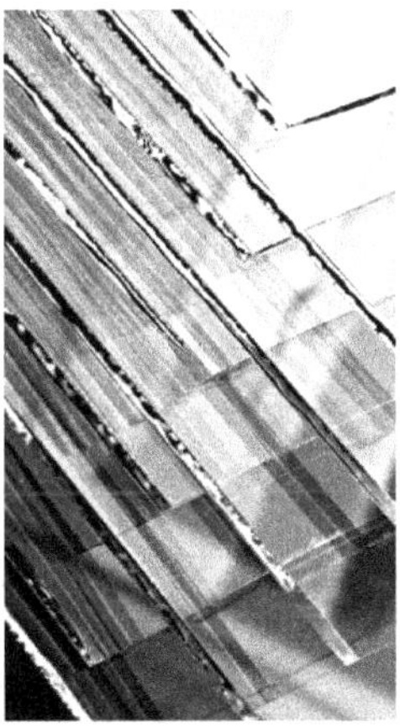

Generalized **'Modern Glass'** types include: flat-drawn glass, rolled glass, float glass, annealed glass, toughened or tempered glass, heat-strengthened glass, laminated glass, safety glass, bulletproof glass and a new family of sometimes flexible aluminosilicate glasses.

Little time will be spent explaining these types because first, hobbyist don't own glass factories to work with and second this is just a little chapter, not a dissertation on the peculiarities of glass.

Anyway, very briefly: **flat-drawn** glass entered the market in 1915, a decade after Emile **Forcault** patented a method for drawing glass. At the center and left side of this sketch is a ceramic trough called a "**debiteuse**". The floating debiteuse lives in the glass furnace and stays warm (1,300 -1,400 °F).

When glass is drawn, the debiteuse is pressed down – which encourages a ribbon of molten glass to be squeezed upward. What is not shown in this image is the "**bait**". A metal bait or anchor (usually a straight piece of steel mesh) is submerged into the bath of molten glass, and is then slowly drawn upwards through the debiteuse as the glass clings to it.

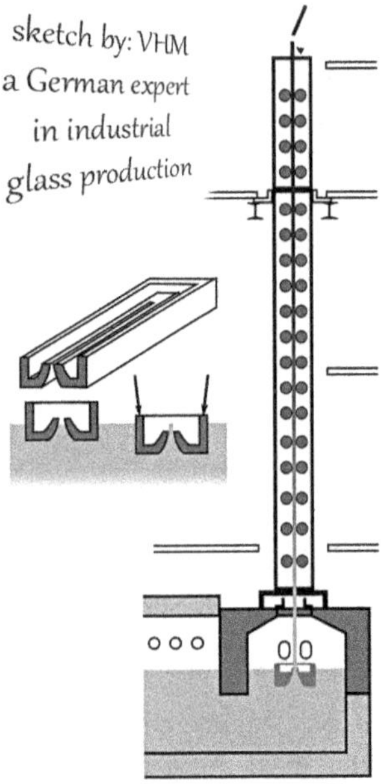

sketch by: VHM a German expert in industrial glass production

https://en.wikipedia.org/wiki/Fourcault_process

Rollers squeezed and helped shape the glass as they also helped draw it up. The glass was cooled and firm enough once it reached a certain elevation in the tower, to be cut and then carried away to be stacked. This process was used for almost 40 years, before being replaced with the float-glass method popular today.

Float-glass begins as molten glass that is poured onto a big tray of molten metal (at first lead was used, now tin). This makes a very flat sheet of glass that can be drawn off the tray, and onto a conveyor-belt like array of rollers. Rollers from top and bottom might squeeze it down to a specific thickness. **Annealed glass** or "standard glass" is basically just float glass that is slowly cooled with blasts of air – on one side. Annealing makes the glass more durable and less prone to breakage. This standard glass is stored away as stock, often to end up being modified by another process.

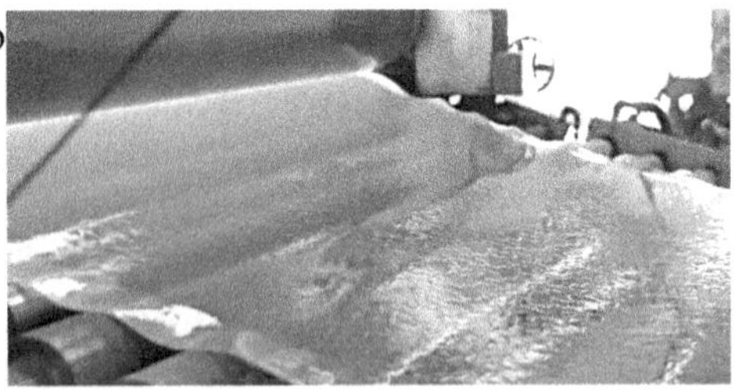

Tempered Glass -also known as toughened glass is what you would find in a regular automobile windshield. This glass might be four times as strong as its float-glass or annealed-glass counterpart, but it can't be cut or drilled. This glass which is also used in shower doors, and large windows in tall buildings will break into many small pieces that are much less jagged and less dangerous than ordinary glass shards. Tempered glass is made by cooling one surface quickly, but the other slowly – so that an internal tension and external compression is created.

Heat-Strengthened Glass is similar to tempered glass, but it is cooled more slowly. The idea is to reduce the amount of tension within the glass. This type is expected to stand up to thermal stress and high-temperature environments.

Laminated glass is the type that might protect a president or pope from an assassin's bullet. This type of "**safety-glass**" is also used in big windows that might face streets and sidewalks or that might frequently be exposed to flying debris from tornadoes or hurricanes. Here the construction might consist of two or more panes of regular annealed glass, but the panes are separated by a sheet of translucent polymer (like polyvinyl butyral) sandwiched between them.

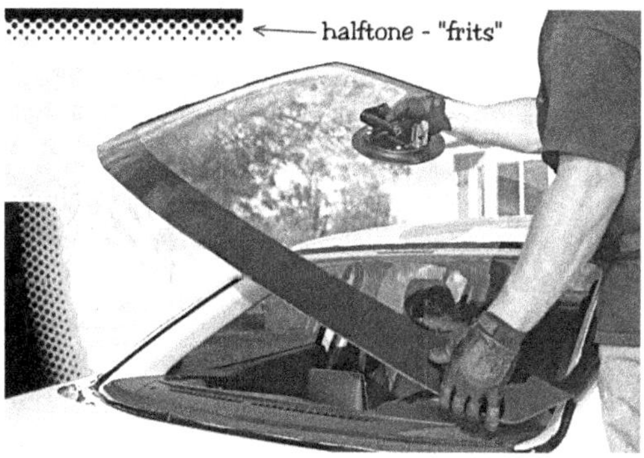

Precision robots conduct most of the modern automobile windshield making process. When a piece of annealed glass rolls in, a robotic arm scores the glass with a cutter. Then another robotic arm with a blowtorch follows the path of the score line, and thermal shock causes the windshield to separate from the scraps. Then a solid strip or a half-tone pattern of black enamel paint is silkscreen-ed on along the windshield's edges. The frit-band or other enamel edge paint will help disperse heat more evenly, but mainly the purpose is to later give urethane adhesives a better grip and weatherproof bond between the glass and windshield frame. The glass (still flat) is then set into a bending jig or mold, where the glass is formed as the whole thing is baked @ about 750 °C. Afterward when cooled, a sheet of vinyl is placed between two sheets of glass – and then pressed carefully with rubber rollers. Still murky perhaps, the laminated windshields are sent to an autoclave (like a big pressure cooker) for an hour or so, for finishing.

Bullet-proof glass is a variation of laminated glass, where multiple layers of glass and plastics are alternated. These laminates usually vary from ¾” to 3 ½”. The first of these **armored** glasses were put to use in some WWII aircraft and armored vehicles (remember that *polymethyl methacrylate* was developed in the 1930's). Because this type of glass can be very heavy, more polycarbonate might be used to help save weight; especially on the inside layer- to prevent spalling.

Aluminosilicate minerals occur naturally, but the development of **aluminosilicate glass** as a distinct material with specific properties began about 20 years ago. These are the drop enduring, scratch-resistant, corrosion resistant, anti-reflective types of glass used in LCDs, touch screens and smart phone displays. (The first i-Phone in 2007 used a plastic screen – but that soon changed).

Typically these **ceramic glasses** contain about 60% SiO_2 (silicone dioxide), 20% Al_2O_3 (alumina), with the remainder composed of CaO (lime), B_2O_3 (boron trioxide) - and finally an element from one of the alkali groups. Currently in about their 9th generation of improvement, these glasses are broadly divided into alkali aluminosilicate glass versions and alkaline earth aluminosilicate glass versions.

These glasses are refereed to as “chemically hardened”. For example - in the “fusion process” glass is immersed in a 400 °C bath of molten potassium salt (typically) that is rich in ions (± charged particles). These large potassium ions diffuse (spread freely) into the surface of the glass. As the glass cools the surface (stuffed with fat, attached ions now) compresses; which causes a stress that substantially strengthens the product. Potassium (K, element #19) is an alkali metal and its use in this instance results in a regular **aluminosilicate** glass. However some of the newest glass ceramic substrates (like “Gorilla Glass Victus” ® : by Corning Inc.) are perhaps 4x better, and these are called “**alkaline earth aluminosilicate** glasses”. Because they incorporate alkaline earth oxides like MgO, CaO, SrO, or BaO.

Back to Thermal Solar Panels

If one wishes to evaluate and compare different glazings for windows or solar panels then a little appreciation for heat energy transportation is called for.

Heat transfer can occur in one of three ways: conduction, convection or radiation.

Convection is the transfer of heat from one place to another due to the movement of fluid.

Conduction is the diffusion of heat (thermal energy) within one material or between materials in contact.

Thermal radiation is the transfer of heat energy by means of photons or electromagnetic waves.

All mater emits at least some thermal radiation. Sunlight is a source of thermal radiation that travels some 93 million miles through the vacuum of space, to reach and sustain the life on this planet. Like space, clear glazings allow radiated thermal energy to pass through them easily. Dirty or opaque glazing frustrates that optimal transfer though, and makes it necessary to create terms like absorptivity, reflectivity and emissivity to explain what is going on.

For any wavelength of light, **absorptivity** and **emissivity** are reciprocals that will have an equilibrium somewhere. For any surface behind clear glazing, a good absorber is also a good emitter, while a poor absorber will also be a poor emitter. So copper pipe or a spray painted beer cans, are considered to be good absorbers that can gather and collect solar radiation well. But these materials then can turn around and emit (conduct) that heat to the air or water that circulates within them.

The emissivity of a surface depends on factors like its chemical composition and surface geometry. For construction of a thermal panel, thick PVC pipe would make a poor choice for absorber and emitter. Thin walled PVC irrigation hose would preform better, but still fall short of copper pipe's capability. While black PVC hose might be a little more absorptive than copper pipe, that copper pipe is more thermally conductive and transfers heat more efficiently. In the great scheme of things PVC hose is far cheaper than copper pipe, and is easier to work with. Sometimes but certainly not always, the cheaper material might be the more practical option.

Emissivity is calculated as the ratio of the thermal radiation emitted by a surface, to the thermal radiation emitted by an ideal surface (a black body) at the same temperature. E' = surface (tr) / ideal (tr). The scale of emissivity runs between 0 and 1.0; therefor 0.952 would be a very high emissivity.

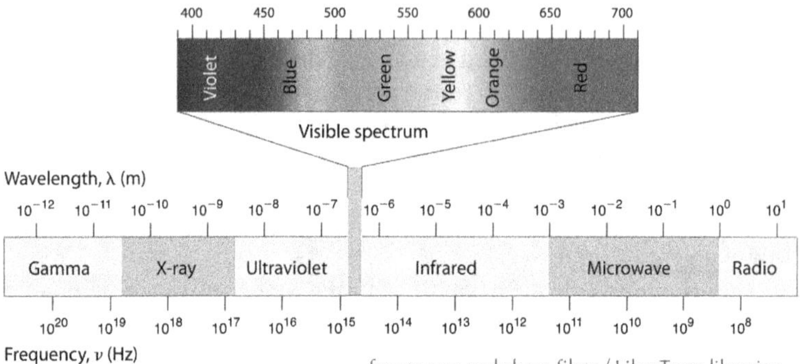

free to use and share filter / LibreTexts libraries

This image was shown before. It is shown again here to emphasize that visible light spectrum falls between invisible **UV** on one side and invisible **IR** on the other. Stars like the sun emit radiation across the entire electromagnetic spectrum.

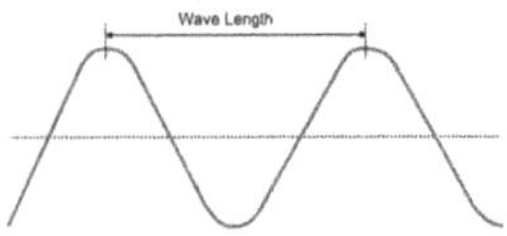

The sun sends out high frequency gamma rays that have wavelengths which are as short as atoms or sub-atomic particles are wide, as well as ELF (Extremely Low Frequency) waves that are 100,000 km or 62,137 miles long, peak to peak. The lion's share of heat arriving from sunlight is in the invisible IR spectrum. But visible light and UV light still contribute somewhat to thermal heat transfer. About 10% of the radiation generated by the sun is UV. The absorption properties of any material in question, will determine how effective UV light will be in its contribution to heating.

UV light is deleterious to many materials; including household interiors, photographs and skin tissue. Too much exposure to UV light can cause wrinkling, premature aging, melanoma (skin cancer) to the skin as well as cornea damage and cataract formation to the eyes. But UV has a few redeeming qualities too. Animals need UV light. Humans need (either vitamin supplements or) some natural UV exposure to make "vitamin D", which is essential for increasing bone density and avoiding the dietary deficiency in children known as "rickets".

But blocking UV out, means loosing a percentage of the total heat that can be extracted from sunlight. Polycarbonate by itself blocks out almost the entire UV spectrum. It absorbs that radiation and does not allow it to pass. Polycarbonate glazing is torn up and deteriorated by exposure to UV light (more so than competitive glazings) and so advertisers brag about UV coatings whenever these might be applied.

But those coatings are just there to reflect the UV and protect the plastic.

Low-E, U-Factor, SHGC and VT

Low-E (*Low-Emissivity)* windows are intentionally designed to **block infrared (IR) and ultra- violet (UV) light.** This is achieved by applying a tint (or retroactively a film) that usually contains some metal oxide (like silver or titanium dioxide perhaps). A Low-E window normally cost at least 15% more than a regular window. Manufactures claim that these windows reduce energy cost by 30% to 50%. The best insulating windows are double-paned or tipple-paned glass with a Low-E coating. However these windows block heat transfer in both directions; and discolor the light that does manage to come through.

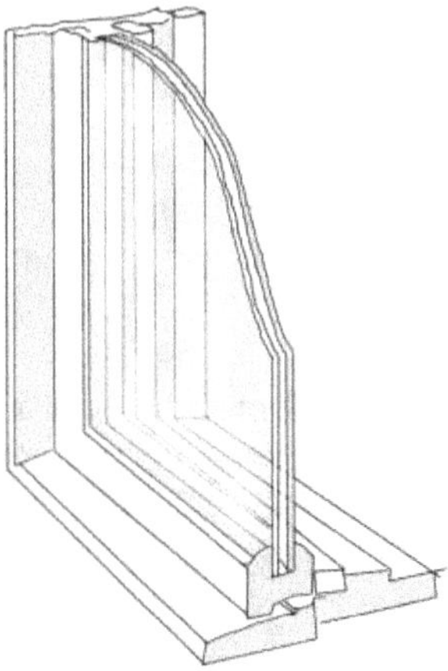

These Low-E windows might be fine for high-rise buildings that are completely climate controlled. HVACs do all the work while the Low-E windows effectively amount to nothing more than slabs of wall insulation that can be seen through. Low-E windows provide some protection against UV exposure.

But this type of glass is completely wrong for a thermal solar panel glazing and wrong headed for a home in the northern latitudes which has many south facing windows. Unless the owner cares little about home utility expense.

With regular glass such homes can and do benefit greatly from sunlight assisting to warm the interior of a house in winter. In summer, antique devices known as "curtains" can be drawn to stop the sunlight.

Quality double-paned windows manufactured 30 years were typically filled with **nitrogen**. Today those windows are most frequently filled with **argon**. Both pure gases are denser than air, and plain air has a tendency to fog up a double paned window. Because air holds water and allows thermal conductivity. (Vacuums don't work either; the seal breaks and air enters). Both nitrogen and argon are **inert gasses** (non reactive as opposed to oxygen which is very reactive). Argon is also a **noble gas** and is a bit denser than nitrogen; meaning that it is less likely to be replaced by air if a window seal should leak. Manufacturing advertisements would have people believe that argon gas is a wonderful energy saving expense worth an elevated price, when in fact its effect is minimal. Argon typically increases the insulating R-value of a window by only half a point[34].

An average insulated-glass (double-pane glass) window has an R-value of around 2.0 to 3.0.

The **U-factor** is a measurement of how much heat flows through a material; lower values indicating better insulation. The number is used to represent how well a window prevents heat from escaping a home in cold weather. (R-value is resistance to heat flow - where higher numbers mean better insulation. R-value is sort of the inverse of U-factor). *"Look for U-values between 0.20 and 1.20, with lower numbers indicating they are better at keeping heat inside".*

The **SHGC** (Solar Heat Gain Coefficient) of a window is a scale that measures how much solar radiation passes through a window. A measure of how much a window blocks heat (an advantage in Low-E treated glass). SHGC values range between 0 and 1. A lower number indicates better blockage (if that is even desired).

VT (Visible Transmittance) measures how much light passes through a window. Expressed as a value between 0 and 1, the higher the number translates to more light entering a space.

^ Light must travel farther if it is redirected. In physics refraction occurs when light enters a new medium and is bent or redirected. The **refractive index** *"is the ratio of the apparent speed of light in the air or vacuum to the speed in the medium".* Light refraction is important in glass optics, but it is of little concern to window glazing.

The Trombe Wall

If you have ever seen livestock or other animals standing next to a wall, outdoors in the sunlight of a cold winter's day, you realize that they are trying to get warm. What you may not have considered is that they are receiving thermal radiation from both sides. Yes, there is less wind or breeze next to the wall, but that is not why they stand there. After sunlight strikes a building in the morning, the sun facing walls begin to absorb heat. Thick masonry

34 https://www.consumerreports.org/cro/news/2008/10/q-a-is-it-a-worth-the-extra-cost-to-buy-windows-filled-with-argon-gas/index.htm

walls are the best at collecting and holding this solar energy. Some of that radiant heat is directed back out, the animals can feel it and that is why they stand there.

In the 1960's a French engineer named Felix Trombe developed a passive solar design that utilizes thermal mass to absorb and store solar energy. "**Passive**" is understood to mean that no mechanical pump, fan or other machinery is required to circulate heat within a building. Trombe wall designs are such that they encourage natural air flow and convection of heat, on their own.

In this image of a Trombe wall, a layer of glazing is understood to be standing on the outside. An insulating air gap separates the glazing from the masonry wall. Cool air will be near the floor inside, because it is denser (heaver) than warm air. As the wall warms up, the air caught between it and the glazing also warms and begins to rise. The warm air has no where to go but to enter the dwelling, through the holes at top. As the warm air exits, it draws in new cool air from the holes near the bottom.

The masonry wall in this instance can also be called a "**heat sink**". A heat sink is a passive heat exchanger. In the world of electronics a heat exchanger is usually made of a high thermal conductivity material – like aluminum or copper.

The next image is of a different type of thermal mass / or heat sink. It's composed of 55 gallon steel drums that are filled with water. Unlike a Trombe wall, this has no one confined or predictable path of air circulation. But heat will convect to the interior of the room eventually. Once warmed, the mass will retain its heat for a long time; certainly for longer that one night.

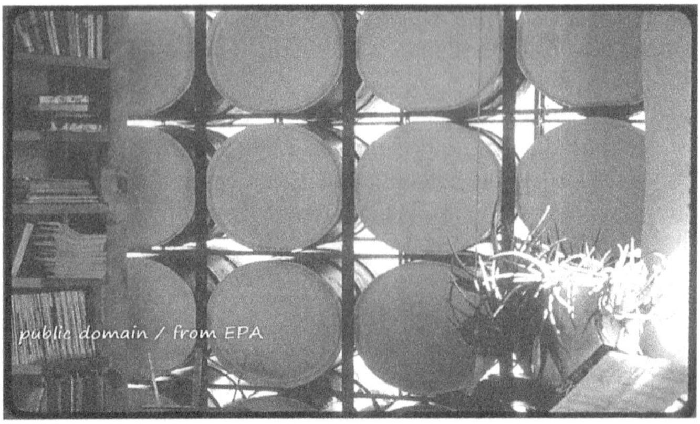

Taken literally, a 'Trombe wall' would seem to imply that as a wall - it should extend from floor to ceiling. A “pony wall” however is a short wall. There is no reason why a shorter wall could not also collect and store thermal energy. In an example like this to the side: a think and dark masonry wall behind glass might serve as a workbench, as a safety barrier from run away cars driven by drunk drivers or potentially as a shield from flying bullets. Just kidding.

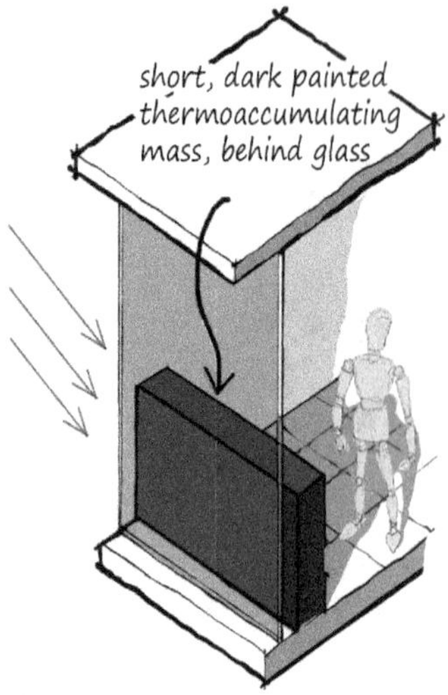

.

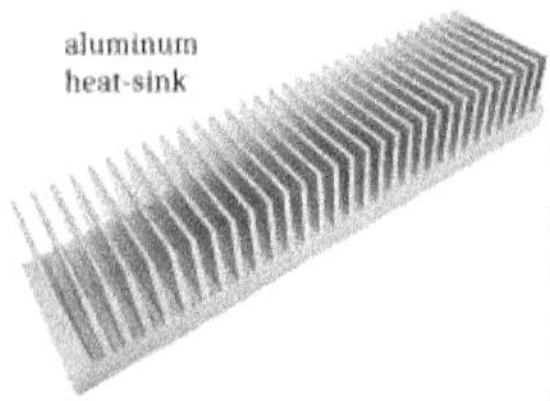

Semiconductor devices like power transistors, computer CPUs or power supplies sometimes need a little help with keeping a regular temperature. The finned heat sink nearby takes excess heat from the device and dissipates it into the circulating air.

Next are two images of the 'Center for Design Research'. An experimental research building constructed in 2011 on the University of Kansas campus. There is barely enough room for a person to pass between the windows and masonry wall. The Trombe like wall has no holes but air can circulate in from either the left or right side. Instruments once recorded the exterior and interior temperature differences year round. Another unique feature was in the unusual glazing used.

free to share and use filter / studio 804 inc
Center for Design Research / University of Kansas

That glazing is described as : "***electrochromic glass***". A glass that employs *"multiple layers of ceramic film that when electrically charged - alter their relationship to each other to tint the glass to varying degrees and manage the sun's light and heat. The glass wall can adapt to up to the minute changes in the building's use or the weather conditions".*

So this electrochromic glass (or "**smart glass**" or dynamic glass) can change its optical properties at the flip of a switch. Smart glass (or film) was invented back in the 1980's so by now there are a few variations. Essentially a polymer film is either sandwiched within glass or glued onto glass, and has PDLC (Polymer-Dispersed Liquid Crystals) nano crystals within. 'Powering on' typically makes the crystals align - creating a transparent window. Powering off would then allow the crystals to fall back into a random order that scatters light; making the window obscure and opaque. Power, a simple low voltage transformer and a couple wires coupled to a switch are all that that are needed to control this type of smart glass.

Thermochromic smart glass contains a phase transition polymer that modifies transparency when temperature changes; automatically, without electricity.

And then we run into colorful LED glass, which is not the same thing as an LED display screen. LED screens like those used for TV's or computer monitors are solid and opaque. But LED glass panels allow light to pass through, making them suitable for architectural purposes. But to explain further would be entering the realm of "off topic" information that has little to do with glass, and even less to do with solar energy.

A **sunspace** is sort of a new word for a sunroom, usually something that has been retrofitted to a house. From about the 1990's the term was first used by people who would enclose an existing porch or deck with glazing (top and sides). This increased the living space and added a bright and comfortable activities room that was useful for at least 3 seasons of the year. Even today some companies offer prefabricated sunspace kits with glazing and aluminum framing that convert outdoor decks into add-on sun rooms.

Concurrently others began to contemplate and build for themselves, **high or low thermal mass sunspaces**. A regular (or hi-) thermal mass room is insulated and constructed with or contains high-density materials which absorb, store and release heat energy slowly. These materials might be concrete, brick, rock or metal cans and barrels filled with water. Materials that possess greater thermal mass than substances like wood and drywall. Such a room is fine for year round gardening or to keep house plants from freezing, but such a plan does little to warm the rest of the house. The thermal mass tends to just radiate heat back into that room, but little further.

A **low-thermal mass sunspace** though, differs in concept[35]. Here the surfaces in the sunroom might be dark floor tile and sunlight absorbing paint on the walls; but the amount of thermal mass is minimal. These surfaces heat up quickly after the sun hits them, but that heat is not stored in a heat sink. Instead that heat is quickly conducted and conveyed by air to the rest of the house; usually forced by a quite, high capacity fan. If well

35 https://www.builditsolar.com/Projects/Sunspace/LTMSSGuide/LTMSSGuide.htm

designed, a low thermal mass sunspace can provide considerable space heating for the house. And cheaply.

There are a few design features to consider. Most **sunrooms** will have at least one door, to separate that room from the rest of the house's environment if need be. But a low thermal mass sunspace **might get too hot** if neglected so an auxiliary vent or two might be a wise safeguard. Naturally one would expect to see a large amount of glazing, and that should face in the right direction for optimal wintertime sun collection. Naturally all non-glazed surfaces should absorb heat well, have low thermal mass but also be well insulated.

Passive Ventilation

The ancient Egyptians were using semi-passive ventilation and cooling systems thousands of years ago. Evidence of this is left behind in artwork on ceramic pottery and tomb walls. Examples across the Sahara and Arabian deserts demonstrate how a simple idea can make life more bearable in a hot desert climate. Wind is caused by the sun and therefore is file-able under the category of solar energy.

Above is an old illustration of **windcatchers** on a roof top. There is much variation to how windcatchers are employed, but generally some face into the prevailing wind and others away. So a draft of fresh air enters one side and pushes stale air out from the other side. A loop of circulating air that cools the building, if there is wind outside.

(Notice the vents' height above ground).

Taller windcatchers catch higher winds. Just like the top tip of a triangular lateen sail on a sailboat, captures more power than the whole rest of the sail. Higher from the surface - winds blow stronger and cooler. Taller windcatchers in the desert will also collect less insects, dust and debris.

Ships at sea have been using this same type of passive ventilation for centuries too. Good ventilation is critical to humans and animals below decks; and can be important for other cargoes. Metal hulls continually precipitate sweat that runs down the inside of a ship's hull. This needs to be removed by bilge pumps and ventilation both.

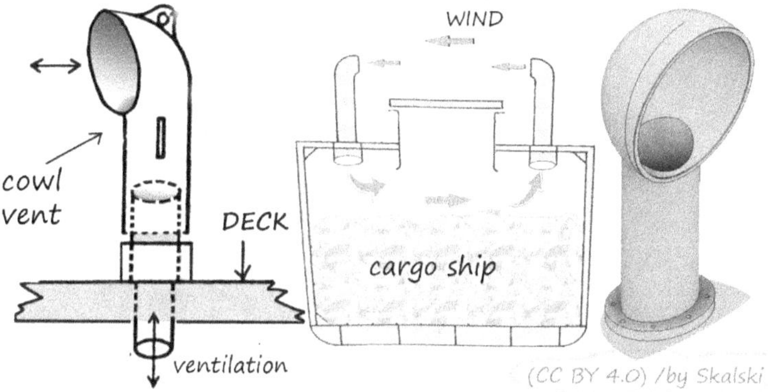

At times there is no wind, but at other times there might be far too much. Traditionally it has been a daily responsibility of the ship's “duty officer” to tend-to and trim a ship's **ventilation cowls**, as needed (even large ones are rotate-able by hand). In normal practice, “*the leeward ventilators are turned towards the wind and the windward ventilators are turned away from the wind. The air enters the compartment through the leeward ventilators, flows in the opposite direction towards the windward ventilators and finally exhausts out of them*”[36].

Roof vents on modern buildings exhaust hot air from attics predominately, but also from underneath the open ceilings of barns and storage sheds perhaps. A turbine roof vent is maintenance free, needs no electricity, is rain water and bird proof. By removing hot air from attics, they can reduce the operating cost of air conditioners and HVAC systems.

Opening the refrigerator door on a hot day **will not** make your house cooler. A refrigerator is cooling the air inside itself, by a heat

36https://knowledgeofsea.com/ventilation-of-cargo-spaces/

exchange with the exterior surrounding air (house air). Refrigerators heat the house air. If the door is left open then the refrigerator must work harder to maintain cool temperature, and so throws off even more heat into the house.

"In dry climates, the ***evaporative cooling*** *effect may be used by placing water at the air intake, such that the draft draws air over water and then into the house".*

From about 8 centuries ago, the originally Moorish palace and fortress complex of **Alhambra**, has exhibited some sophisticated passive climate control.

Sitting atop a plateau, Alhambra was a self-contained city, separate from the rest of Granada (in Andalusia, Spain) below. Twenty-three sultans of the Nasrid dynasty ruled the 'Emirate of Granada' (southern Spain) from there, until the end of the "Reconquista". It took about 780 years of bloodshed for Christians to finally drive invading Muslims from the Iberian peninsula (Spain and Portugal). The Muslims finally capitulated there and Christopher Columbus's first voyage of discovery was financed there at Alhambra, both in the year 1492. Back to the point; the first earnest construction on this UNESCO World Heritage Site, began in 1238. It is possibly the most famous monument of Islamic architecture that exist. Its appeal is enhanced by open courtyards with water pools, multiple gardens with air cooling (evaporative cooling) water fountains that are all fed by water that is conducted to the plateau by aqueducts.

In dry arid climates "swamp coolers" (evaporative coolers) are simple machines that are very effective at making the environment more comfortable. They work by evaporating moisture into the air. They're cheap to build and install. The only materials needed to make them are a blower fan, a pump, a filter pad, some water and a sheet metal box. As environments differ and may become more humid however, evaporative coolers become less effective. Air-conditioners take over when the dew point of air grows. Air-conditioners (and HVACs) remove both heat and humidity from the air by using compressed gas refrigerants. Air-conditioners might be more versatile because they can work in both humid and dry atmospheres. But air conditioners and HVACs are far more complicated and expensive machines than simple swamp coolers.

Winding this topic up

At the beginning of this chapter about 80 pages back, was an introductory image of a "solar window". It attempted to depict both the apparent daily and yearly vacillations of the sun's path. The sun's altitude is governed by the month or season, but the sun's azimuth by the hour of the day.

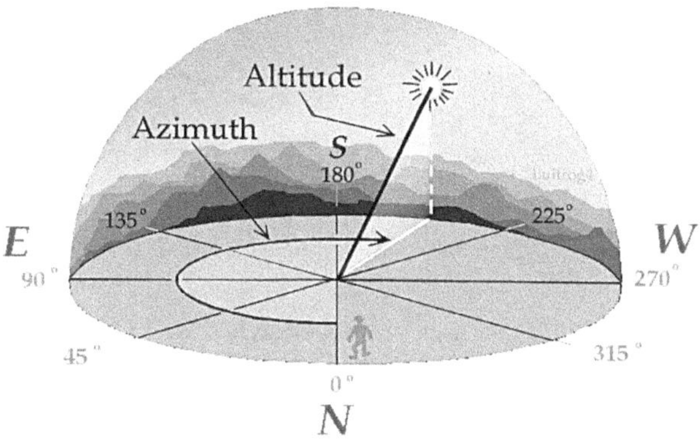

For those living in the Northern hemisphere, pointing a solar panel towards an azimuth of due south (not magnetic south) is by far the most common choice. However a growing number of grid tied solar photovoltaic panel owners, are choosing a more southwesterly azimuth.

Ideally the plane of a panel's surface should be as perpendicular to the sun as is possible. But since the sun appears to roam around so much, no one fixed position can be ideal for very long.

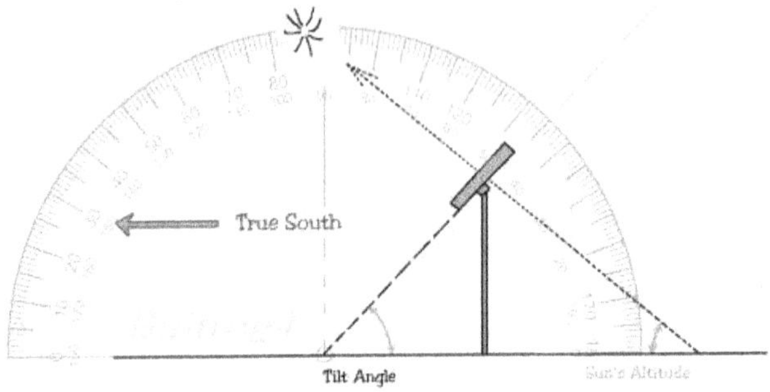

One common compromise is to choose a **fixed angle** for the panel that matches to the local latitude. (*The angle of incidence refers to the angle at which sunlight strikes the Earth's surface*). On an equinox at the equator, the sun is directly overhead at noon. In other places, the angle of sunlight's incidence is very close to matching the local latitude. Choosing an azimuth of due south and a tilt angle that matches the latitude should return a maximum average result. Of sunlight gathered – through out the year, for a fixed panel.

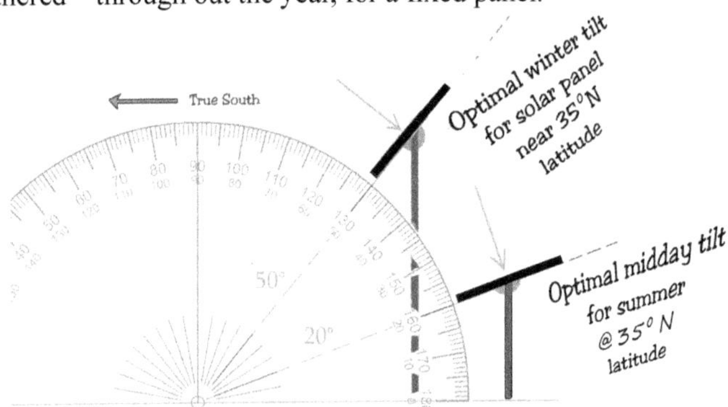

Typical tilt compensation for seasonal sunlight change. (Latitude +/- 15°).

Institutions or people that are not too concerned with cost, might purchase **solar trackers** which employ servo or stepper motors to keep photovoltaic panels aligned with the sun.

Others with the time and energy might install **adjustable solar panel mounts**. That means probably that twice a year they intend to go out there and perhaps climb a roof with a pair of wrenches to manually adjust the panel tilt. A broad or general rule of thumb used to maximize solar panel efficiency, is to increase the average angle tilt by 15° in the winter, or decrease average tilt by 15° in the summer. (Several other little calculations on the Internet differ from one another, but the resulting tilt angles are fairly inconsequential).

Not all solar panel types benefit from readjustment every 6 months. **Solar thermal** panels that circulate just air for instance, need only be optimized for wintertime performance. (Heating a home is the largest and costliest energy expenditure, remember).

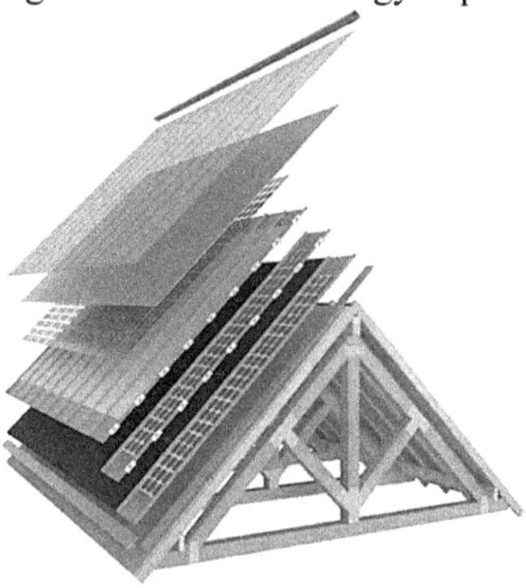

In some instances where new construction is involved, it might be possible to align a building's roof pitch and direction, for optimum performance with solar panels. This would help with the installation of traditional panels, but would greatly improve the effectiveness of newer "Thin-Film" solar technologies.

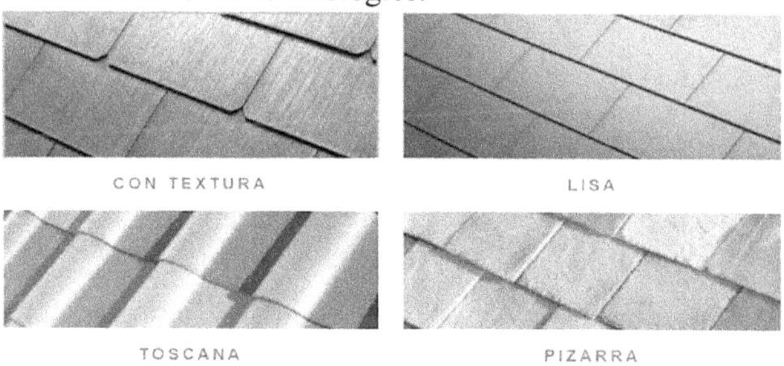

Thin film photovoltaics create less power for the same surface area as monocrystalline PV panels, but they can be far more attractive when they mimic regular roofing that covers an entire surface. These new photovoltaic roofing tiles are being made in different colors and with different textures. In the continental United States, only the roof pitches between 6:12 (27°) and 12:12 (45°) will be perpendicular to sunlight (when considering the best averaged tilt angles, chosen for fixed solar panels).

A "12:12 roof pitch" (45 degree) is about as steep as anyone builds a roof these days. In N. America the latitude where a 12:12 roof pitch would be ideal for average year long panel tilt - would be up around Montreal, Canada; Minneapolis, MN; the border between Wyoming and Montana and Salem, OR.

The western border between Canada and the U.S. runs along the 49th parallel. In those cold environs, wintertime photovoltaic panel tilts, thermal panels and of course Trombe walls can just stand vertical.

This image is borrowed from the book Vol I, and is used here to remind someone that a common sundial like this one, is pointed in the opposite direction of a solar panel attempting to collect light. The gnomon and style should both be pointed towards the South Pole, if this sundial were located south of the equator. For accuracy, the angle of the gnomon should match the site's latitude also.

Skylights allow natural light into a building. They can improve light efficiency by augmenting artificial light and can even increase heat energy efficiency by warming a thermal mass, like floor tiles for instance. Skylights can enhance the aesthetic appeal of a home, provide useful ventilation (if they are of a type that opens) and probably improve the resale value of a home. Unfortunately they might also go omitted as an expensive luxury option, by new home builders wishing to cut cost. In the center of the image below, the cylinders are like large diameter stainless steel stove pipes that have a mirror surface on the inside. Light enters the dome at the top, bounces around as it travels down the pipe, and then is dispersed widely by a prismatic lenses at the bottom. These tunnel or tube skylights are appropriate for homes with tall attics, where there is a long distance between roof and ceiling.

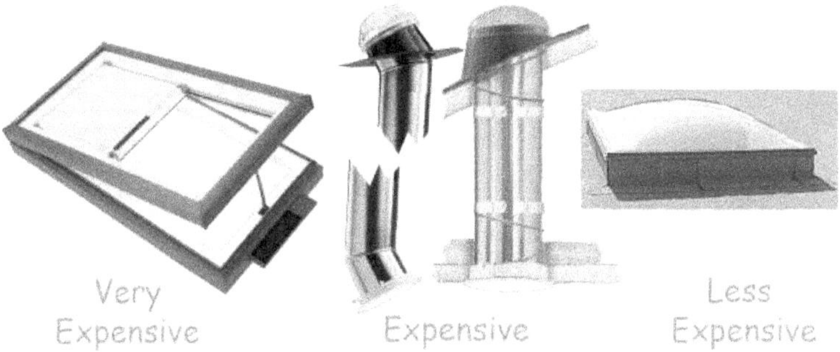

Below: someone somewhere invented a very economical type of skylight that simply uses a plastic soda bottle filled with water. A whole is cut into the roofing and the bottle is glued into place with rainproof silicone caulk. Naturally this was intended for use in the shacks and shanties of the poor, living in warm tropical climates. But in temperate climates the light emitting idea could be extended for use in the non-insulated roofs of storage sheds, hay barns and chicken houses.

Back in the 1960's a pair of PhD's working in a soil hydrology laboratory for the USDA invented a **solar evaporation still** that could suck useful drinking water out of the ground. Even in the arid desert around that lab in Tucson, Az., they realized that below the surface, soil entrapped useful moisture. Their solar still was made by digging a pit in the ground, placing a collection pot in the bottom and covering the hole with a sheet of plastic. A pebble was dropped in the center of the tarp. Additional moisture could even be gathered by placing green vegetation under such a vapor trap.

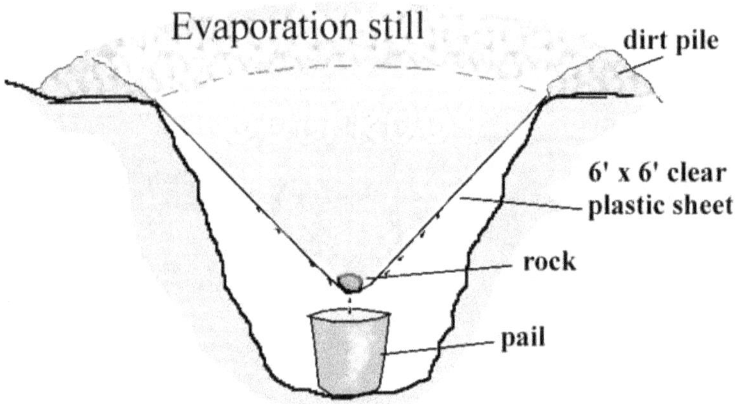

Apparently the first solar stills ever made were constructed in the harshly arid Chilean nitrate mining community of Las Salinas, in 1872. A Swedish engineer evidently constructed a series of stills that were eventually capable of purifying an impressive 23,000 liters (6,076 gallons) of valuable clean water, from brackish water, per day.

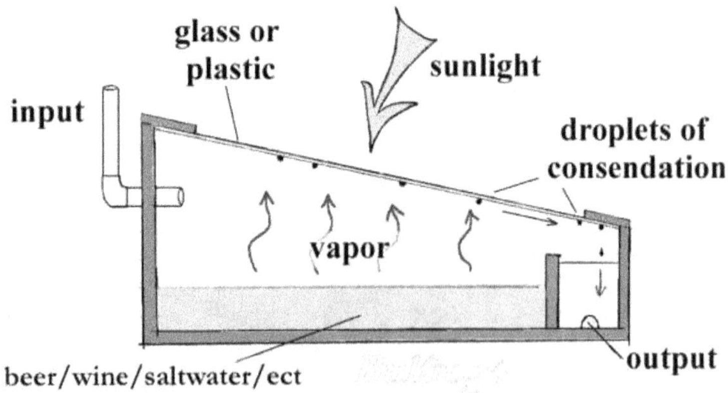

This is a generalized cross section of a simple “basin still”. It's easy to construct, effective and is used in many parts of the world for basic drinking water purification. Brackish water can build up mineral deposits inside and sea water can cause corrosion after extended use.

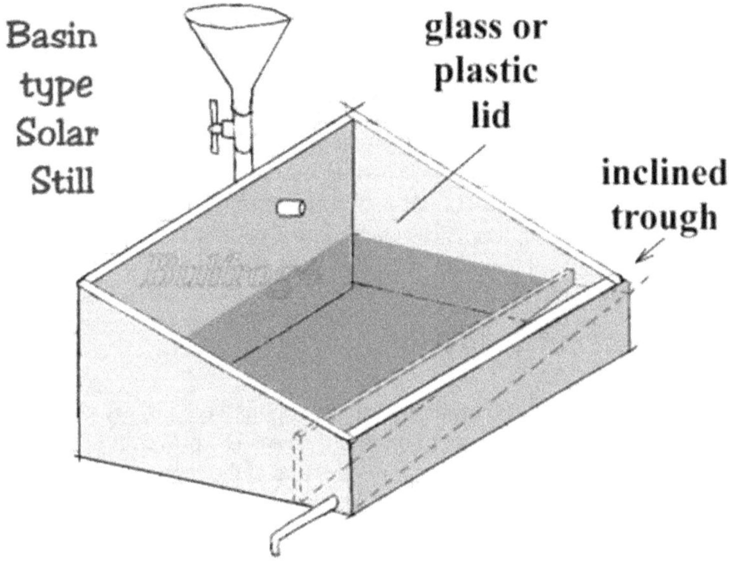

Nowadays one might find a desalination device in the life rafts or escape modules of ships and offshore oil rigs. One like pictured on the left below is inflatable and made of plastic. In the course of a day, one might produce enough fresh water to keep several people alive.

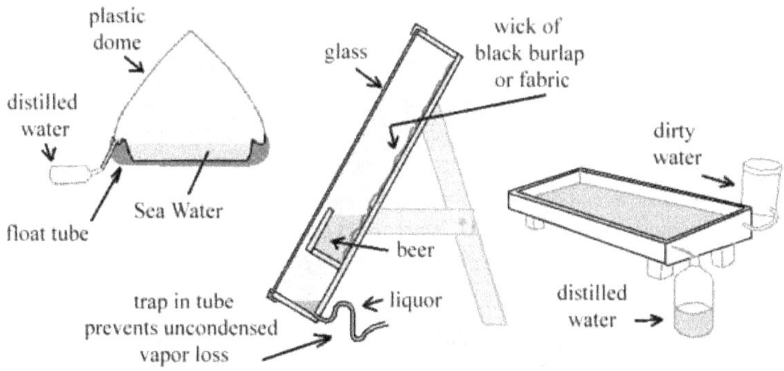

The center and right images above are appropriate for different latitudes. On the right the basin still, gathers sunlight from straight up; an attitude appropriate for locations close to the equator. The still in the center is intended to be a "**tilted wick**" design; a steeper pitch for a much higher latitude. Moisture is intended to soak into a coarse fabric like burlap and then climb the cloth before it eventually evaporates into vapor. Then the beads of sweat roll down the glass and glide to the bottom, missing the tray.

A more thorough investigation of " Solar energy / Solar Power" should include the topic of **energy storage**. Photovoltaics were mentioned earlier in this chapter but electricity has been ignored until now. Unfortunately the end's of several chapters have been trimmed and information culled, to keep this volume constrained to a target size of 500 pages. The world of battery chemistry and details pertaining to the sensible construction of one's own DYI solar power system are examples sadly, of information that must be omitted – due to lack of room. But a few high points can be hit upon, within the next few pages.

The Battery Sink

Solar PV panels, dynamos and alternators powered by wind, water or steam power, need a place to put their excess DC power. For the last couple of centuries, batteries with anodes and cathodes separated by an electrolyte - have traditionally provided that energy storage. *(Capacitors can store electricity for short duration's - micro seconds/ to minutes).*

Nowadays newer ideas for energy storage systems include lithium-ion batteries, fuel cells and supercapacitors. The first **fuel cell** was invented almost two centuries ago (well, 1839) and the main difference between one and a regular battery is that the chemicals must be circulated and refreshed, whereas in a battery the chemicals are self contained. But fuel cells and supercapacitors are presently a bit too complicated and rarefied for small home solar power storage solutions.

> Over the years there have been numerous commercial battery compositions. Consider this partial list of battery chemistry types: ***zinc-carbon, zinc chloride, manganese oxide, ammonium chloride, potassium hydroxide, aluminum-ion, chromic acid, nickel oxyhydroxide, silver-oxide battery, zinc air, mercuric oxide, nickel–cadmium (NiCd), nickel–metal hydride (NiMH), lithium-ion polymer (LiPo) and lithium nickel manganese cobalt oxides.***

Twenty, thirty years ago most batteries used for solar were led-acid deep cycle types. Deep cycle golf cart batteries were available and popular. For domestic solar applications multiple batteries were connected in either series or parallel configuration; and then called battery "**banks**" or battery "**sinks**". These banks needed insulation and protection from the elements, but also isolation from living quarters because they produce flammable "off-gas" (diatomic O_2 & H_2). But how those battery sinks are connected makes a big difference. Most commonly the industry peddles components for 12V, 24V and 48V systems. At first the 12V systems seem attractive because the components are cheaper. But the amperage in a 12V system necessitates the use of thinker, heavier wire; and copper is expensive. Eventually it might become evident that a 24V system is more efficient and economical in the long run. Larger homes requiring off-GRID solar power might adopt a more capable but also more expensive 48V system.

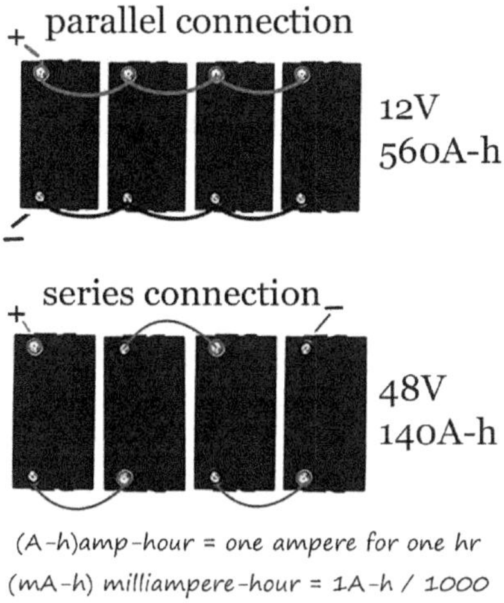

Connecting all the positive battery poles together and separately all the negative poles together of multiple batteries - creates a **parallel connected** battery group. Assuming the multiple batteries are nearly identical, this creates a battery sink where the voltage does not change, but the reservoir of available current (amperage) expands. Conversely, hooking the batteries up in **series** - increases the potential (voltage) but decreases the reservoir of available current (amps). Hooking batteries up in parallel is not a very desirable idea. In parallel configuration a weak battery can draw down the stronger ones. The solution in this case is to match or balance the individuals together as closely as possible. (Don't mix and match different brands, capacities or discharge specs).

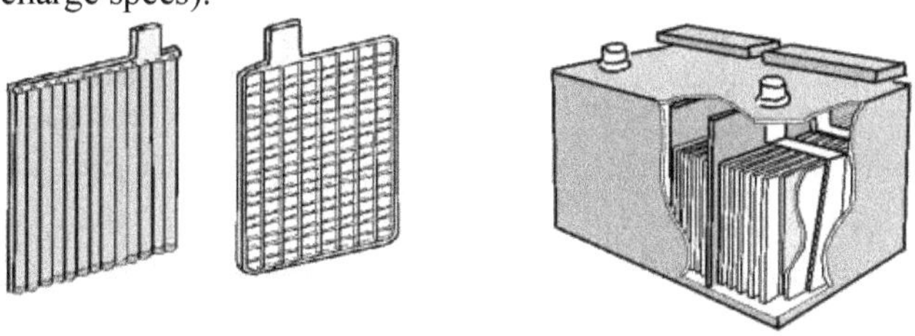

Rechargeable batteries

Typical old fashioned zinc-chloride flashlight batteries were disposed of after being used up. These were examples of non-rechargeable "**primary batteries**". When the electrochemical reaction in a battery's cell is reversible, it's called a "**secondary** cell", A **rechargeable** battery. (Incidentally, a battery charger functions by passing a current through the cell in a direction opposite to that of the flow of electricity during discharge).

Automotive lead-acid batteries have been used in cars for over a century now; since the 1912 Cadillac first sported the electric ignition. Although having a low "energy density" for it's weight, the old Pb/$PbSO_4$ (lead -acid) still out-shines all other battery chemistries in one department – the demanding task of starting a car on a cold morning. A chore that any affordable deep-cycle battery is unable to do.

Also it is almost impossible to freeze "oil of vitriol" (H_2SO_4 / sulfuric acid) which is why other electrolytes like hydrated potassium aluminum sulfate (alum) are seldom ever used.

Automotive batteries have traditionally had two useful ratings stamped on them; CCA (**Cold Cranking Amps**) and RC (**Reserve Capacity**). *[CCA = the number of amps a battery can produce –for 30 seconds @ freezing temperature. Reserve Capacity (RC) is the number of minutes that a battery can deliver 25 amps at or above a 10.5 volt threshold].*

Reserve capacity and **discharge tolerance** makes deep cycle batteries preferable to automotive types, when it comes to off-grid electrical storage. Any battery will last longer if it is not allowed to discharge to a great degree. A deep-cycle battery discharged to 50% every day will last about twice as long as one cycled to 80% of capacity. For less strain and increased longevity, deep cycle batteries should probably be drained no more than 10% on a daily basis.

Small car's with small motors might suffice with an automotive battery rated at 400 CCA , but a large car or truck might need at least 750 CCA. Diesels even more.

Beware of $LiFePO_4$ battery marketeer's hype when it comes to cold cranking amps. They cheat by being inconsistent, and don't test using the honest original (SAE) criteria for measuring CCAs. Actually, where the first "**C**" in CCA should stand for "**Cold**", lithium battery makers would change that to mean "**Continuous**" (cranking amps). Seems dishonest. Also most batteries will now have an "Ah" (**Amp Hours**) rating. This is a capacity rating (similar to a gas can holding so many liters of fuel). Between 40-60 Ah is typical for most car batteries.

$LiFePO_4$ Batteries

The new **lithium** $LiFePO_4$ batteries are quickly becoming the darlings of the present, off GRID energy storage conundrum. These are a little less energy dense than other lithium chemistries (like $LiCoO_2$, LiMn2O_4, NMC) and so are a little larger as a result. Although big, they are very light if compared to a lead-acid battery however. While prices have declined in recent years, some $LiFePO_4$ batteries - the most desirable ones, are still substantially expensive.

$LiFePO_4$ batteries have significantly more recharge life cycles than other batteries. They are tough in terms of being drawn down deeply and yet still being able to recuperate. They recharge **quickly** and in the meantime have a low self-discharge rate. But they won't start a car on a cold morning – without a fortune being spent. Here a model must be able to heat itself or several must be hooked in parallel.

Build Your Own DIY Solar Power System

A **galvanic cell** generates electricity through a chemical reaction (but an **electrolytic cell** uses electricity to drive a chemical reaction). To the right are some popular galvanic cells used as scientific

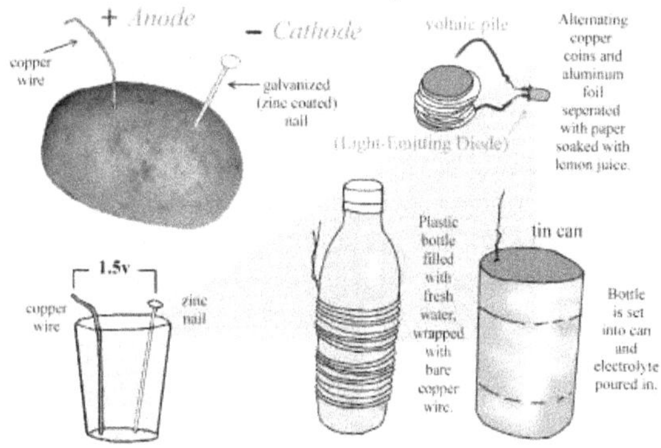

experiments for school students. Sticking dissimilar metals like copper and zinc into a potato - creates a battery.

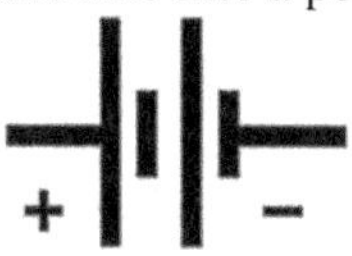

The stylistic symbol for a battery imitates a "**voltaic pile**"; the first battery. Back in 1771 Alessandro Volta teased his professorial contemporary (Luigi Galvani), by naming a creation after him (the galvanic cell). The **nominal voltage** (standardized reference point) for both a galvanic cell and an alkaline cell, is 1.5 volts. Different cell chemistries have differing nominal voltages. For instance a nickel–metal hydride (NiMH) cell is rated at 1.2 volts. A LiFePO4 cell has approximately 3.2 volts, but a NMC cell (lithium Nickel Manganese Cobalt oxide) is typically between 3.6V and 3.7V. Lead–acid battery's rate 2.1 volts for each cell. For a single cell in a car battery the voltage can range from 1.8 V (loaded at full discharge), to 2.10 V in an open circuit at (full charge). That's why the whole battery's voltage might seem to fluctuate between 10.5V and 14.7V. *[12V lead acid batteries come with 6 cells, but 12V $LiFePO_4$ batteries only need 4 cells].*

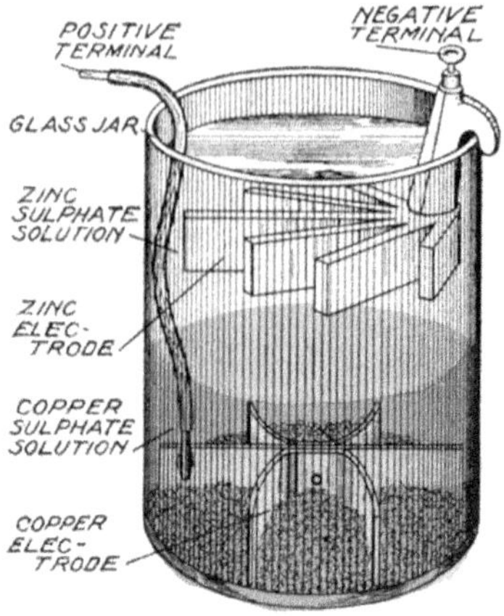

Fig. 64. Gravity Cell

from the Cyclopedia of Telegraphy and Telephony / 1919

From 1837 the electromechanical telegraph sounder used electromagnets and a single wire to send messages huge distances. The ground provided a return path for the electric current. The "**crow's foot**" gravity cell was the battery used by telegraph systems in the Old West. The negative electrode was made of zinc and was shaped in a way to maximize its surface area. The star shaped positive electrode was made of copper. Distilled water was poured in over blue **copper sulfate** crystals the then: presto, a 1.5 battery cell was born. As current was drawn, a (clear) layer of zinc sulfate solution formed at the top. This top layer was kept separate from the bottom copper sulfate layer by its lower density and by the polarity of the cell. Hence the name "**gravity cell**".

Inconveniently, this type of cell must be continually drawn from, to keep the two solutions from mixing by diffusion. The health or disposition of the crow's foot battery could be judged at a glance. After a month or two, the copper sulfate would break down. Periodic maintenance consisted of removing some of the zinc sulfate and replacing it with water and a few more blue crystals. Telegraph stations combined several gravity cells together into one battery bank.

Copper(II) sulfate ($CuSO_4$) is sold as a fungicide and algaecide; but it is also trivial to make by using electrolysis - with copper electrodes, sulfuric acid and DC current. (Making one's own oil of vitriol is not a particularly complicated chore, either).

No matter what the source of off-GRID power (wind turbine, micro water turbine, steam engine or solar panel), a **charge controller** is needed between that source and the battery sink. A charge controller regulates the voltage and current to the batteries, preventing overcharging and extending battery life. There are many models now with a wide price range. The better, pricier ones will feature protection against overvoltage, reverse current, and deep discharge.

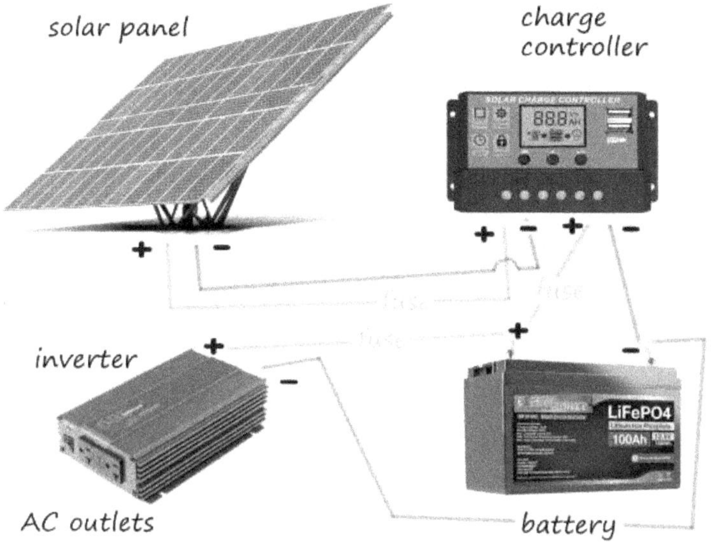

An **inverter** changes DC current into AC current. At first this task may seem straight forward, but it is not. Chopping AC into DC is simple to do with a "converter"; but inverting flat DC into the smoothly **transitioning (+/-)** sine wave of AC is difficult. Cheap inverters have destroyed many tools and appliances over the years. More components are used in a true sine wave inverter, hence the increased cost.

An "inverter" converts DC current into AC current; hard to do.

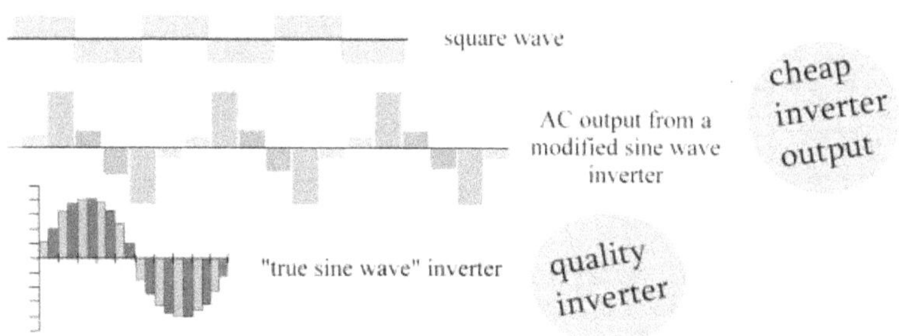

Running Gear - of typical late model Studebaker wagon

The tongue is hinged on the queen bolt, between the two front hounds.

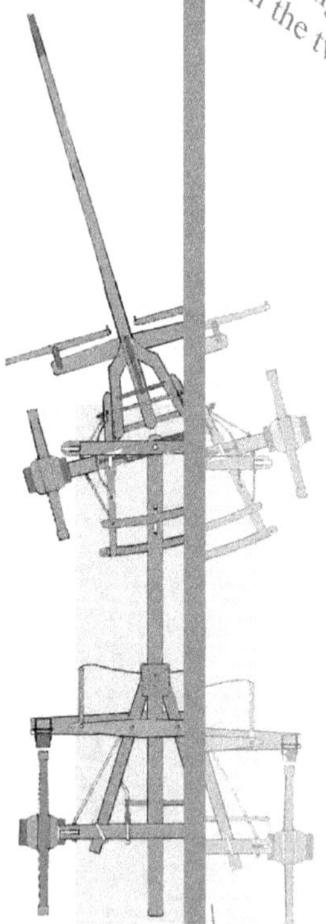

the Wagon Box fastens to 4 upright bolsters

<skip me page>

Ch. 5, Roman Builder

Segovia Aqueduct, Spain. Photo by Juan Laurent, c. 1856-1867

This is a section devoted to construction; to building things. The greatest builders in antiquity were the Romans, although they undoubtedly borrowed good ideas and techniques from other civilizations. The Romans may have not constructed the largest singular monuments, but theirs was an ambition to affect nature - to improve upon it. They surveyed, terraformed and landscaped outward from the center of the metropolis. Far and wide across Europe, many

Roman structures have withstood the savage, deleterious test of time.

Considering both the old republic plus the eastern empire, then 2,000 years of 'time' would be an average mean estimate, for the age of things Roman. Any building constructed today will be very lucky, to last for a *fraction* of that time. So a few images, a little history and some examples of Roman achievement are presented here as an modest introduction. But eventually the conversation will turn to functional ideas to use, should a reader wish to construct his own home.

"Engineering" can sometimes be used as a dirty word, to imply cheapness. A significant secondary reason for studying materials science is to reduce the amount of materials needed to complete a project. If one is manufacturing consumer items, it is sensible to use only as much material as is necessary, to manufacture and sell an object. That is commonsense economics.

Oftentimes though, the final product is weakened. All too commonly, over-manipulated designs are adequate for the time being, but are weakened to a point where they will eventually fail the tortuous test of time. Usually it is the first example of a bright new product that endures to become an antique, not the later examples that were deliberately **engineered down** and constructed with less material. We might consider that a Roman aqueduct or the mayor's new expensive house is "overbuilt". But is that true, if they last?

"The building is only so tall as the foundation is strong enough to build on". These few historic achievements to be introduced as examples, were multifaceted undertakings that required planning, surveying, engineering and massive amounts of organized labor.

In contrast, what is offered later in this chapter, are little ideas for small construction projects.

The Romans worked from the ground up. They used cranes, winches, water mills and pile drivers. They invented or used surveying tools like the groma, chorobates & dioptra.

Their legacy and possibly most important contribution, was in their great passion for building in the first place.

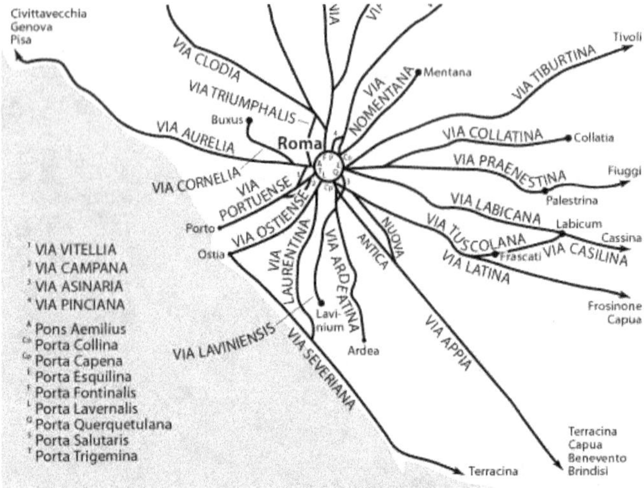

Creative Commons map | derived by Maximilian Dörrbecker

There is an old idiom that states: "**All roads lead to Rome**". While the expression can have multiple meanings, primarily it is an observation that Rome was the hub of a sprawling road system. The Romans built these quality, hard packed paved roads for sensible reasons; to promote commerce, to allow their legions to travel swiftly throughout the empire and also to occupy or employ their legions, whom might have gone idle or have grown less cohesive, otherwise.

At its peak Rome had 113 provinces, 372 roads interconnecting those provinces and 29 great military highways radiating from the capitol itself. In fact the Romans constructed some **50,000 miles** (80,000km) of superior hard packed roads throughout their empire.

Their best paved highways were called "via munita". The Romans built less expensive gravel roads too, called "via glareata"; most of which are paved over now. Sometimes the Romans would even inlay luminescent or fluorescent paving rocks down the middle of a road. Invisible ultraviolet light from distant stars would then excite the luminescence and so enable travelers to move even on very dark nights.

Pont du Gard, France
photo- courtesy of Benh LIEU SONG

An even greater achievement by the Romans possibly, is in the quality and quantity of aqueducts that they built. An "**aqueduct**" is a long distance waterway that might incorporate ditches, canals, arcades, tunnels, siphons, pipes, and bridges. The previous image is of the tallest aqueduct bridge (@48.8 m or 160 ft. high) named 'Point du Gard'. This aqueduct was built in the 1st century AD and it was engineered with a gradient that drops only 1 inch - in 1,500 feet.

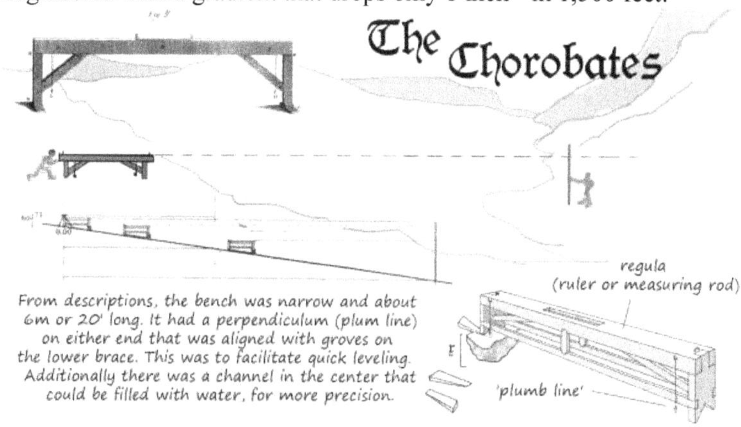

* TEXT ENLARGED* (from image) *From descriptions, the bench was narrow and about 6m or 20' long. It had a perpendiculum (plum line) on either end that was aligned with groves on the lower brace. This was to facilitate quick leveling. Additionally there was a channel or trough in the center that could be filled with water, for more precise leveling.*

By 226 AD, no less than 11 aqueducts brought fresh water into the city of Rome. *Quote: "The aqueducts delivered a staggering amount of water to Rome. Estimates suggest that the total daily flow of water through the aqueducts could reach* ***over 300 million gallons****".*

Moreover, the Romans literally built hundreds of these lengthy watercourses throughout their empire. Prominent aqueducts were built (by Romans) in France, Germany, Spain and Serbia. One Roman aqueduct in Carthage, Tunisia - was 132 km (82 miles) long. Try hiking 82 miles sometime; (probably would take 3 days).

Beyond mere agriculture, an aqueduct's continuous supply of fresh water was significant because it facilitated public hygiene, health and prosperity. This water flowed into drinking fountains, public baths, public latrines and even into some private homes. A few of these aqueducts (the ones maintained over the centuries) are still conducting water today. Some of this construction longevity might be attributable to the resilience of the unique cement that the Romans used.

The Groma

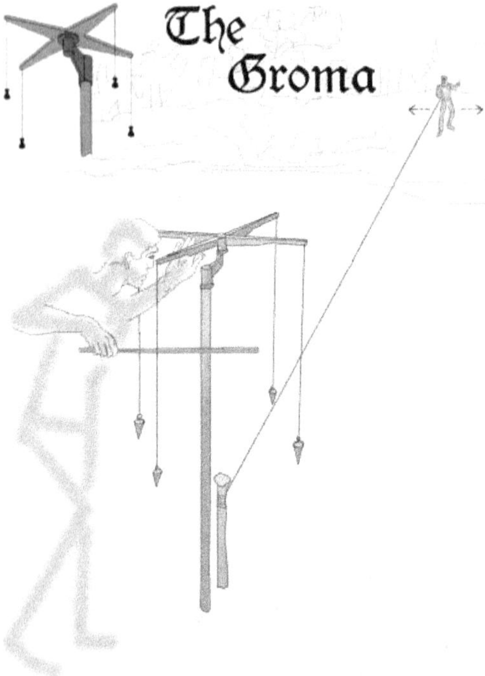

** A **viaduct** may look like an aqueduct from a distance. It's still a bridge that consists of a series of arches, piers or columns connected to each other. But the first stone viaducts were built in the 19th century to conduct steam trains across valleys; maintaining relative elevation. Trains can't climb seep gradients.*

The Roman **groma** was a tool used for laying out towns, army camps or surveying agricultural land. Its arms were perpendicular or "square" with each other. Plumb lines (lead weights or "bobs" tied to strings) draped down. Then beneath, horizontal string lines were adjusted until they ran directly under these plumb bobs.

The Roman **stade** (length of a stadium) may have influenced the length of a **furlong** (which in turn influenced the length of a mile, the area of an acre and so on). But by the Middle Ages another standard of measurement called the "**rod**" was needed to describe the furlong.

This illustration is taken from a woodcut, published in Frankfurt, c. 1536. It depicts 16 random people (coming out of a church service)

lining up their left foot – to determine the length of a **rod**. Someone else may have used an "ox goad" (a stick to poke draft cows with) that was exactly this length for a reason. It might have been useful for measuring one's plot or acreage, in a field shared with other farmers. Later the surveyor's rod (or perch) was standardized at being 16.5 feet, or 5.5 yards long. A **surveyor's chain** was, and remains - 4 rods (22 yards or 66 feet long).

16th century woodcut by Swiss-German artist - Jost Amman

So a rod like the woman above is poking the cow with, was used to define the larger **furlong**; (from two old English words furh-lang meaning furrow-long). It was the distance a team of oxen could plow without resting; about **660 feet or 1/8th of a mile** (or 0.125 mi). The dimensions of the original **acre** (4 rods x 1 furlong) might seem narrow and elongated, but for a reason. A plow is narrow. It is time consuming and arduous to turn a draft team with plow around, to then get them lined back up for a return. Also, some of the best soil for farming would have been in narrow strips alongside stream and river banks.

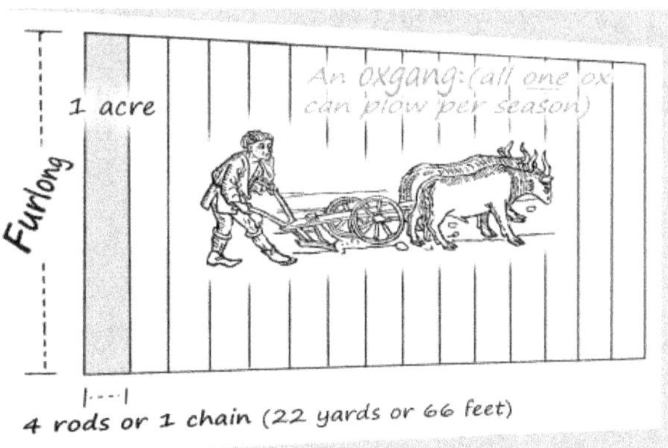

An **acre** was the amount of land that a man and one ox could till in one day. A thousand years later our acres come in many shapes, but possess the same area (43,560 sq. ft.).

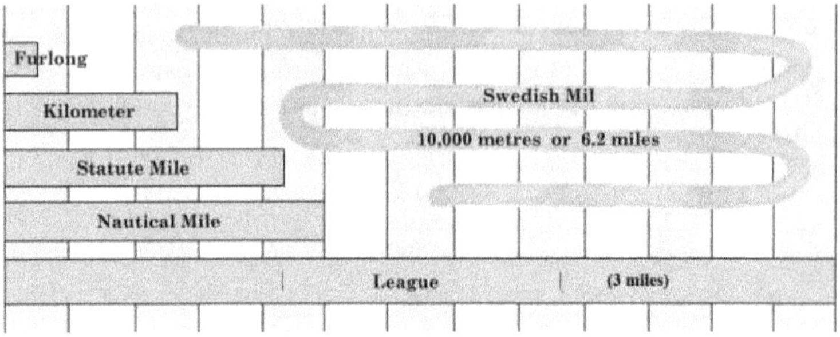

– **5** furlongs (5 x 660') just happen to be very closely equivalent to **1 kilometer** (1.00584 km exactly).

- A track for Quarter Horses (whom are sprinters) is **2 furlongs** (¼ mile) long. A typical racetrack for horses at a fairground is usually **8 furlongs** (1 mile) long.

- But for a Thoroughbred to compete for the the prestigious "**American Triple Crown**", three differing track lengths are encountered. The **Preakness Stakes** is just short of 9.5 furlongs (1 + 3/16 miles / 1.9 km), the **Kentucky Derby** is 10 furlongs (1.25 mi / 2.012 km) and the **Belmont Stakes** is the longest at 12 furlongs (1.5 miles or ½ league).

– *One **league** = the distance a Roman Army could comfortably march in one hour (**3 miles**, 24 furlongs or 960 rods).*

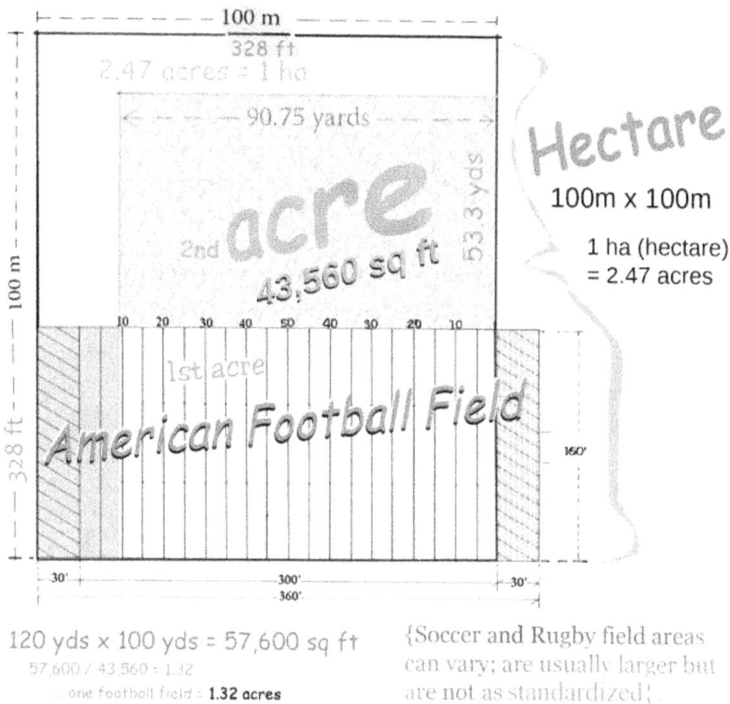

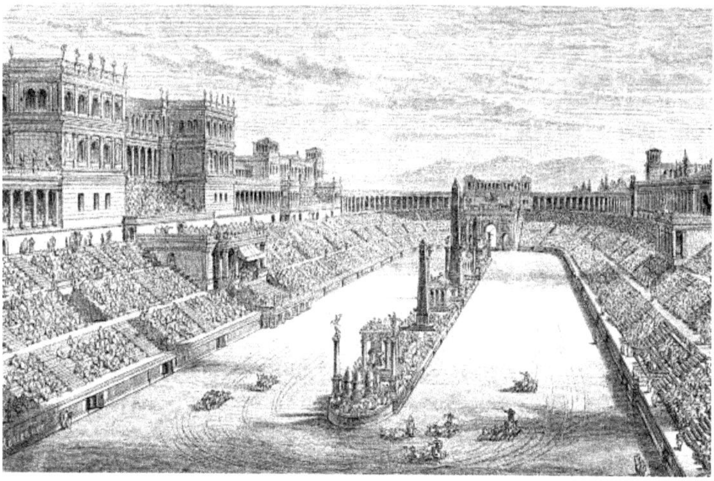

Above is an engraving that reconstructs what the 'Circus Maximus' hippodrome in Rome should have looked like. (From: Ward and Lock's Illustrated History of the World, published c.1882 / artist – Rehlender).

Horse racing, but especially **chariot racing** was the most popular sport in ancient Rome, appealing to all social classes. The word hippodrome comes from combining the Greek "hippos" (horse) and "dromos" (path or way)". Almost every Greek or Roman city of any importance had a track, circus or hippodrome to entertain the populace. Much money was bet on the outcome of these races.

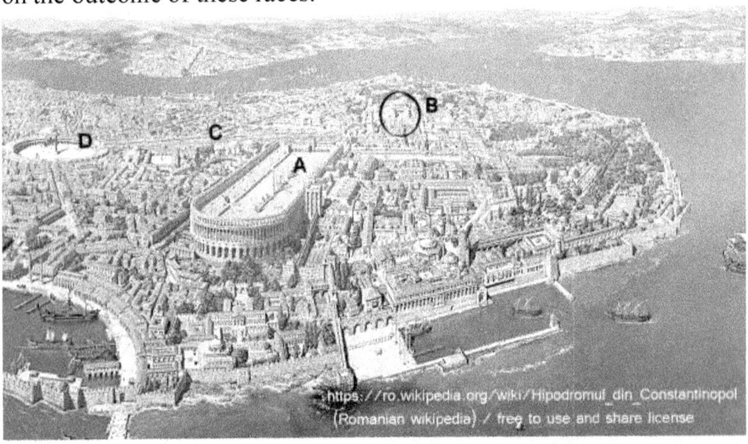

(A) Hippodrome of Constantinople, (B) Hagia Sophia, (C) Argyroprateia, (D) Forum of Constantine

One entertainment venue that rivaled Circus Maximus was the great hippodrome in Constantinople (a city renamed **Istanbul** in **1930)**. Above, the underlying image (by contemporary artist Antoine Helbert) shows what the Hippodrome of Constantinople should have looked like in the 10th century AD. The race track and seating for somewhere between 50,000 and 100,000 spectators are all gone and built over today, but the hippodrome's skeleton can still be distinguished from the air. The area labeled "**C**" was the "argyroprateia" (translated: "place of the silver sellers") where moneylenders and silver merchants hung out.

Constantinople was an epic city by almost any standard of measurement. A bastion of western civilization for 1,000 years, it was the largest, most powerful, most influential city in the western world, for about 800 of those years. The rest of the time its population was decimated by warfare, riots, plagues and numerous earthquakes. (Today the population of Istanbul is about 16,237,000).

The Chariot Race in the Hippodrome / 1882 painting by Alexander von Wagner

The original Rome did not become a proper "empire" until about 146 BC; only after subjugating both Greece (at the Battle of Corinth) and Carthage - with the last of the "Punic Wars". Some four centuries later in 286 AD. Emperor Diocletian then divided the far reaching empire into two halves. A tetrarchy of two emperors and two designated successors was established. The administration and imperial court of the Western Roman Empire was moved to **Milan**, Italy, while the capital of the Eastern half of the empire was eventually built atop the ancient city of **Byzantium**. But this division probably contributed to Western Rome's ultimate demise.

Over time the two widely separated halves began to drift apart. *"East and West failed to adequately work together to combat outside threats, and the two often squabbled over resources and military aid".* By the 4th century AD the Western Empire half of the empire was showing signs of decline; from internal corruption, economic troubles and frequent invasions by strong barbarian tribes. By the 5th century (476 AD.) the western half of the Roman Empire would collapse.

Considering Logistics

The distance between Rome and Constantinople is about 1,356 miles (2,195 km) or 900 miles 'as the crow flies' (by air and in a straight line). An average backpacker today or a traveler back in the Middle Ages can or could have comfortably covered 15 – 30 miles in a day. But that's thirty miles in a day, only if the terrain is relatively level, the weather fair and the person is both in good physical shape and intent on covering long distances. Wagons would have been lucky to cover from 10 to 20 miles in a day. Horses with a rider might cover about 40 miles in a day; but horses need food, water and rest like anything else. Under urgent circumstances professional couriers might have covered 50 – 60 miles in a day. But not continuously, day after day for great distances. *(On foot, indigenous western American Indians like the*

Hopi, Navajo, Mojave, Yaqui and the Apaches could easily out distance horses, when it came to long distance travel. That might have taken a couple of days to prove though).

If a traveler averaged 20 miles a day on a trek between Rome and Constantinople (again 1,356 miles apart), the trip would take a minimum of 68 days – or **more than nine weeks**. Even if they were efficient travelers that could consistently travel 30 miles per day, they'd still need to stop regularly to **acquire food** somehow. You can't carry 1.5 to 2 months of food on your back. A second or third horse might carry enough food for the long trip – but the horses themselves need to stop and eat and sleep too ya know. The inevitable point being made, is that the two halves of the Roman Empire were **very far apart** from each other.

In the 4th century AD the Hippodrome of Constantine, the famous Hagia Sophia cathedral and indeed the whole new city of Constantinople were refurbished and significantly enlarged by Constantine the Great. He and his Roman entourage (including tens of thousands of slaves and Goth prisoners of war) moved in and remodeled what before was known as the Greek city of **Byzantium**; an old city – dating back probably to 667 BC. What they built and were prepared to rename "Nova Roma", was destined to outshine the old western Rome in many ways. Especially in terms of centralization, strategically defensible location and in terms of wealth because of its lucrative trade. As the old Rome decayed and declined, Constantinople steadily grew more wealthy and powerful. Mainly because the city was smack dab in the middle of ancient, established trade routes. East and West (Asia and Europe) met here; the location was accessible by both land and by sea.

Constantine was born in what is now Serbia and he was the first 'Christian emperor'. As emperor between 306 to 337 AD, he had a lot on his plate – so to speak. He successfully triumphed over several foreign and domestic military threats. Before moving to Byzantium to create Nova Roma he (in 318 AD) ordered the construction of the Old St. Peter's Basilica, (located on Vatican Hill in old Rome).

That building took about 40 years to complete, but some 1,100 years later in the 15th century, the church was falling into ruin. So the "Old" St. Peter's was demolished in 1505 and the construction of the new cathedral or basilica that sits in it's place now (in Vatican City) was begun, but not finished until 1626. * The basilica has been there all this time, but the <u>Vatican city state was only recently established (1929)</u>. Vatican City is the smallest independent country in the world. * "**Basilica**" and "**cathedral**" are <u>not synonymous</u>. A cathedral is where a bishop hangs his hat; his home, center of jurisdiction (diocese). St. Peter's (in the Vatican) is not the cathedral of Rome; St. John Lateran is. (In terms of hierarchy an "archbishop" is higher ranking than a bishop and a cardinal carries more prestige than either).

St. Peter's Basilica / 1757 / Giovanni Paolo Panini

Getting back to Constantine and his construction projects, not only did he build (or re-build) St. Peters Basilica in Rome, but also the far away Hagia Sophia in Constantinople; in the same capital city that was later named after himself for more than 15 centuries.

Several Roman emperors (like Augustus, Vespasian and Hadrian) had a penchant for building impressive, long lasting structures with public (tax-payer) money; while aggrandizing their own name in the meantime. That's what many inflated politicians aspire to do today, too.

Even today the two churches rank among the world's most impressive buildings. Both churches currently standing are in at least their 3rd iteration. Both were also built over the remains of pre-existing pagan places of worship. For many centuries the Hagia Sophia contained the most internal space of any building in the world. St. Peter's would surpass its size in the 16th century and today, Hagia Sophia would rank about 9th down the list (of largest churches).

Constantine's version of Hagia Sophia – completed and consecrated in 360 AD, actually caught fire and burned down in 404

AD. The cathedral was rebuilt under Emperor Theodosius, but it also was burned again to the ground, 128 years later (in 532 AD) during the infamous "**Nika Revolt**".

The 3rd version of the Hagia Sophia, the one standing today, which preforms as a museum and as a mosque, was rebuilt from the ashes, by emperor Justinian I. (Between 532 and 537 AD).

This 'Nika uprising' apparently began after two chariot racers were arrested for murder and were sentenced to execution. Initially the emperor Justinian I refused to pardon them.

*Again chariot racing was immensely or unusually popular in Constantinople. Racers were celebrities. Originally there were four zealous rival chariot racing (or circus) factions: the Greens, the Whites, the Blues, and the Reds. Over the years the four rival teams were reduced to just two factions – the Greens and the Blues. These popular teams were bitter opponents and the general population grew to be polarized in support for one or the other faction. These two circus factions garnished considerable political power and even controlled separate neighborhoods within the city. But at the time of this revolt, they joined together in contempt for authorities and 'the powers that be'.

"*After the botched execution of two circus faction members, arrested for murder, both the Blues and the Greens appealed to Emperor Justinian for mercy. Justinian commuted the sentence to imprisonment but did not pardon the accused*".[37]

The crowd turned into an angry mob and began chanting (from inside the hippodrome) **"Nika"** (Gk: conquer or victory / also the name of the Greek goddess of victory in battle. **Nike** is an alternate spelling.) repeatedly. They began to riot. Members of the aristocracy

37 https://www.thecollector.com/what-was-the-nika-riot/

who were displeased by new taxes that Justinian had recently imposed upon the rich elite, used this uprising as an opportunity to add to the unrest and attempted to overthrow the low-born emperor. The situation turned into a full blown rebellion that lasted for five days. A great fire ravaged the city. Some impressive buildings like the Senate House and the church of Hagia Sophia were destroyed. Justinian was prepared to flee the city but his wife Empress Theodora, persuaded him to stop and grow a backbone. *(Theodora had been a courtesan and also came from low-birth background; unseemly origins for a future empress. But she had an acumen that many other rulers lacked).* Finally Justinian and the army managed to capture most of the mob within the hippodrome, sealed off the exits and began massacring the miscreants. By the end of the day, more than 30,000 rioters were put to the sword. During the chaos, some of Justinian's more rancorous political opponents were conveniently dispatched as well.

postcard from 1922 / the Hippodrome and Hagia Sophia

Above, the field of view in the camera shot (taken by Solita Solano over a century ago) shows the floor of the missing hippodrome in the foreground. Minarets were added to the church after the city fell to the Ottomans in 1453. Somehow this building is still standing after almost 15 centuries of intermittent earthquakes. An unusual architectural feature might be responsible for lending the structure strength.

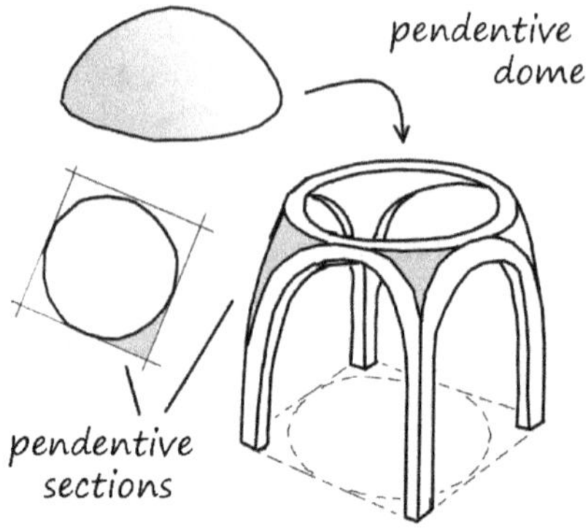

A pendentive is three-sided, curved section of vaulting. It allows an elliptical dome to rest over a rectangular room.

The first **pendentive** domes were experimented with in 3rd century AD Rome, but were perfected or finally trusted in the Hagia Sophia.

The weight of the dome is carried by the pendentive sections and the load is transferred to arches, who's legs are combined and run down to the four corners of a room, with floor-space that is now square or rectangular.

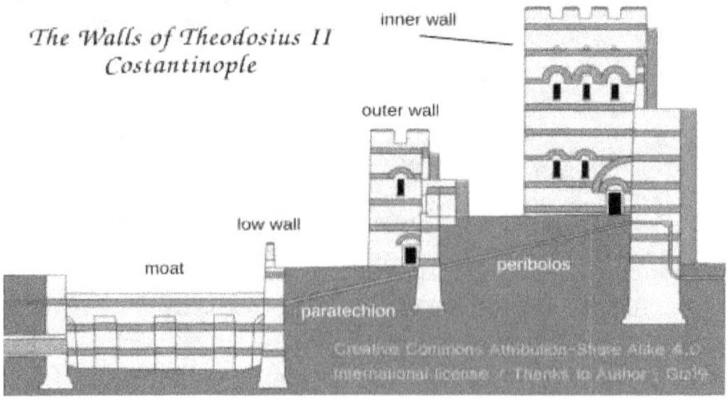

What really made Constantinople so resilient and able to last so long were the city's world class defenses. The city was besieged on numerous occasions by various armies, but the defenses of Constantinople proved impenetrable for nearly nine hundred years.

First the city was surrounded by water on three sides. The shores were lined with about 18 km of defensive walls. On the land side however stood the formidable 3.5 mile long "**Theodosian Wall**", representing defenses nearly 200 feet thick. Shown in cross–section in the previous image, this barrier was actually three walls and a moat. The moat was 20 m away from the "outer" wall, 20 m (65') wide and 10 m (32.8') deep in places. The outer wall was 2 m thick at the base and 8.5 m (28') tall. The outer wall also consisted of 90 raised towers. These were offset with the 96 larger towers of the "inner" wall – to provide archers with overlapping fields of fire. All walls were crenellated (crenels are gaps or indentations in a parapet).

A raised terrace 50' wide separated the outer wall from the inner wall. The inner (or main) wall was 5 m (16.4') thick and 12 m (39') high. The 96 towers built into the inner wall ranged from 15–20 m (49-65.6') tall. Interestingly, during peaceful times these towers were partially privatized. For the cost of upkeep and maintaining a tower a tenant got to live in some valuable property – with great views in every direction. This arrangement occasionally came in handy because after an irregular earthquake, both towers and walls came tumbling down. The city could ill afford to repair all the damage to defenses – in a timely fashion (without help). The city was struck with three earthquakes between 437 and 448 AD. The worst one happened in 447, where it destroyed over 50 towers and large stretches of wall. Attila the Hun and his massive horde were prowling the Balkan neighborhood at the time, so citizens of Constantinople were very keen to re-stack the blocks of their fortress.

The Theodosian Wall also had nine especially reinforced military gates. The strongest of these was the strategically placed "Golden Gate and the Castle of Seven Towers". (The Golden Gate bridge at San Francisco Bay, CA. takes its name from this star shaped gate/fort). But it was at another "gate" (the Gate of Saint Romanus) which was so weakened or demolished by artillery fire that the Ottoman Muslims finally breached the legendary walls and entered the city (May 29th, 1453).

The "Triumphal Quadriga" / Creative Commons Attribution 3.0 Unported license: photo by Tteske

These famous, life sized, hollow, beautifully cast chariot horses were stolen by Napoleon in 1797 and carted off to Paris. After the Battle of Waterloo (1815) however, Napoleon was exiled while the horses were returned to St. Mark's Basilica; to the Venetians who stole them in the first place. The statues, possibly cast in the 2nd century AD, had been displayed at the hippodrome for close to 9 centuries before the Venetians purloined them. To get the statues stored away in ships – the Venetians sawed the heads off. The collars seen in the picture are not original, but are there to cover up the cut marks. The statues are not real bronze but almost pure soft copper; and are fragile. The were mounted outside and above the entrance of St Mark's Basilica for about 800 years, before they were replaced with copies and brought inside.

The **Latin Conquest** or "**Sack of Constantinople**" in 1204 was a dark chapter in history. But it was preceded by an equally treacherous event named the "**Massacre of the Latins**" (by the city's Greeks in 1182). There was a palatable enmity between the eastern (largely Greek) Orthodox Catholic church and the western (Latin) Roman Catholic church. The religious dichotomy was real, but not the only reason for the latent hatred between Greeks and Italians.

Briefly; by the 11th and 12th centuries AD the growing presence and financial power of Latins in Constantinople was beginning to annoy the rest of the population, which was mostly of Greek ancestry. Merchants from Venice (in the Adriatic Sea) and from other city states like Pisa and Genoa (on the Mediterranean side of the Italian peninsula) had more ships and steadily began taking business away from native Byzantine merchants; a trade dominance that the latter had enjoyed for a long, long time. Many Latin merchants immigrated to Constantinople, became the biggest importers and exporters, bought the largest estates and became landed aristocracy. Resentment rose to a point where in 1182, the native **Greek Byzantines murdered some 60,000 Latins** (not to be called Italians yet) and sold the rest (some 4,000 others) into Turkish slavery. This treachery would not be forgotten.

< Timeline: **Crusade I – IV** >

- The **1st Crusade** (to free Jerusalem from the Muslims) began in 1099 AD and was launched from Consatntinople. It was the only crusade (of 4) to succeed, insofar as it captured Jerusalem, Edessa, Antioch and Tripoli. However the Crusaders massacred unarmed Muslims and Jews in Jerusalem. They also alienated the Byzantines when they dishonored their agreement with its leader (Alexios) who had promised them aid in exchange for the return of Byzantine territory. Alexios did not deliver as well as expected, and the Crusaders did not return the seaport and Principality of Antioch back to the Byzantines as promised.
- The **2nd Crusade** (1147,49 AD) was misguided militarily flop. The Crusaders passed through Constantinople to reach the Middle East. (It should be noted that the Eastern Orthodox Byzantines were not nearly as committed to recapturing the Holy Lands as the Roman Catholics were).
- The '**Massacre of the Latins'** occurred in 1182.
- **Jerusalem fell** to Saladin's army on October 2, 1187.
- The **3rd Crusade** (1189, 1192 AD) took a route that circumvented Constantinople, some participants taking long sea routes. Although not as embarrassing as the 2nd Crusade, this 3rd one also failed to recapture Jerusalem.
- The **4th Crusade** (1203, 04 AD) was an embarrassment and a failure, where most of the participants never even made it to Jerusalem. Actions by greedy Venetians and freelance Crusaders caused the Pope to excommunicate both, for the pillage of Zara (Croatian Christian city). Later these same merchant Venetians and pseudo pious Crusaders circumvented the great chain that closed the Golden Horn waterway and attacked Constantinople from behind. After capturing the city the Crusaders re-installed a previously imprisoned ruler and

then withdrew to wait (4 weeks) for the agreed upon payments to be made.

When the payment of dept was refuted and ignored by the Greeks, the Crusaders and Venetians invaded and recaptured the city a second time. **The Sack of Constantinople** was not a massacre (although there was much brutality and bloodletting) but a pillage of all the valuables and considerable wealth belonging to the citizenry. Much of the city was destroyed and it never again yielded the military power and political influence that it once had.

- The **Black Death** (1346 – 1353) was a monumental pestilence which decimated perhaps 1/6th of the world's population by killing about **75 million people**. Because Constantinople was a terminus of the 4,000 (6,400 km) long Silk Road across Asia, the plague probably spread to the rest of Europe from there. Spread by mice and rats taken aboard merchant ships, as well as by any other warm blooded mammals that carried fleas infected by the *Yersinia pestis* bacterium. This bacillus is endemic to the rodents of Central Asia – where caravans along the Silk Road passed through.

  (Plague had also caused a prior great dye-off in Constantinople (in 541 AD). The **"Plague of Justinian"** was the first historically noteworthy outbreak of this same bacillus; smaller outbreaks spawned about a generation apart / every 20-25 years. Noteworthy, because ancient Constantinople was by far the largest city in Europe at the time and as many as **5,000 people a day died in the streets**. Emperor Justinian himself (mentioned before in association with Hagia Sophia and the Nike Revolt) succumbed to this plague).

- In 1453 AD and after 1,000 years, this anemic Christian empire fell to Sultan Mehmed II and the Ottomans.

This "siege cannon" or "super sized bombard" did not participate in the Muslim **'Siege of Constantinople'** (1453) but was made 11 years later. After 343 years, possibly the first and only time this "Dardanelles Gun"was ever used was against the British navy – during the 1807 "Dardanelles operation". (The British were attempting to prevent Ottoman Turkey from joining France, during the Napoleonic Wars). Much later this very obsolete cannon was presented to Queen Victoria in 1866, as a gift from the contemporary Ottoman Sultan.

The Dardanelles Gun is 17 feet long, 16.8 tons heavy, 3.5 feet in diameter and could shoot a half ton stone ball about a mile and a half. The previous picture shows the barrel (left) and propellant chamber (right) unscrewed and ready for reloading. The slotted indentations were to hold the levers used to twist the two haves apart or back together again when reloading. The original gun this one was patterned after, was created by a Hungarian engineer and was even more ungainly. Named "**Basilica**", it was 10 feet longer and several tons heavier. It took 42 days, at least 60 oxen and nearly 400 men to drag Basilica to Constantinople (some 140 miles) from where it was cast. Two other big siege bombards and about 62 lesser cannons required tedious conveyance also.

(Ottomans had no monopoly on cannons; big ones were being

cast all over Europe in the 15th century. The Byzantines too had some old, very obsolete cannons but after being bankrupted by the Latins, could afford no modern, efficient, smaller ones to place upon their defensive walls).

By and large the effectiveness of these large cannons during their participation in the siege – is overstated or exaggerated. These were status symbols that had more psychological effect than destructive potential. For example once fired, it took 3 hours to reload and re-aim the gun, amounting to 3 shots a day (or no more than 7). Projectiles were scarce. Smaller cannons maintained the din between the big shots. Due to its tremendous recoil, the Basilica cannon also killed many of its operators. It began cracking after just two days of firing and had to be reinforced. The gun needed to be cleaned frequently; its barrel soaked with large amounts of *"warm olive oil to prevent cold air from penetrating and enlarging the fissures"*. Basilica finally exploded, killing everyone standing nearby.

The pitifully few defenders made repairs to the wall if they could. *(Estimates vary, but some enumerate 7,000 initial Byzantine defenders against 80,000 regular and 20,000 irregular Ottoman attackers).* Finally, after **53 days of concentrated cannon fire**, the weakened structure on both sides of the St. Romanus gate collapsed and the Muslims rushed in, to murder and rape the city with enthusiasm.

While some history books might cite this Siege of Constantinople as being ultimate evidence of gunpowder and cannon reigning supreme over traditional stone fortifications, it still took an embarrassingly long time against a hapless enemy that could barely respond, before the results were realized.

Concrete is the most common building material.

Some of the strongest concrete in existence today was mixed some two thousand years ago by Roman laborers. This has to do with a "**pozzolanic**" activity between lime and a siliceous - aluminous material (from a special volcanic ash) that the Romans discovered by accident, and then improved. By comparison, modern day Portland cement begins to degrade soon after it is set.

The problem with standard everyday modern Portland cement - is that not all the ***calcium hydroxide*** (slacked lime /Ca $(OH)_2$) is bound up. The important binder in cement, at the molecular level, is ***Calcium Silicate Hydroxide*** (CSH). There is much to explain and much to know about this topic. But not at the moment.

The Romans used cement in large amounts for building, but they were hardly the first to make concrete. (*Archaeologists have found a type of concrete dating to 6500 B.C in Syria, 5600 B.C. in Yugoslavia, 3000 B.C. in China, 2500 B.C. in Egypt and 300 B.C. in Greece).*

By volume, mass, price or by most any other metric, concrete is the most common building material in today's world. To make concrete one needs only sand, gravel and cement. One can make his own cement too, although it might be impractical to do so. Once called "poor man's cement", pioneers and farmers of previous generations knew how to make their own lime cement. (Not encouraging anyone to do so now). To make cement, one needs to **calcinate** calcium. This implies having a source of calcium rich mineral to begin with and then cooking it. The best sources of calcium normally come from rocks, crystals or minerals that are either <u>carbonates or sulfates of calcium</u>;

like **limestone, marble, travertine, chalk, gypsum, alabaster, caliche, marl** or **dolomite.** Many of these rocks are sedimentary, having been formed at the bottom of ancient oceans, from skeletal debris left by marine organisms that were able to extract calcium from the seawater. (Sea shells are a fairly pure source of calcium too).

> A pure chalk or pure limestone is called a "calcite" or "calcium carbonate" and has the formula $CaCO_3$. "**Calcination**" is derived from a Roman word for the burning of lime. Calcination is the decomposition of calcium carbonate into calcium oxide (quicklime) and carbon dioxide [$CaCO_3$ > + heat > $CaO + CO_2$].

So for thousands of years people have been crushing calcareous rocks, mixing them with a fuel like wood or coal and then burning that mix (at about 900 °C //1,650 °F) to drive off the carbon dioxide. The hygroscopic product (**calcium <u>oxide</u>** or "**quicklime**") must undergo **carbonation** before it can ever harden again. **Calcium <u>hydroxide</u>** or "**slacked lime**" differs, by being produced once hygroscopic quicklime is mixed with water. Packaged slacked lime can look perfectly dry, but it still has chemically bonded water. Less caustic and easier to work with, slacked lime is only part way through the lime-cycle. It can harden back into a rock like mineral by absorbing CO_2 again.

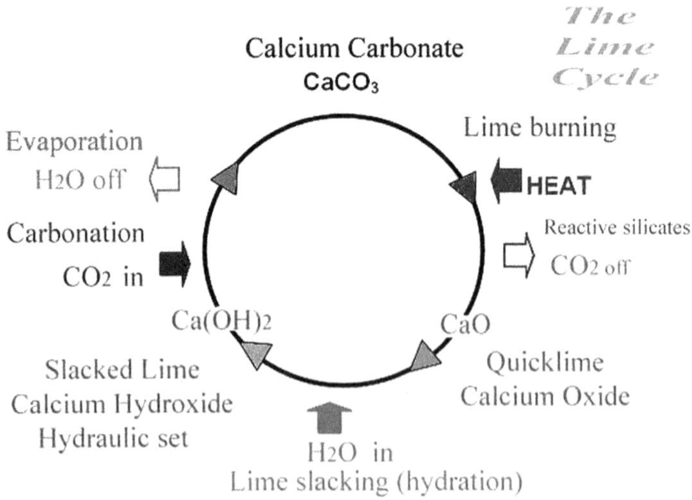

Limestone "rick" (rocks stacked in firewood). (*Unrelated - a rickhouse – is a barn for stacking barreled spirits, while they age*).

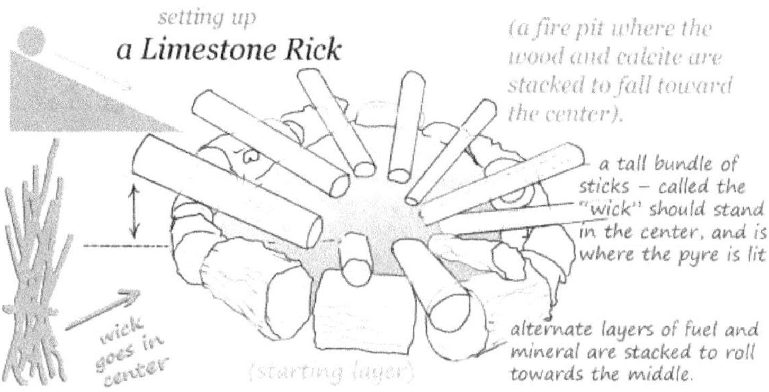

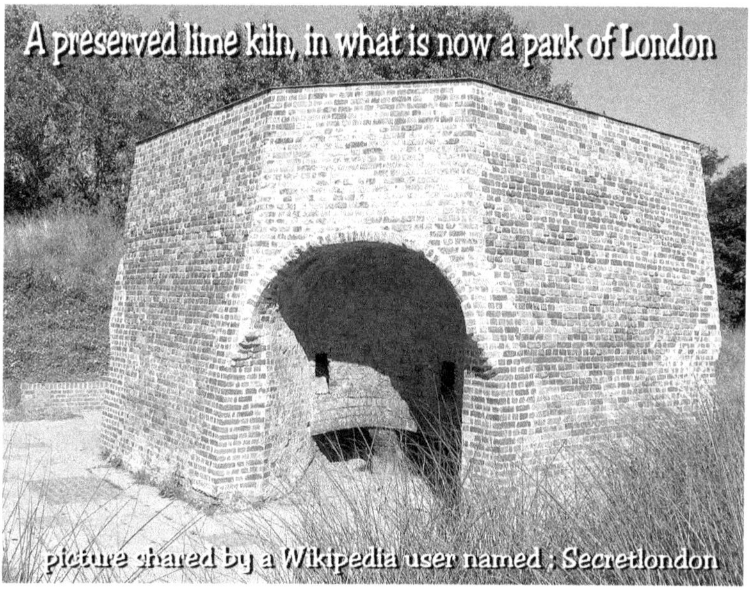

Above is a typical **lime kiln** derelict, a type of which can be found all over the world. This one is of ancient design, although it seems to have been built relatively recently (*1816 in Burgess Park, London)* and was used up until the 1960's. Inside it has an inverted cone shaped bowl, and it could substitute for a **bloomery** (metal furnace) if need be. Some ancient lime kilns were nothing more than a pit in the ground.

Lime

Lest one forget: lime has many important uses. **Lime** is a popular interior **wall plaster.** We don't physically see it in our glassware or in our refined metals but it was there during their manufacture, being used as a purifying **flux**. Industry uses quicklime as a **catalytic agent** to scrub sulfur from smoke emissions too. Lime is used to make both **fiberglass** and **paper**. We use it in our modern **antacid tablets**, in our swimming pools to adjust **pH balance**, in our **blackboard chalk**, in our **soaps**, as a filler in our **plastics** or as an extender in our **paints**. Agricultural lime is applied as a **fertilizer** to acidic soils or to improve the arability and stability of clay rich soils. Underneath many new highways and large buildings, lime is often mixed into the supporting soils to harden them and increase load bearing properties. Lime is often used in hot-mix asphalt where it can considerably improve the longevity of that road pavement. Our water and sewage plants use lime. Lime is the beginning basis for pretty much **all** construction **concretes** ancient or modern.

END OF ORGANIZED TEXT

BEGIN 'HAIL Mary' TEXT.

The subject remains focused upon construction, but will now switch to small projects that human individuals can accomplish by themselves, or with minimal help.

- A Hail Mary play in an American football game, is where a pass is thrown forward and high. The offensive team rushes forward to where the ball is expected to drop; where hopefully a designated receiver (there are 6) will catch it. Sort of a play of desperation – hence the reference to the Catholic "Hail Mary" prayer. First to use the play were Knute Rockey's Notre Dame, Fighting Irish football team, in the 1930's .
- The football play referenced here, is intended to describe the disorganized nature of the next material to follow. Images will be tossed – not into the air, but soon into these next pages. Explanations in text will be squeezed in-between, wherever they can fit. The prayer part comes with the expectation that by applying Gestalt principles of design, this visual information will be perceived in a way that enhances understanding and usefulness.

A picture of someone attempting to build a low income shelter. So far it looks as if 26 boards and 9 sheets of plywood have been used; the shack's floor-space should be about 200 square feet, when finished. A whole lot is missing in this structure. But one can imagine that with more rafters, end walls, a door, a window, insulation and a foundation being added, that this little A-Frame could turn into a cozy little cabin.

Young people in the world today are facing a **housing problem**. The cost of rentals are staggering today. The banks and the bureaucracy are not there to help you. Renting one's lodging is akin to throwing money away; it's never to be seen again. If at all possible, one should invest in a small property and then live on it. Thereby re-direct that otherwise lost rental money, into personal equity. Yes, easier said than done.

After WWII there was another serious housing shortage. Millions of American GI s returning from war got married and had families, but soon discovered that there were not enough existing dwellings to go around. In 1947 some 6.5 million families were living in temporary housing or with older relatives. That soon changed - as the passage of the **G.I. Bill of Rights** (1944) allowed the purchase homes with little or no down payment. Soon huge neighborhoods of smallish "tract homes" sprang up. *"Approximately 200,000 new homes were constructed each year during the post-war decade to accommodate returning veterans and their families".*

(base photo) U.S. pilots in front of some Dymaxion Deployment Units
North Africa, 1944 / Library of Congress Photo

Buckminster Fuller (of geodesic dome fame) was forever looking for low-cost housing solutions. His was a pursuit of building design that achieved maximum performance with minimum material. Driving down a rural farm road one day in the 1940's he encountered several small grain silos on the side of the road. He approached the company that made the metal silos and together they allied to turn out a couple hundred of these shelters (above) for the Army Signal Corps. Looking very much like a metal version of a Mongolian yurt, this Dymaxion Deployment Unit (DDU) was a smaller version of his older Dymaxion House. For a short time these were sold (for $1,250), as kit homes from Montgomery Ward.. From 1941 the DDU civilian kit came with an insulated wallboard finish, portholes (windows), door, kerosene-powered icebox and stove. But metal shortages stifled production during the war. After the war ended, the portable Dymaxion House was considered a little austere when compared to a more permanent tract home, which could be purchased with the help of a guaranteed loan.

A modern day, mail order yurt, made in Mongolia. These rarely have floors because they are genuinely transportable. "Western made" yurt kits usually come with a floor that is leveled on a temporary foundation.

Diamaxion (Dymaxion) house, built by Butler Brothers, KCK / Library of Congress

[If ($A = \Pi r^2$) and dia = 20', then floor area for that hut is 314.159 sq.ft.]

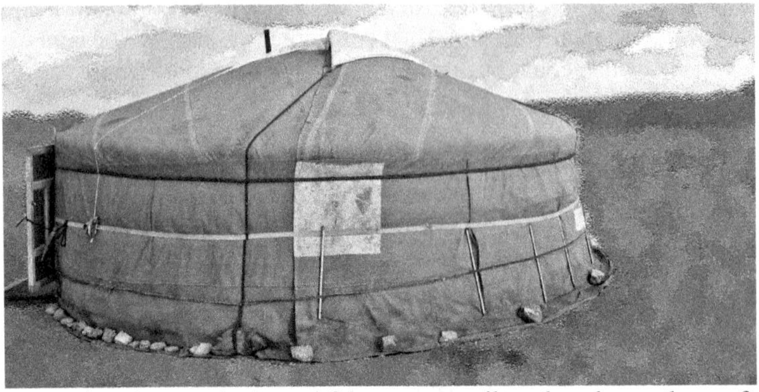

Authentic yurts had wool felt mats or quilts, placed over the roof and lattice walls for insulation. Animal skins or fabric tarps and ropes held the quilts down.

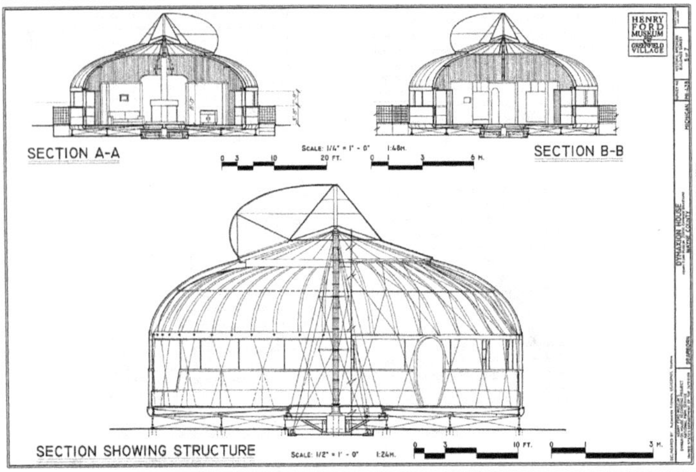

This picture from one of Buckminster Fuller's patents, shows a version of the Dymaxion House with a **wind vane.** It looks as though the vane is mounted on a section of the roof that rest on or is attached to a <u>compression ring</u>, and can rotate toward the incoming wind. Air vents could be opened or closed to control air circulation; opening vents that faced the airflow would force a cooling air draft into the shelter.

Buckminster Fuller did not design and build the first geodesic dome (Walther Bauersfeld did) nor did he invent the first round house or the first metal skinned shelter. Like anyone else – he took ideas he liked and developed and improved upon them. Just like the ancient

Mongolian yurt was not dreamed up one night and built the next morning. The yurt's origins are probably not Mongolian at all, but might be better credited to the Scythians or some other ancient nomadic tribe that inhabited the vast Asian steppes; long before the Mongolians ever appeared. (Above) the typical Mongolian yurt skeleton also had a type of compression ring, called the crown-wheel. Vents and skylights, of either shaved leather or modern day plastic might be installed. This particular crown wheel also has a stovepipe collar, which supports and isolates a hot stovepipe.

In WWI, Canadian-American-British engineer and inventor Major Peter Norman Nissen (on the left above / 1917) fairly much invented the first metal skinned, affordable housing. There was a desperate need for such shelters at the time. The "Nissen hut" was economical in its use of materials, portable, prefabricated, could be erected by six men in four hours. At least 100,000 were produced in the First World War. Nissen huts will be revisited shortly.

Mongolian Yurt

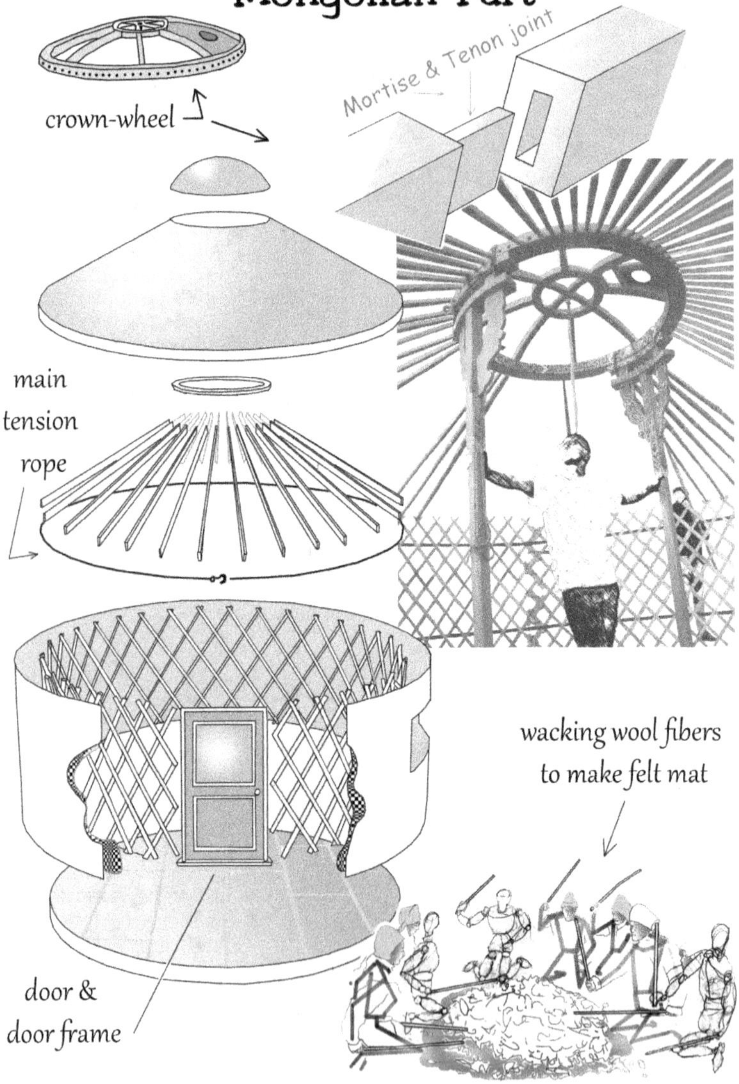

So the thin rafters of a yurt are carved and shaped to attach to the crown wheel, with a mortise and tenon joint. The rafters' other end will rest upon and be tied to a rope or cable that itself rest atop the lattice. The ends of the lattice wall, anchor to the door frame. At lower right women (usually) made wool felt blankets by beating a pile of wet wool with sticks. Once beaten and flatted to a pulp, the felt was then rolled up inside a wet tarp; and spooled around a pole. This drum or cylinder of wool felt was then rolled for hours, being pulled all over the prairie by kids riding horses.

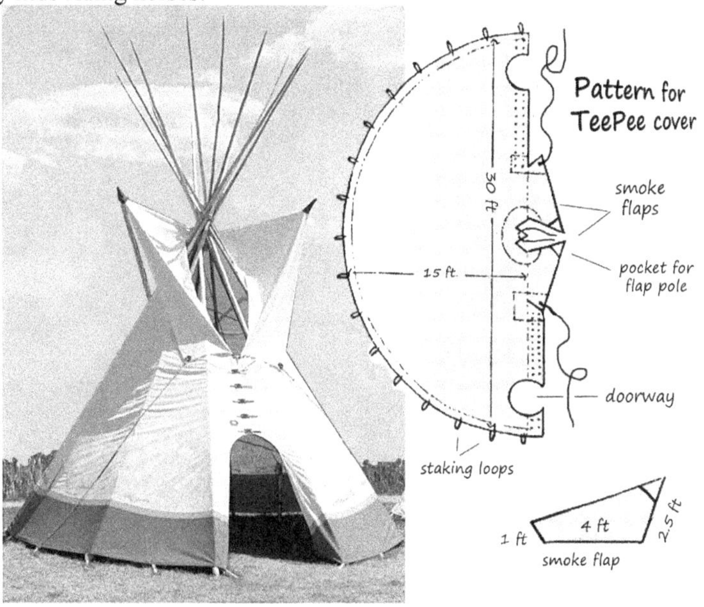

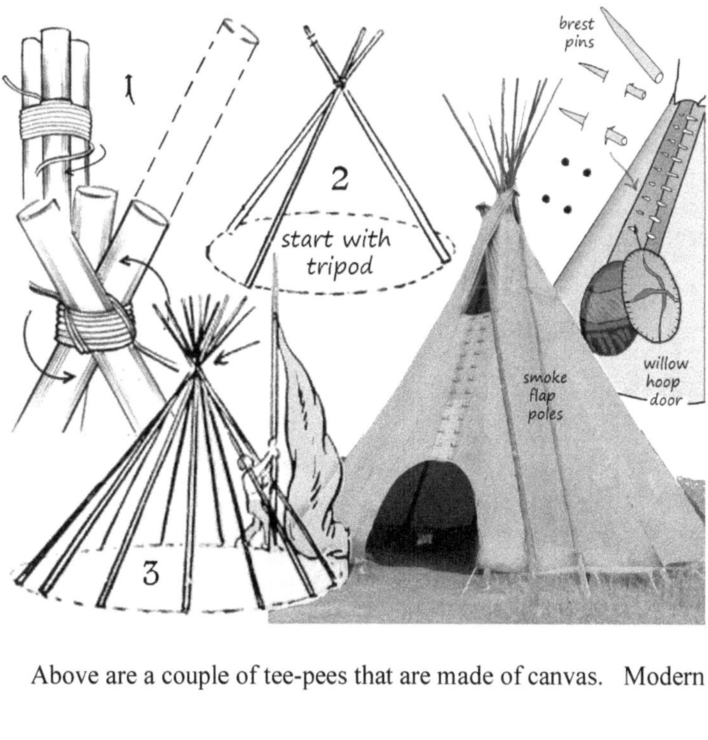

Above are a couple of tee-pees that are made of canvas. Modern tents like these can be purchased in kit form. The Teepee (or tipi) is an ancient shelter, but is it believed that it's recent form was influenced in some ways by the arrival of horses (those escaping and breeding after the arrival of the Spanish in the 16 century). The plains Indians perfected the design for a nomadic lifestyle, where the tent was quick to erect or to take down and subdivided into parcels that were easy to transport.

Source: Library of Congress, Skin tepees, Shoshone Indians

Although the tanned buffalo hides stitched together were heavy, separate sections of the hide tarpaulin could be divided between pack animals. Two long "smoke flap" poles that reached to the top of the tipi, could be adjusted to control airflow; (wide open to allow hot air to vent away during the summer, closed to hold warm air in during winter, opened a crack to let smoke from a cooking fire escape).

An average tipi might have required as many as 28 Buffalo hides to adequately enclose the shelter. The size of plains Indian teepee villages could vary greatly, from just a few, to an encampment of over 1,200. A large gathering consisting of several tribes occurred just before the Battle of Little Big Horn for example.

"In cold months, an inner lining would be constructed from animal hides, blankets, or strips of fabric. It would be hung along the lower portion of the inner wall, creating an extra layer of insulation to keep the inhabitants warm while they slept. Grass or brush could be placed between the outer wall and the lining to add even more insulation".

* It could be noted that the Roman Legions lived in leather tents too. Each "contuberniam"(the smallest organized unit of soldiers in the Roman Army / about 8 -10 men) had its own individual waterproof tent. Usually made of goatskins, the squarish box-tents were contracted for and purchased by the Army's equivalent of a "purser". Whenever a Legion marched to a new location, each soldier carried all his gear and personal effects himself. However, following the Legion was a pack train of pack animals and wagons bringing up each contuberniam's heavier items like the tent and cookware. A legionaries food ration was largely dispensed as raw whole grain. The soldiers did their own cooking; using a small stone gristmill - called a quern-stone and a 2-piece oven called a clibanus.

Above is the inside of a Navajo hogan. This one is a roadside museum display, not some actual person's home. Made of pine and juniper logs, usually octagonal in shape. Outside the walls and roof are chinked and packed with mud. The door of a hogan always points east, to welcome the morning sun. Notice the heater/stove, made from a steel oil drum.

Two different Apache wikiups - 19th century

The words wikiup and wigwam are fairly synonymous. Structurally they are both made in a similar way (green or flexible poles are bent over and tied to other poles to make a dome shaped lattice or skeleton. "**Wikiup**"sort of implies a small, crude, single family dwelling used by desert dwelling Indians of the southwest. These were built on site with local materials, but not moved. Sometimes a wikiup was turned into a sweat lodge (or sauna) by covering it with hides, throwing in hot rocks and then sealing the entrance.

Free to Share and Use photo /Plimoth Patuxet is a complex of living history museums in Plymouth, Massachusetts founded in 1947, formerly Plimoth Plantation.

"**Wigwams**" are best associated with the eastern /Atlantic coast tribes of the Iroquoian-speaking confederacy or others from Ontario and Quebec. Wigwams came in different sizes and were generally communal in nature. The larger wigwams could hold several families, which would divide off their private spaces with leather curtains or reed screens. There may have been more than one cooking fire in a longhouse.

A wooden pole lattice, covered by woven reed mats that are then covered by large tree-bark shingles – seems to be the typical configuration. Archaeological excavations of many longhouses in New York state prove that they could range between 40 and 400 ft long.

More yurt pics

Above, a close up of the crown wheel, showing how one end of the skinny rafters fit with mortise and tenon joints. (A tenon fits into a mortise). The following picture on the next page shows how the same rafters rest upon and are tied to the lattice wall of a traditional yurt.

Commercial yurts are manufactured in Canada, America and in Europe. (Below) 2"x4" rafters are notched so that their weight can be carried by a steel cable. Such units also usually come with floors.

Yurts can be built by ones-self or purchased in many sizes. When they are double walled and well insulated, they can be quite comfortable in the coldness of winter.

Duro-Last® is an established PVC roofing material. But some yurt makers use this leather like membrane for the outer shell; it comes with a 15 year warranty.

Quote: *"The Duro-Last® membrane is a proprietary thermoplastic formulation consisting of resins and plasticizers, stabilizers, and flame retardant / UV absorbents. This membrane is then laminated to a weft insertion knitted scrim that is produced by over-stitching, a non-woven web".*

If there was ever a culture that has adapted to high winds and miserably cold weather - it's been the Mongolians.

Below the etching shows how in antiquity the larger yurts (of Khans or generals) were moved across the steeps – on to better grazing for the livestock. Turning wasn't necessary. The wheels were easily removed.

(source) Internet Archive\ "The book of Ser Marco Polo" (1824) \ pg 255

More Nissen Huts

Above men are assembling some early Nissen huts, on the Western Front, in 1916. It can be seen that these units will have floors.

On the Western Front - 1916

(Below) more British troops assembling Nissen huts, but this time in 1939; a different war. On the left, the crew is attaching purlings to ribs using "hook bolts". On the right, two soldiers are standing on a purling as they fasten an interior corrugated metal panel.

The first thing a crew would do was to lay 4x4" wooden beams (or "bearers" as they were called) on the ground.

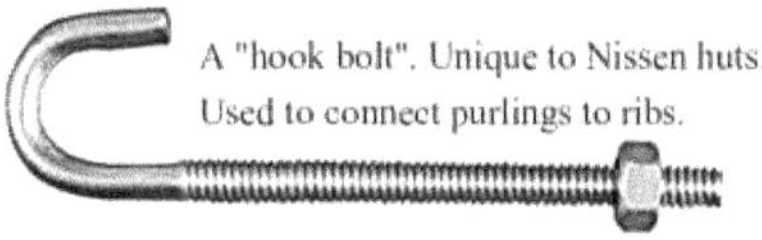
A "hook bolt". Unique to Nissen huts. Used to connect purlings to ribs.

These bearers were spaced, squared and then leveled with bricks or blocks and then later perhaps, back-filled with dirt. The ribs and floor joist screwed into the bearers. Wood tongue & grove floorboards complemented most huts, but occasionally the floor was just dirt. The two ends of the hut were wood framed and clapboard covered.

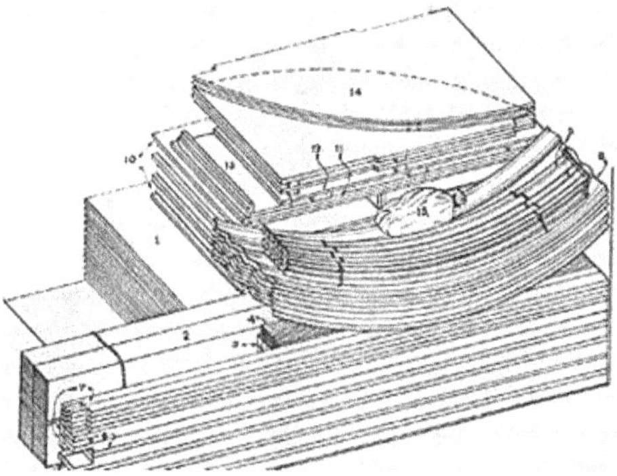

The building kit could be carried by one wagon.

The kit contained: *Wooden bearers that were placed on level ground / Steel ribs bolted onto those bearers / Wooden joists that were secured to the bearers to support flooring / An inner lining of light corrugations and an outer layer of corrugations and covered ends.*

Nissen's huts lacked insulation, but they initially came with a "matchboard" lining (tongue & grove paneling) on the inside that created an air pocket or barrier. Cold and wet troops would rip this off to fuel the stove.

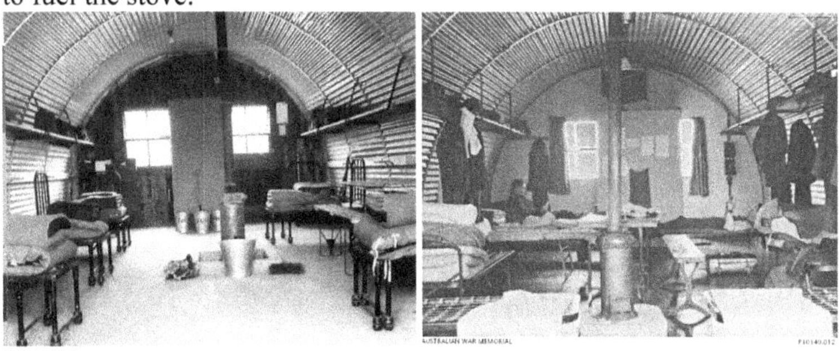

Army huts were delegated to different uses. On the left above, is an Army 'School of Hygiene' interior; on the right is an officer's barracks. Some 10,000 Nissen huts were hospital versions that provided about 240,000 beds.

The Quonset Hut

An elegantly simple, cheap and versatile plan of construction.

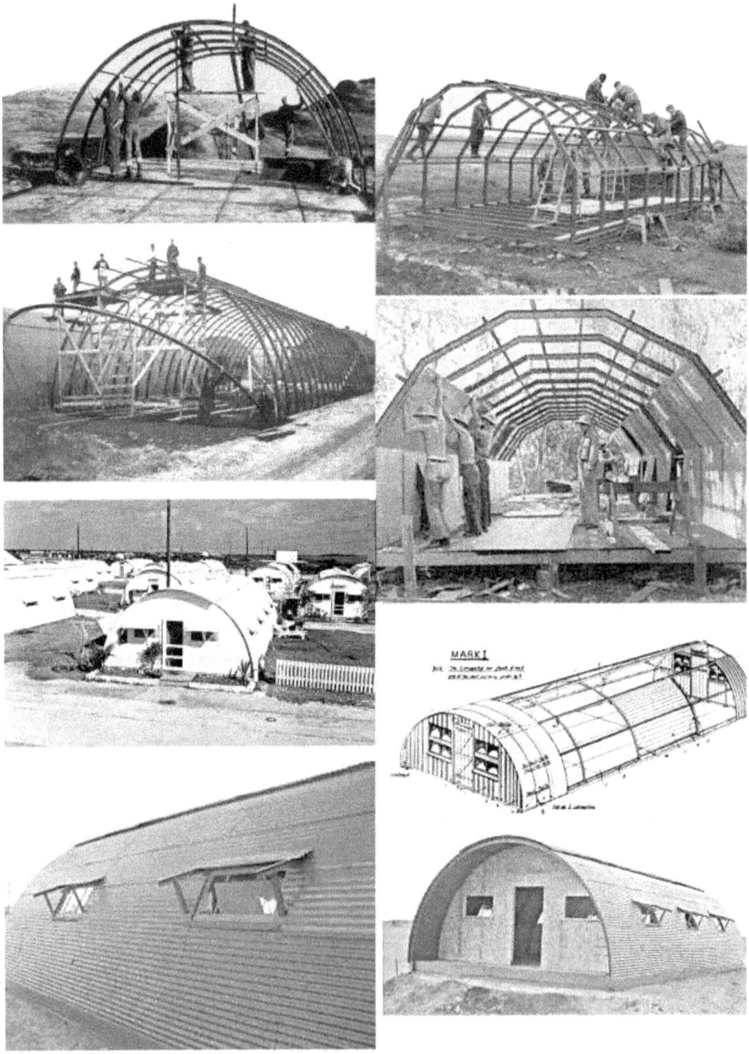

Nissen's design was patented in 1916 and other companies had to pay a royalty to built them. However when the next conflict started (1939), his company waived its patent rights – for benefit of widespread wartime production. The American “Quonset huts” originated from this development. Ultimately Nissen huts were fabricated in three standard sizes, based upon their internal spans (actually the diameter for half a circle) – 16', 24' & 30'. The longitudinal length of a bay was flexible, but in some multiple of 6 feet.

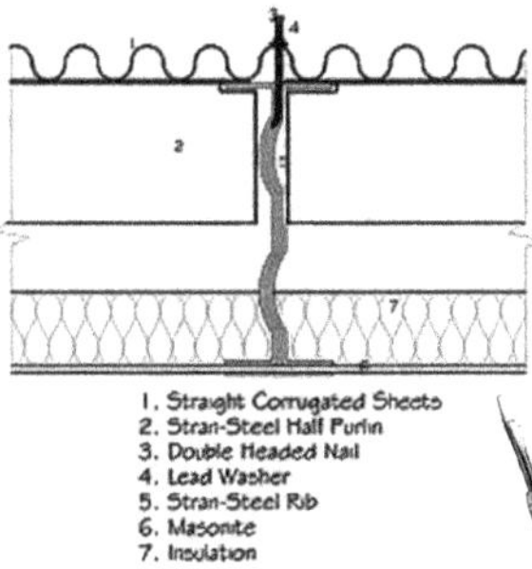

Quonset Hut
Stran-Steel Model
Wall Detail

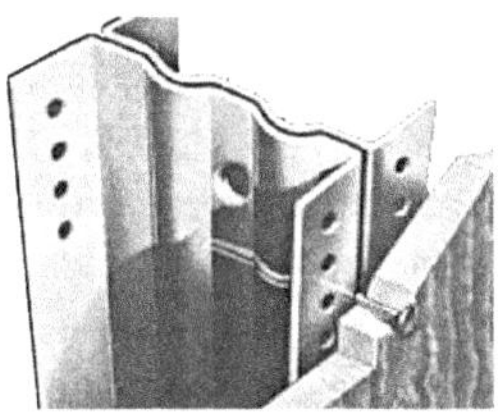

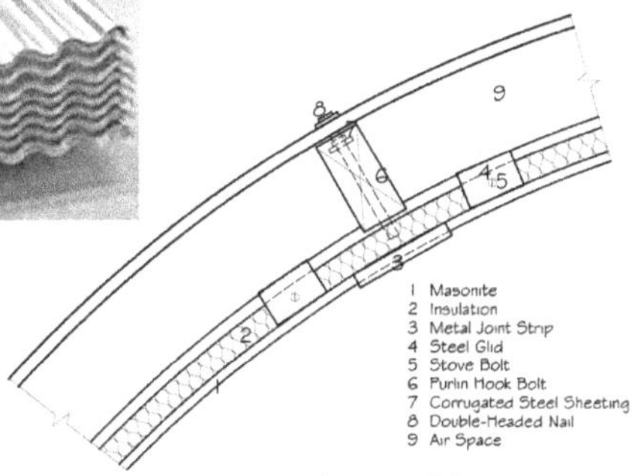

Quonset Hut T-Rib Model Wall Detail

When the first Quonset huts appeared they came with a few small improvements. They could be “sided"” on the outside using hammers and lead headed nails, they came supplied with a Masonite ® interior paneling and they had a modicum of insulation, in the form of paper or wood fiber.

Between 1941 and 1946 the American military erected 150,000 to 160,000 prefabricated “Tin Cans” (a popular nickname for Quonset huts). The demand for these structures was very high, so soon a half dozen companies were needed to make them. The dimensions of the original Quonset (named the T-Rib) closely mimicked the Nissen huts, at a size of 16 ft. x 36 ft..

Naval Air Station Barbers Point, Ewa, Honolulu County, HI /Library of Congress

Quonsets

During WW2 the Quonset design evolved through five different models, each one becoming cheaper, more adaptable and easier to erect. At the end of the war though, there were only two basic sizes of Quonsets being produced; the "20" and the "40". The '20' generally measured twenty feet by forty-eight feet, and the '40' (or "Elephant hut") stood forty feet wide by one-hundred feet long.

The name "**Quonset**"comes from a native American Indian word for a peninsula that juts into Narragansett Bay, Rhode Island. It meant 'a small, long place'. Quonset Point became one of the finest **deep water ports** on the East Coast. Shrunken in importance now, the area once contained a large United States Navy Base, a Naval Air Station, a shipyard for nuclear submarines and Camp Endicott. Camp Endicott (part of Davisville Naval **Construction Battalion** Center) was the home to Atlantic based seebees from 1942 until 1994. Seebees (**CBs** = Construction Battalion) – are enlisted personnel that build things, like airfields. The first American factory to fabricate these huts, was located near the Quonset Base and seebees were soon trained to erect them. By the time America entered the war on Dec 8, 1941, some 8,200 new

T-Rib Quonset huts had been produced. A few Quonset huts sent to Iceland quickly proved their resilience in their first winter of use. The owner of the company that first made Quonset tin cans (George A. Fuller) once bragged about them thusly, following an Icelandic storm[38].

> *"A night gale of hurricane proportion that wrecked shipping in the harbor, tossed crumpled PBYs on the beach like paper hats, and ripped the covering completely off of many British Nissen huts, left the Quonset huts practically undamaged."*

(Same building above and below) Kaneohe Bay, Golf equipment and repair shop, Marine Corps Base Hawaii (MCBH).

38 https://www.quonsethuts.org/huts/

US, Naval Base, Storage facility, Pearl Harbor

Both the Navy and the Army bought Quonset huts. But the Navy erected the most by far and many of those are still standing, but scattered across the world.

Aside from the regular barracks these huts were used for offices, hospitals, mess halls, barber shops, latrines and storage sheds.

A NSF (National Science Foundation) Public Access Repository image

Someone took the time to determine that Quonset huts served at least 86 different official uses. All wood "screen huts" that did not rust, were used in the South Pacific and canvas covered insulated huts were built for the cold in Alaska, Northern Europe and Antarctica.

In the 1950's and also during the "Cold War", the harsh continent of the Antarctic was being invaded and investigated by several separate nationalities. The Americans launched *"by far the largest Antarctic expedition, with more than 4,700 naval and marine personnel, 44 observers, 13 ships, and a number of aircraft"*– which settled in and set up shop at McMurdo Station, in 1956. At first, the regular metal and masonite Quonset huts were used, but soon a special type of Quonset hut known as the Jamesway (developed a few years earlier) was deployed.

Optimized for cold, the Jamesway Hut was more of a tent than it was a stiff walled hut; but it looked like a Quonset from a distance. The Jamesway was developed for the Army Air Corps to erect and use quickly in cold Arctic regions. The first time they were used was in the Aleutian Islands Campaign, after the Japanese invaded the tip of Alaska in 1942.

Message Center headquarters, at Camp Denali, near Fort Richardson,and Anchorage, Alaska During Exercise Little Bear War Games / 1959 / Public Domain image

The Jamesway hut was novel in that it incorporated thick "R7" mats or blankets of fiberglass insulation and these were faced with muslin and enclosed in plastic-treated cotton. The huts used curved laminated plywood arches, with centers that folded in quarters on metal hinges with pins. The packing crates that a hut came in, doubled as insulated floor sections for the hut.

The Aleutian Islands Campaign took a year to resolve, mainly due to great distances, bad weather and the extreme harsh environment. The combined American and Canadian troops committed to the campaign amounted to 140,000, while the number of Japanese numbered 8,500. There was a little fighting on Attu Island, but by the time the Allies got around to reclaiming Kiska Island – the Japanese had slipped off into the night. Probably very glad to be leaving the muddy, wet and miserably cold place. The only troops that likely slept in any comfort during the campaign – were the Air Corp personnel lucky enough to be housed in Jamesway huts. These huts also proved their worth in Korea, when temperatures dropped to -54° F (-47.7° C) during the "Battle of Chosin Reservoir".

The Davisville Naval Construction Battalion Center was a United States Navy Seabee base located in Davisville, Rhode Island. Operational from 1942 until 1994. Library of Congress photos.

The huts above reside in Rhode Island, where the first Quonsets were built. The top two photos show diagonally spanning straps that act as **shear bracing**. This sway bracing acts to improve the longitudinal rigidity of a structure. At the least, such reinforcement prevents stud walls, roof rafters or metal arches (in this case), from collapsing like a stack of dominoes during their erection. Shear bracing is needed to initially stabilize a skeletal structure long enough for roofing or sheathing to be attached. This type of bracing remains in place.

V-2 Rocket Facilities / White Sands Missile Range / components of 2 huts joined

In 1941 the standard 20 by 48-foot general purpose sized Quonset hut cost about $800. After the war a civilian could purchase a military surplus hut for $1,000. That's a semi insulated building with a floor in some cases, for less that **$1.00 per square foot**. Still, a barn; not a house. Some very important amenities were missing in that price, like electricity, central heating, sewage and plumbing.

The shape of the Quonset is still with us today, but the prefabricated kits are different. Without trusses and purlings, the posture of the modern Quonset is supplied by the shape and rigidity of corrugated **arched panels.** These are simply bolted together. Since there are no floor joist to hold the spanning arches together at the base, a good concrete footer with anchor bolts is needed to hold the hut up. An interior concrete floor might be added later.

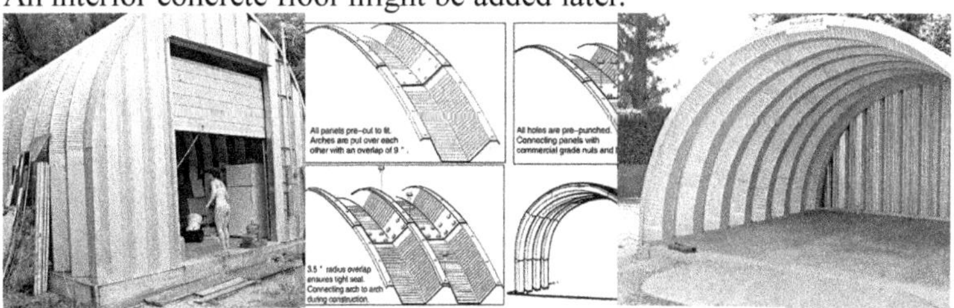

From Flicker / using Google Free to Share and Use Commercially, filter

Today (2025), a **pre-assembled-modular home** cost anywhere between $100 to $200 per square foot to buy. **Custom luxury homes** can cost $300 to $400+ per square foot.

A little money might be saved if trying to adapt a modern Quonset shaped steel hull into a livable home – but expect the cost to be comparable to installing a modular. *(A modular is transportable; usually in two sections, like a high quality trailer house that is set on a real foundation after the wheels are removed. Trailer homes and campers depreciate in value. But modular homes (prefabricated) and custom homes (built on site) ordinarily increase in value).*

While most of these new Quonset shaped prefabricated steel buildings are used as barns or storage sheds, some are actually turned into living abodes. The **issues** of converting a steel building into a comfortable house are numerous. Like a dome, the curvature of the Quonset shape, decidedly complicates custom construction.

Modern interiors / From Flicker / Google Free to Share and Use Commercially, filter

Proper insulation is one of those issues that must be surmounted. Common fiberglass batt insulation can not be directly applied to these metal buildings. Blown foam insulation might adhere but it would be thick in the groves but too thin in the lands section. A "fur-down" with wood will work best, but would be labor-some and expensive.

Above is R-19 batt insulation being installed in a 2"x6" wall with studs 2 ft. OC (on center). Notice the "x" bracing for shear strength. Behind the brace is (soft, weak subsiding) foam board; not dense like plywood, but adding probably another R2 (@ ½") in insulation.

Above, insulation is being sprayed between wood rafters in a vaulted ceiling. The excess is shaved off. Where the narrow inward face of the rafters is exposed to the living area, wood makes an acceptable insulator. But the inward facing lands of a corrugated steel panel are very wide and would present much more surface area for heat to escape from, should the excess foam insulation be shaved off.

Spray foam insulation would need to be covered up, because it is ugly. And grows uglier with time, as the surface oxidizes and turns yellow, and then to brown. Perhaps a reflective foil faced, bubble backed or closed cell polyethylene foam might be used.

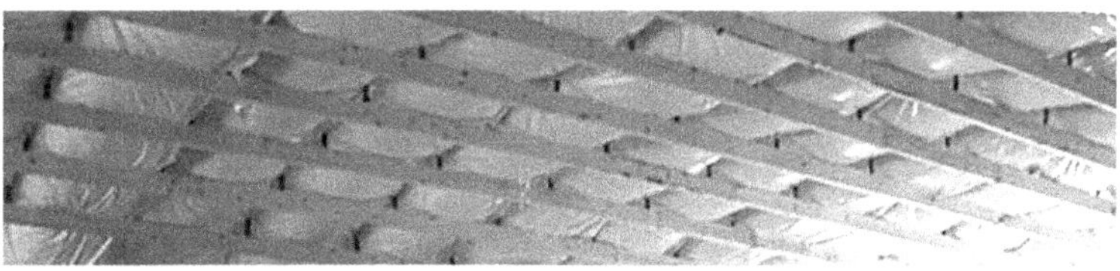

Lackluster R-factor might be a problem with bubble wrap though. Is there even an adhesive that will work between bubble-wrap and steel, without peeling off due to the metal's sweating when the weather changes? A **fur-down** would seem to be the best insulation solution for such steel huts. *A fur down ceiling, is a lowered section of the ceiling that is often used to conceal plumbing, electrical wiring, or ductwork. (External fur-downs are called **soffits**).*

Potential **drawbacks** of the Quonset hut type home range from lack of aesthetic appeal, to poor resale value, zoning restrictions and odd or annoying acoustics.

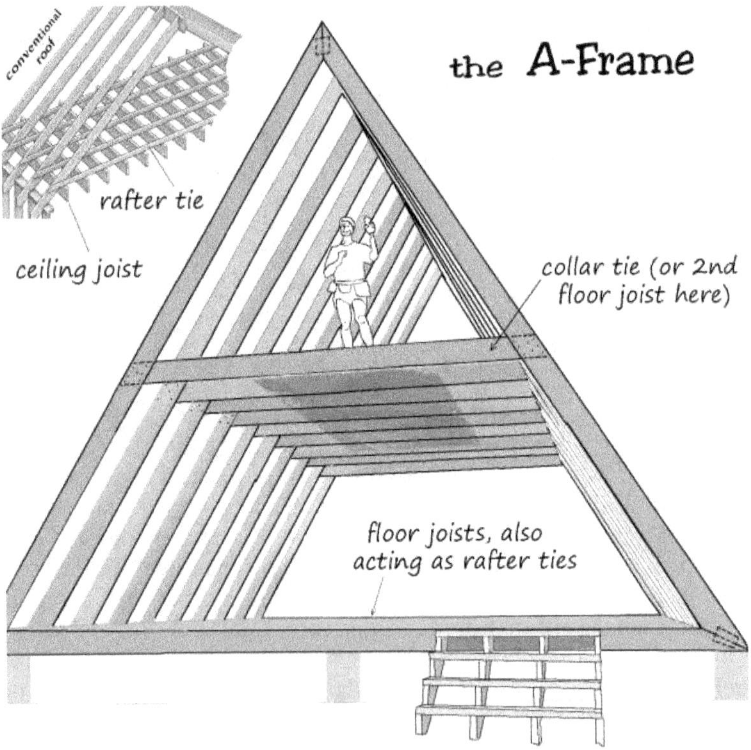

By convention, both rafter ties and collar ties are "tension ties" that are found in a typical (cut) wood framed roof. A **collar tie** is a structural reinforcement that prevents rafters from separating at the ridge. They belong in the upper third of a roof; might not be attached to every pair of rafters but are uniformly found on "gable" ends.

A **rafter tie** holds the bottoms of a pair of rafters together, so that the rafters do not spread walls under weight. The lowest chord of a simple triangular roof truss, preforms the same function as a rafter tie.

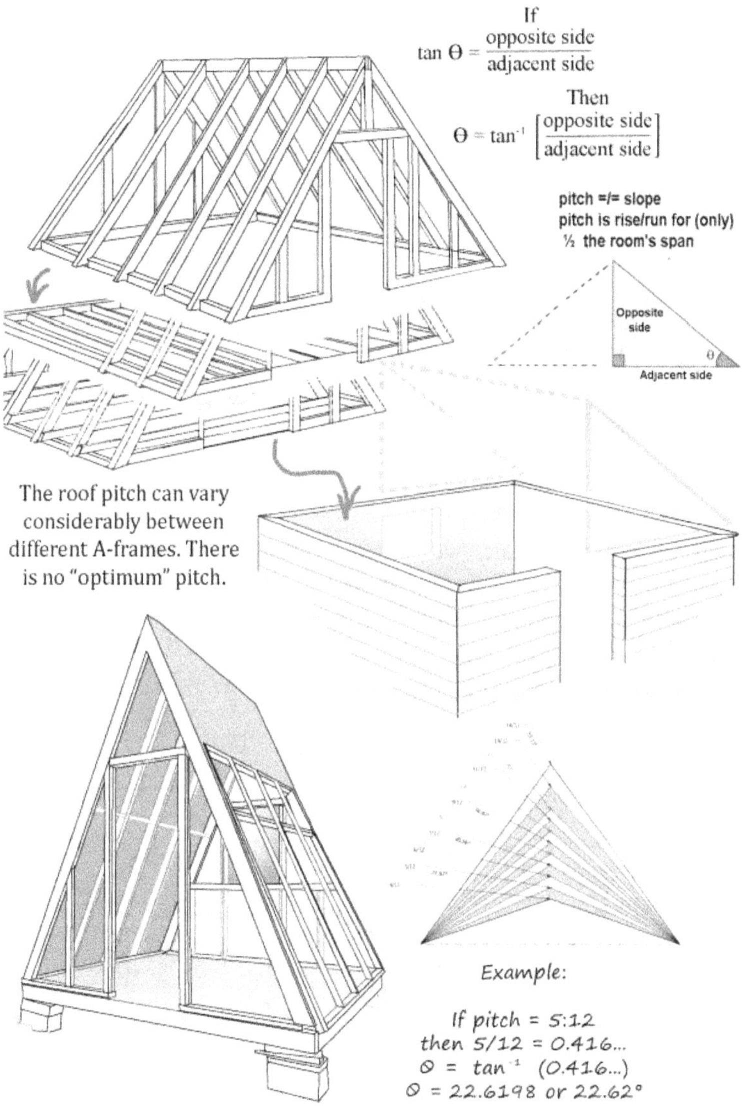

The roof pitch can vary considerably between different A-frames. There is no "optimum" pitch.

The image above shows a slight inconsistency. Pitch and slope are two differently calculated expressions of roof incline. **Pitch** is an older historic expression that describes the ratio of the vertical rise to the entire horizontal span; (not just half the span of a room). **Slope** on the other hand is the ratio of the rise to the (half span) run and is typically expressed in inches per foot. Pitch becomes a fraction, often reducible, whereas slope remains a ratio (inches of rise per / standard 12" run).

Some people now also describe roof incline in terms of degrees, using the formula $\Theta =$ archtan (rise / run). Others might attempt to express the steepness of a roof in terms of percentage; as in, Pitch Percentage = [(rise / run) x 100].

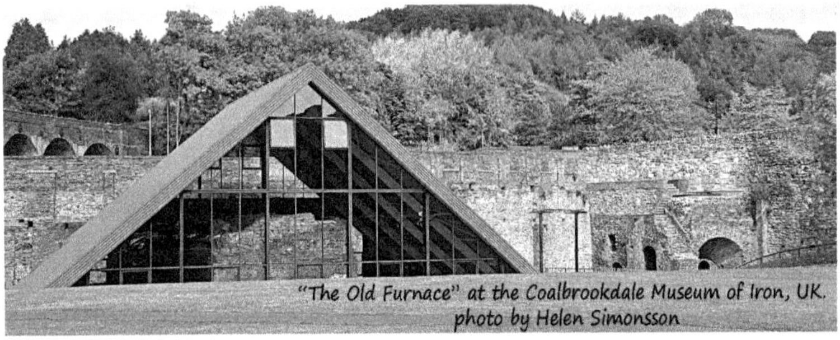
"The Old Furnace" at the Coalbrookdale Museum of Iron, UK.
photo by Helen Simonsson

Above, a low and wide metal A-frame shed protects the remnants of a very old blast furnace. Similar large structures commonly protect archaeological digs, which would otherwise re-exposed to the ravages of nature's elements. Below is a scene from inside the same shed. Notice the sway bracing or shear bracing on the left.

A 3.5 century old blast furnace. Famous for being the first to reduce the sulfur content in iron, by burining coke instead of coal. The Ironbridge Gorge Museum Trust

The "Darby Furnace" is where Abraham Darby (from 1705) revolutionized the iron industry by introducing **coal** as a substitute for **charcoal** in the smelting process. His approach involved **"coking" coal**, which effectively removed the sulfur from coal. Iron quality improved, which facilitated the manufacture of better machine tools and advanced machinery.[39] After studying how brass pots were cast in sand molds (in the Netherlands), Darby went home, experimented with and patented lidded **cast iron pots**, which he named "Dutch ovens".

A-frames look good in alpine settings and they blend in well with the tall coniferous trees in such domains. A main benefit from the shape is the increased snow load that such a roof incline provides. In mountainous alpine regions that routinely experience snow, a heavy wet snow can literally collapse a flatter roof. With an A-frame the snow generally slides off. The shape veritably eliminates most of the downward weight of a snow load.

A-frames are simple uncomplicated structures to build. Large, luxurious custom A-frames have been built, but smaller cabin like ones are much more numerous. These make good vacation or rental homes. The charming ergonomics might wear thin after a while though. Much space is wasted within an A-frame. Not only empty space in the lofted ceilings but on the floor too, when you bump your head while walking down a hallway or must get down on hands and knees to retrieve something down close to the juncture of floor and

39 https://www.ebsco.com/research-starters/history/invention-coke-smelting

wall.

Senior Homes

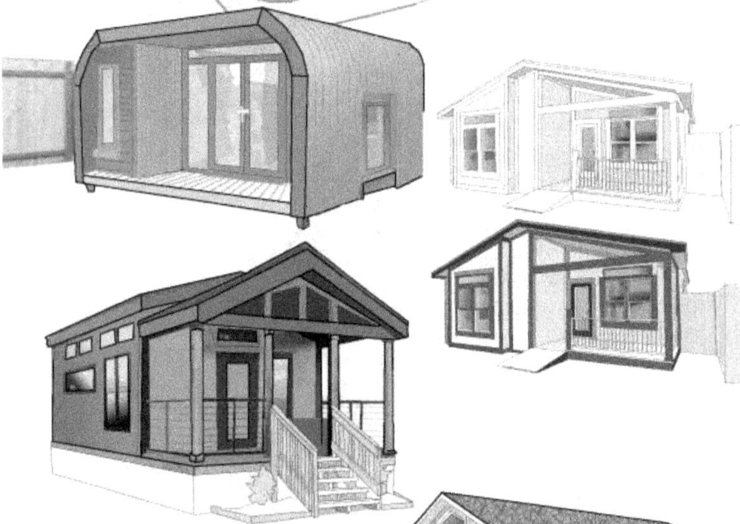

Little houses (smart und simple)

<u>The most essential room of any home is the bathroom</u>. Yes people can live in a cave or in a tent. They can dig a hole in the ground and put an outhouse over it. But a dwelling won't feel like a real home until it has a warm room with hot water to bathe or shower in. Some tiny custom homes for the elderly have been whittled down to the bare necessities, yet they are still comfortable. Bedroom, bathroom,

kitchen, perhaps a closet. Seniors that buy these homes, have disengaged themselves from collecting possessions. They need little else from a home, except perhaps a porch or shade canopy outside, from where they can wave to or gossip with neighbors. Usually these little homes are nestled together in private communities. Lawn or landscaping services are contracted to maintain the neighborhood's exterior decorum.

The point being made though is, that a tiny clean abode can be entirely adequate, in the right setting. Whether for a single individual, couple or small family – a tiny house will cost less, be easier to build and be easier to heat. The obligatory plumbing, sewage, and electrical should cost less also. But actual work, hobbies and play can be conducted outside, or beneath the shelter of other buildings.

Trailers

Creative Commons Attribution-NonCommercial-ShareAlike

In 2012, seven years after hurricane Katrina wrecked the Gulf Coast, a federal judge awarded a \$42.6 million class-action settlement to people who had already benefited from the free temporary housing provided by FEMA. Following Katrina (2005), FEMA was rushed into purchasing more than 12,000 travel trailers and 9,800 mobile homes, to house disaster victims along the Gulf Coast. Later a complaining lawsuit claimed that the materials used in the trailers, were substandard and contained unusually high levels of formaldehyde (which is probably untrue; the same wood was available in every lumber store across the country at the time). A federal judge and FEMA, managed to 'pass the buck' and the penalties onto the backs of (insured) companies that built the units.

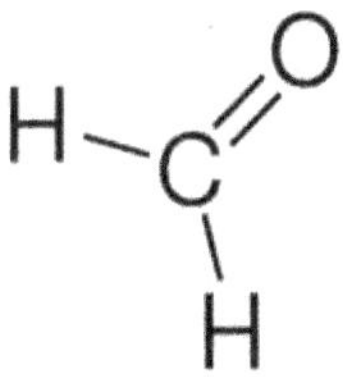

In 1984 the EPA classified **formaldehyde** as a probable human carcinogen, and established a permissible exposure limit 0.75 parts per million (ppm) averaged over an 8-hour workday. (Very high by today's standard). Comprehensive regulations were lacking until the Toxic Substances Control Act amendments in 2010 established stricter limits. Formaldehyde (CH_2O

but with structure $H_2C=O$) would seem to be a simple molecule, but it's actually a tricky compound that comes in several different forms. Traces of the organic (because it has carbon) molecule are found in tobacco smoke, automobile exhaust, forest fires and in outer space. The name was coined – or it became a generic trademark back in 1893. At the time it was used as an antiseptic and preservative for biological specimens. Counter to what many may have been assumed, formaldehyde was not used as an embalming fluid during the Civil War. (The main embalming fluid was a mixture of arsenic and soap).

- *Embalming was not common before that war, but soon enterprising individuals descended onto the battlefields like vultures, to collect and preserve the corpses (which soon began to smell badly otherwise). Families that may have made the trip to a battlefield to reclaim a body, often had to pay a $100 ransom to get a body back that was embalmed without permission. Anyway, embalming soon became a lucrative business.*

Formaldehyde can be made several ways. It's an intermediate in the oxidation (or combustion) of methane. Industrially it is made by using a catalyst to cause the oxidization of methanol. (Carbon monoxide and hydrogen will react over a catalyst to produce methanol). Sunlight and oxygen can react with methane and other hydrocarbons, to produce CH_2O in the atmosphere. But after being formed, formaldehyde does not accumulate or persist in the environment, because that same sunlight or bacteria in the soil or water, will break it back down. The human body converts it into formic acid.

*(**Formic acid** is smelly and is produced naturally by ants, stinging nettle, some fruits & some vegetables. Globally it has many uses: production of rubber, dyeing textiles, limescale remover, toilet bowl cleaner, leather tanning and as a preservative in livestock feed).*

Being a common precursor to more complex compounds, formaldehyde is indispensable in the formation of consumer goods (resins, plastics, textiles, hair relaxers, permanent adhesives, foam insulation, molded products, paints, explosives, antimicrobials in vaccines and animal feeds, disinfectants and embalming agents). Formaldehyde **adhesives** (for wood products like plywood and laminated beams at first) entered the market in the 1940's.

Labeled a carcinogen now, this chemical's utility is still too important to disregard. The EU sets a maximum limit of 0.2% concentration of formaldehyde in any finished product; and a label for anything containing more than 0.05%. As of 2010, Congress passed a bill that restricts allowable formaldehyde emissions from wood products to be less than 0.09 ppm. *<info on treated umber forthcoming>*

In 2023 wildfires destroyed the homes of many in Lahaina, a

Hawaiian town on the island of Maui. Instead of shipping trailers as temporary housing, FEMA contracted a company to build 109 experimental, more robust portable housing units, which had the potential to be turned into affordable long-term housing, following their temporary use. These homes (designed by Liv-connected) were filled with light, had large windows, wood-paneled walls and high ceilings[40]. They were fabricated quickly in Colorado and Idaho, from quality materials; then shipped 2,000 miles by barge. Importantly, the units were prefabricated but could be knocked-down into panels (except for the pre-fabed bathroom and kitchen) that then stack easily on a flatbed truck. A feature that can substantially reduce shipping cost.

Purchasable today, these houses are modular, within the limits of adding to the length or adding a second story. Plans range from a 480-square-foot one-bedroom unit to a 980-square-foot three-bedroom home. (2025 Price currently is high; about $300 / sq. ft.)

Many people (some retired) have chosen to live in recreational **motor homes** and **camp trailers**. These can be moved to a new location easily. If not parked on private land however, these are still obliged to pay rental fees for parking and utility hook-up.

Moving a **trailer home** on the other hand can amount to a significant expense. First, the 'tax man' has his hand out – selling a certificate from the county treasurer confirming that all property taxes on the trailer are paid. A 'Certificate of Title' will also be needed. (A trailer built before 1974 may not even be eligible for relocation due to federal regulations). Then, you cannot move the trailer house yourself. You must hire a licensed, bonded, and insured professional moving company. That can be costly. Prices will vary, but for distances of 50 miles or less - expect charges of $5,000 - $8,000 (for single-wide) to $10,000 - $14,000 (for a double-wide trailer house).

This chapter began with things Roman, its ambition is pointed towards building on solid foundations. One could build a robust shelter around a trailer home or camper core, but in the long run that decision would be neither practical or very economical. Most

40 https://www.fastcompany.com/91400481/mauis-new-temporary-homes-show-what-better-disaster-housing-can-look-like?utm_source=firefox-newtab-en-us

contractors won't try to remodel trailers because these are flimsy and are doomed to depreciate in value no mater what you do.

Modulars

Separate notions can be implied when interpreting the term "modular". Some might consider Quonset huts to be modular, because they can be expanded easily. Other people might call homes made from shipping containers modular, because the boxes can be stacked and arranged like LEGO toy bricks.

But the term "**modular house**" now connotes "*a prefabricated home built in sections in a factory, which are then transported to a site for assembly on a permanent foundation. Unlike mobile homes, modular homes are constructed to the same building codes as traditional homes and are not designed to be moved after installation*". This type of modular house is not planed for the ease of future modifications. Neither is a **modular look** - the same as a **modular design;** a principle that subdivides a system into smaller parts. Separate individual modules can be interchanged, moved around and still join or interface with each other. A building plan with provisions for future expansion in mind.

While many farmers and ranchers might use **railroad boxcars** as barns or storage sheds, other people have succeeded in turning boxcars into comfortable homes. But to do so is neither easy or cheap. Boxcar production is in decline, loosing favor to the newer **intermodal shipping container** concept (containers that can be used across different modes of transport – such as back and forth from ships to trains to trucks).

Boxcars must be loaded and unloaded from a "spur line" (side track). Shipping containers however, can be delivered to almost any location and parked, where the contents can be processed more leisurely. Boxcars are stronger and more robust than **intermodal shipping containers**. Another difference is that the doors on a boxcar are located mid-span of the unit and both sides have slide open doors. Shipping containers generally have swing open doors, on only one end.

The first intermodal shipping containers were invented by the U.S. military and used during the Korean War. By 1967 and the Vietnam War, over 200,000 of these shipping containers had been procured by the military. It needed tough steel boxes to protect supplies. Not tied to rails, not needing to be packed and repacked for an ocean voyage, these containers meant fast delivery. Their name "**Container Express**," was eventually shortened to "**Conex**". About 75% of those containers were shipped only once. They made useful storage supply units and shelters once set down, and so they generally stayed put. Many steel shipping containers today get transported only once. The cost of manufacture might the same or less than the cost and trouble of shipping an empty container back. And they are not well made. The roof's steel is so thin that a person can leave crinkles or dents, just by walking over it.

The four corners are strong and that is where the weight is transferred to whenever the containers might be stacked. Otherwise these boxes are as soft and crush-able as a huge soda can would be. Once a door or window opening is cut out of a corrugated side panel, that entire section has lost load carrying strength – and therefore must be reinforced.

The floor joist are "C" shaped metal 2"x 6" size, and the floor cover is usually 1.125" marine plywood. Cantilevering these boxes (they usually come built in 10', 20' or 40' lengths today) is impractical without the help of expensive beams, or both post and beams from underneath.

Ninety five percent of the intermodal shipping containers today are made in China, by three companies. The containers all have 8 ft. widths (sized to fit railroad flatcars). Standard height is 8 ½ feet but seen more frequently now are versions that are precisely 1 foot taller.

The maximum expected service life of a shipping container is somewhere between 18 and 22 years. There is high turn over rate however. Because many become "single use containers", when a deposit has been payed and a customer chooses not to return them. From a shipping company's viewpoint, the transportation expense of returning one might be so high that it's easier to sell just sell it at a loss and buy a new one.

Containers returned are inspected - before further shipping use or potential sell off to the public. Inspected containers come in three ratings: CW (**Cargo Worthy**), WWT (**Wind and Watertight** / but no longer certifiable for sea travel) an "**As Is**" which implies poor condition. Some containers may have been contaminated by transporting fumigants, pesticides or other toxic chemicals.

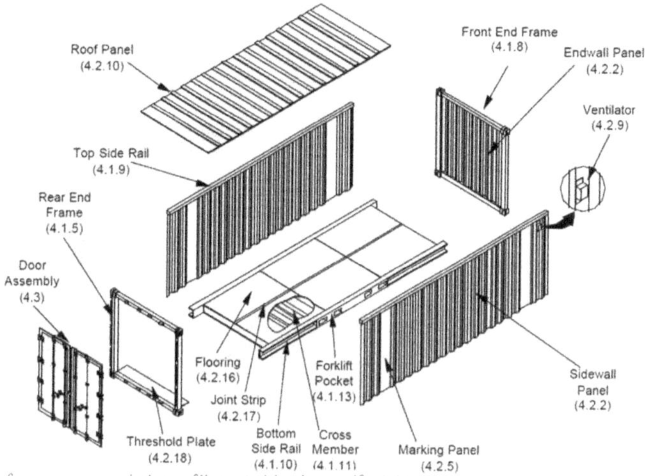

free to use and share filter / shippingandfreightresource.com

If **converting one to a living space**, it is a fallacy to expect that one of these containers can be stuck out in the wind and cantilevered without concern. They are not that strong to begin with and require support to be safe. Again, cutting openings for windows or doors significantly compromises their strength.

Insulation is an issue, and here again a "fur-down" with 2 x 4s stuffed with batts of fiberglass wool, seems to be the best solution. Insulating the floor from underneath will be no problem. Inside the ceiling will be dropped a few inches. But these already narrow (@ 8' wide) boxes will loose another foot of internal width once left and right fur-down walls with sheet-rock or paneling are installed.

Condensation can occur in any home if ventilation is inadequate. But metal roofing and metal clad walls can amplify condensation, where beads of sweat accumulate and run down walls or soak into insulation. This can lead to metal corrosion or create mold and mildew, which might foul indoor air quality.

> *"Condensation under a metal roof occurs when warm, humid air inside a building comes into contact with the cooler surface of the metal, causing moisture to form".*

During cold weather, while the interior of the building is heated, a metal roof or metal wall will remain cold. This temperature contrast promotes condensation. Regions with high humidity are more prone to condensation issues, but no mater where the area, the atmosphere retains some water vapor at all times. Good ventilation is the best remedy for this issue. **Ventilation** is critical for the crawl spaces and attics in wood framed homes as well.

Kit homes

This add was published in Popular Mechanics magazine, in 1908. The advertisement stipulates that all the boards are milled and cut to length.

Being "knocked down" implies that all the boards in a wall for example were cut to fit and then tacked (nailed lightly) in place as it was being assembled. Once complete each wall was dissembled; the boards (perhaps numbered) were gently knocked apart and nails pulled, before the lumber of that section was bundled up for shipment.

The add goes on to claim that this measure saves 30% of the labor cost – should a contractor perform that work on the job site. Remember, the electric powered "skill saw" was yet to be invented. It takes authentic "elbow grease" to cut a board with a handsaw, and some boards required multiple cuts.

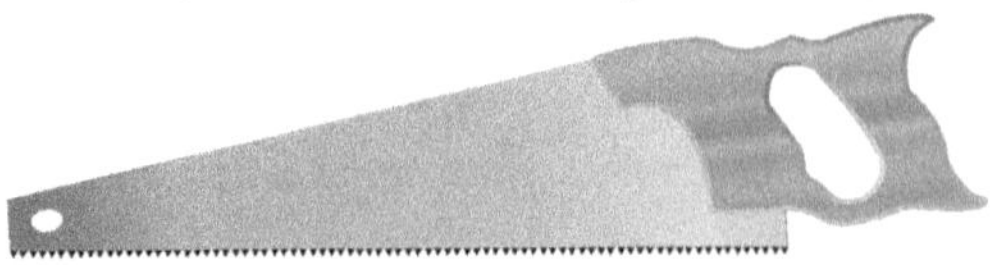

The Aladdin Company was a pioneer in the pre-cut, mail order (kit) home industry. Where all the doors, windows, flooring, lath & plaster, tar paper, paint and hardware were included in the kit, for one price. Aladdin had its own catalog but it also subcontracted thousands of house kits through mail order companies like Montgomery Ward and Sears, Roebuck and Co.

Some industries built entire **company towns** around new factories, mines and mills; where kit homes provided the employee housing.

FOUR ROOMS AND BATHROOM

Honor Bilt

The Fairy

No. 3217 "Already Cut" and Fitted.

$965.00

See Description of "Honor Bilt" Houses on page 9.

THE FAIRY is a comfortable bungalow home, with shingle siding. It has many features that will appeal to the housewife. There is quality in every foot of material. Yet, we will furnish all the material to build this bungalow at the very low price quoted. The splendid value is made possible by our modern and successful system of "Honor Bilt" homes.

The Living and Dining Room. The porch affords a pleasant place for warm summer evenings. The half glazed front door enables the housewife to see the caller before opening the door.

Having entered you find the living room combines with dining room. Hence the most economical use of floor space, without sacrifice of appearance. This combination living and dining room is about 12 feet wide by 15½ feet long. There is room for a piano between two windows on the outside wall. Your dining table and chairs may be set at the end of room nearest the kitchen and hidden by a screen if so desired. There is ample space for other furniture as shown on plan.

The Kitchen. A swinging door leads from the dining and living room into the kitchen. Here is a well planned kitchen, room to place cabinet, table, sink and range to make work easy which reduces the housewife's work and saves many steps each day. Two windows give plenty of light and air.

A door connects the kitchen with the rear entry, which has space for ice box, that can be iced without tracking dirt into the kitchen. From the rear entry, stairs lead to the outside and down to basement.

The Bedrooms. The front bedroom opens from the living room. It has a fair sized clothes closet with shelf.

A small hall is open from the living and dining room and gives privacy to rear bedroom and bath. Bathroom is just the right size and is conveniently located.

Basement. Excavated under entire house. Cement floor.

Height of Ceilings. Basement is 7 feet from floor to ceiling. Main floor rooms are 8 feet 6 inches from floor to ceiling.

What Our Price Includes.

At the price quoted, we will furnish all the material to build this four-room house, consisting of:

Lumber, Red Cedar Shingles for Siding; **Lath;**
Fire-Chief Shingle Roll Roofing, guaranteed for 17 years;
Framing Lumber, No. 1 Yellow Pine;
Flooring Interior Floors, Clear Yellow Pine; Porch Floor, Clear Edge Grain Fir;
Porch Pergola;
Finishing Lumber;
High Grade Mill Work (see pages 86 and 87);
Interior Doors of White Pine with Two Cross Panels of Fir;
Trim, Clear Yellow Pine, choice grain and color;
Windows of California Clear White Pine;
Medicine Cabinet; Building Paper; Sash Weights;
Eaves Trough and Down Spout;
Stratford Design Hardware (see page 118);
Paint for Three Coats for Outside Trim;
Stain for Two Coats, for Shingles on Walls;
Varnish and Filler, for Inside Trim and Doors;
Complete Plans and Specifications.

Built on concrete foundation, cement blocks above grade.

We guarantee enough material to build this house. Price does not include cement, brick, concrete blocks or plaster.

Can Be Built on a Lot 28 Feet Wide.

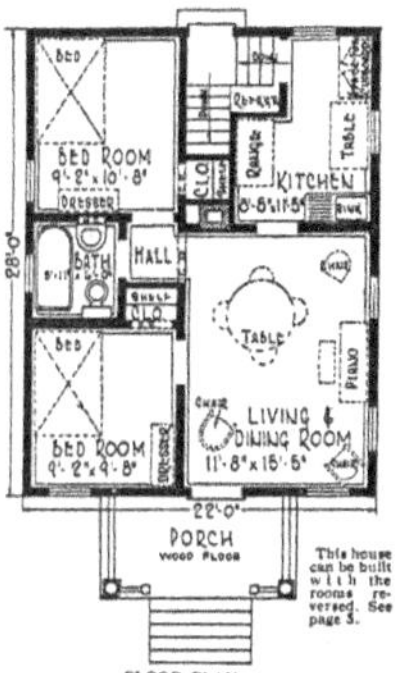

This house can be built with the rooms reversed. See page 5.

FLOOR PLAN.

OPTIONS.

Sheet Plaster and Plaster Finish to take the place of wood lath and plaster, $90.00 extra. See page 88.

Storm Doors and Windows, $48.00 extra.

Screen Doors and Windows, galvanized wire, $29.00 extra.

Oriental Slate Surfaced Shingles instead of Fire-Chief Roofing, $30.00 extra.

For prices of Plumbing, Heating, Wiring, Electric Fixtures and Shades see page 95.

Page 24 *SEARS, ROEBUCK AND CO.*

This page is taken from a **1926 Sears, Roebuck and Co**. catalog. This house was intended to be built over a concrete floored, 7' high, cement block walled basement. Dimensions are 22' x 28' on the outside., 11'8" x 15'5" for the Living & Dining room combo. The framing wood, trim and cabinetry provided were clear #1 Yellow Pine. Rain gutters and downspouts were included. The kit came with Red Cedar shingles for siding and "Fire Chief - Shingle Roll Roofing" (which had a 17 year guarantee). Slate shingles were optional for $30.00 extra. Concrete blocks, cement and plaster were not included.

Adjusted for inflation, that $965.00 (in 1926) price tag becomes $17,967.63 (in 2025). Not counting the basement, the floor space of this house (22' x 28') is 616 square feet. In 1926 the materials cost to buy this shell of a house was $1.56 per sq'. The cost to buy this kit in 2025 after inflation, **should be only** $29.16 per sq '.

Sears started selling kit homes in 1908. Montgomery Ward began selling books filled with house plans in 1909, but not "ready-cut" kits (branded as "Wardway Homes") until about 1918. Sears began offering financing (mortgages) on their kit homes in 1917. That was a big step that Montgomery Ward would not take until 1925. The Great Depression hit in 1929 and Montgomery Ward would pull out of the kit home business in altogether 1931. Sears would last it out until 1987. About 30,000 'Wardway Homes' were sold, while Sears sold about twice that many kit homes.

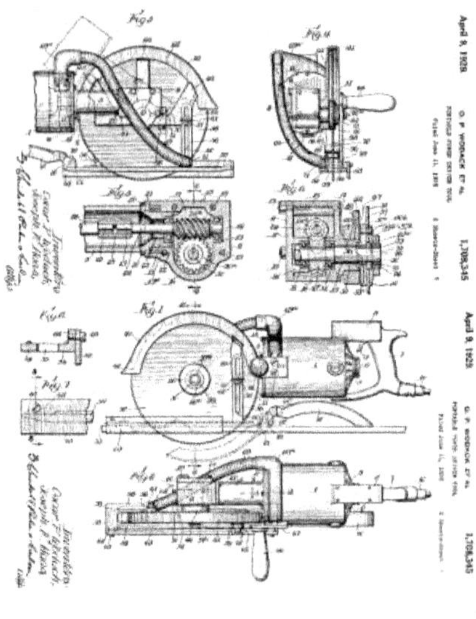

The first proper **electrical circular saw** was produced by the Electric Hand Saw Co.- which later became **Skilsaw** Inc. It was a "worm drive" because of its worm gearing – shown in this 1929 patent application. About a year later the Porter Cable company created the direct drive circular saw (nicknamed a "sidewinder") to circumvent Skilsaw's patent. Hardly anyone had either one of these new tools because they were too expensive; invented during the **Depression** (when workers of every trade and every level of education, were impoverished and standing in **bread lines**; waiting for the sustenance necessary to survive). It wasn't until well after WWII, that the typical contractor could afford the luxury of using an electrically powered hand saw.

Domes – a quick crash course

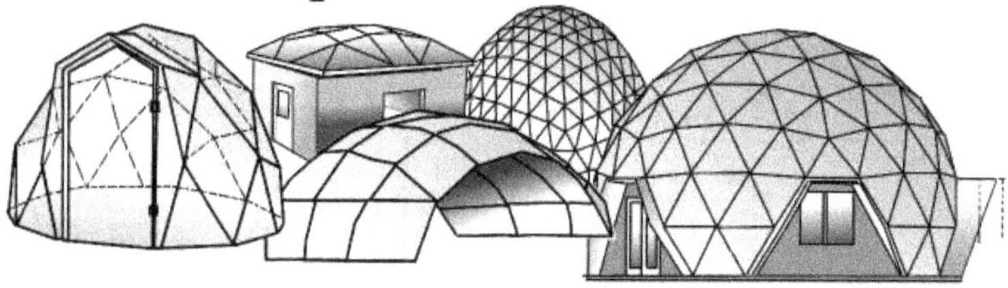

"A **geodesic** is the shortest line segment between two points on a **sphere** or other **curved** surface". **Polygons** are 2-D (two dimensional) shapes. **Polyhedrons** are 3-D solid shapes or representations thereof.

A **polytope** is just some 3-D geometric model with flat sides, used whenever modifying the dimensions of a polyhedron. **Polyhedra** with regular polygonal faces turn out to be especially important. That's where the influential Platonic Solids come from. The process of tessellating one of these choice polyhedra and then expanding its components onto the face of a sphere, is how differing geodesic domes are calculated. Imagine forcing air into a round balloon, which is itself placed inside an imaginary straight edged, but springy and flexible Platonic Solid. Imagine the long edges bending and increasing in number as the shape itself swells and becomes a more uniform sphere.

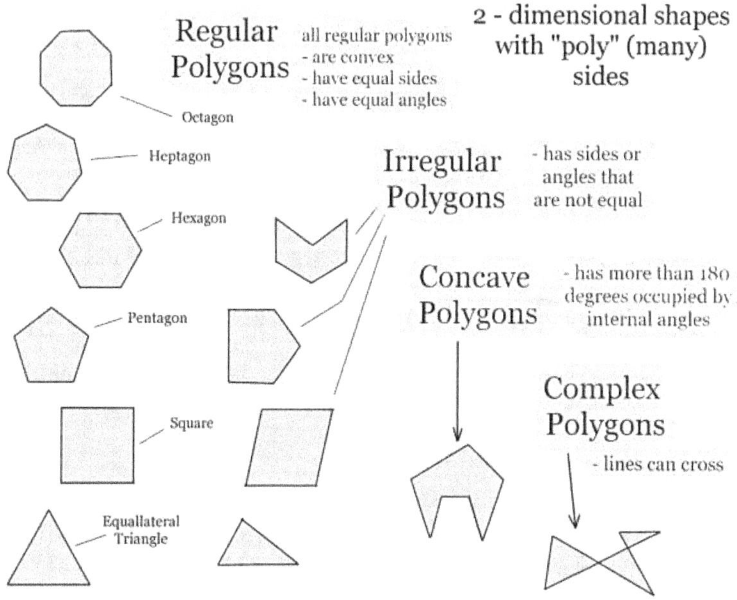

*"**Platonic solids** are three-dimensional shapes where each face is the same regular polygon, and the same number of faces meet at each vertex. There are **exactly five types**: tetrahedron, cube, octahedron, dodecahedron, and icosahedron".*

Tetrahedron	Hexahedron	Octahedron	Dodecahedron	Icosahedron
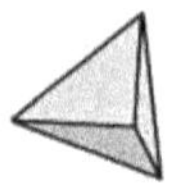	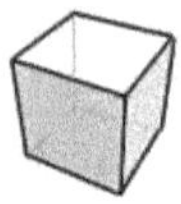	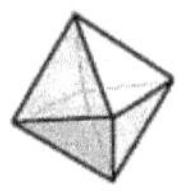	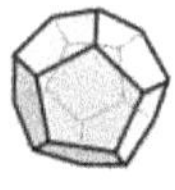	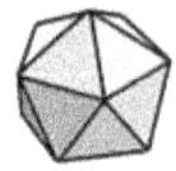
4-faced	6-faced	8-faced	12-faced	20-faced

> *"In geometry, every polyhedron is associated with a **second dual structure**, wherein the vertices of one correspond to the faces of the other and the edges between pairs of vertices of one correspond to the edges between pairs of faces of the other".*

If there are only 5 Platonic Solids, then there should be only 5 duals. But the dual of a tetrahedron is another tetrahedron. The dual of the hexahedron is the octahedron and vice versa; the dual of a dodecahedron is the icosahedron and vice versa. You are still left with only 5 unique convex regular polyhedra. There exist uncountable irregular polyhedra, convex polyhedra (both regular and irregular), concave polyhedra and stellated polyhedra. Each with a dual.

> * **Duals** are created by truncation. "*In geometry, a truncation is an operation in any dimension that cuts polytope vertices, creating a new facet in place of each vertex*". For example: the cube (a hexahedron) has 6 faces, 12 edges and 8 vertices. Truncating or lobbing off the tips (vertices) about 1/3rd down each edge, creates new plane surfaces that now look like equilateral triangles. An octahedron is created; with the same number of edges, but now with 8 faces instead of 6 and 6 vertices instead of 8). Icosahedrons are the same way: truncation causes a morph into a dodecahedron, with all new pentagonal (five cornered) faces. To plan a dome there is no reason that one cannot begin with a dodecahedron, subdivide the 12 pentagon faces into triangles and push the vertices out to the inscribing sphere.

Geodesic polyhedra

Domes are essentially just geodesic spheres cut in half. Most domes are configured from one of the platonic solids. Sometimes the tetrahedron or octahedron, but usually the icosahedron is chosen as the starting point. Spheres hold a lot of volume for their amount of surface area. Triangles make very sturdy structures. Geodesics mix both.

There are a great number of mathematical ways to for the architect or engineer to vary the arrangement of cross-members in the structure of a geodesic dome. But the larger domes get complicated fast.

Octagons Don't Tessellate

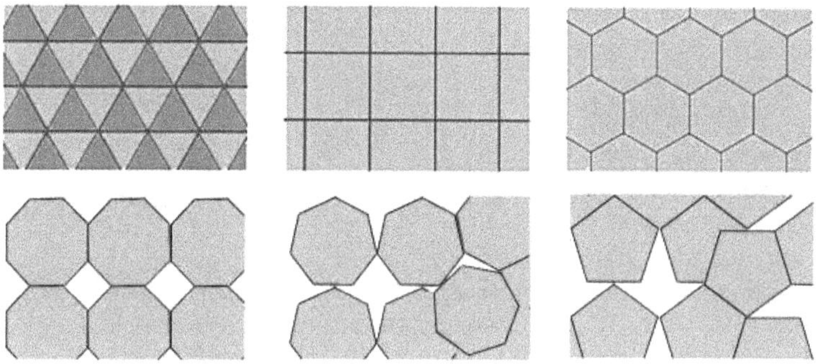

A "**regular tessellation**" is a pattern made by repeating a "regular" polygon. (Rectangles aren't <u>regular</u> polygons when their sides differ in length. Squares are regular polygons).

> *"Tessellation is a pattern formed by covering a surface with one or more geometric shapes, called tiles,* ***without any gaps or overlaps****. This concept can be seen in various forms, such as floor tiles, mosaics, and even in nature, like honeycombs".*

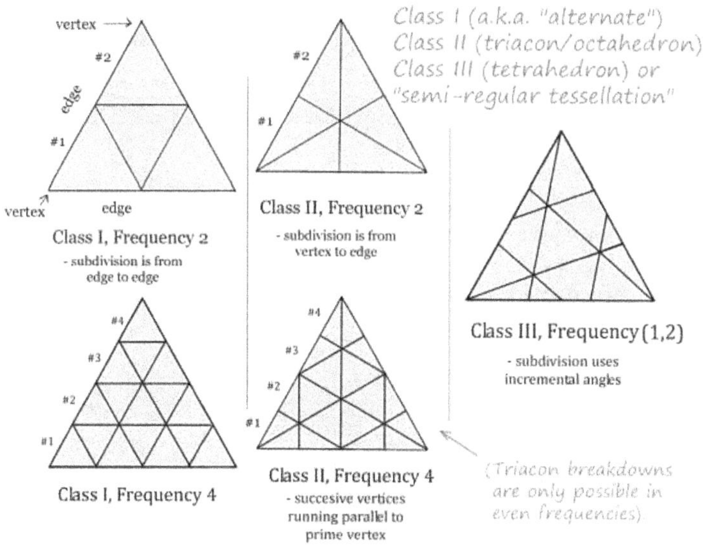

"Geodesic polyhedra are defined by the equilateral triangles of the primary face (i.e., the face of the underlying tetrahedron, octahedron, or icosahedron) laid out on a ***60° grid*** *so that their vertices always align with grid crossings. This produces three classes of tiling or tessellation. The edges of the primary triangular face in Class 1 are parallel to the grid lines. The edges in Class 2 are perpendicular to the grid lines. Those in Class 3 are neither parallel nor perpendicular to the grid lines".*[41]

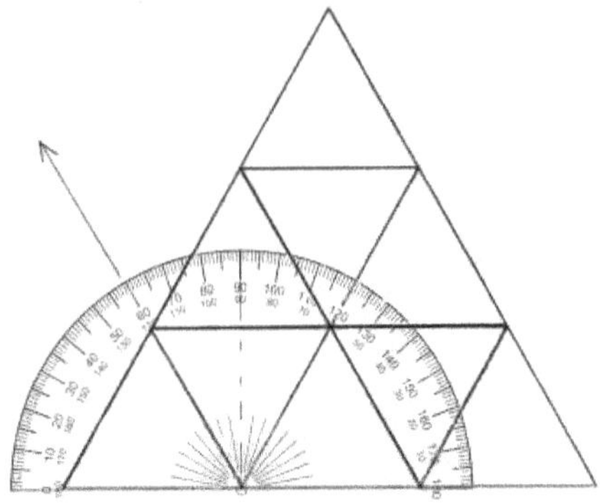

It becomes evident that most shapes won't properly tessellate. In fact only triangles, squares and hexagons will. Triangles become the most popular polygons to tessellate domes with. And the icosahedron is the most nearly spherical Platonic shape to begin with.

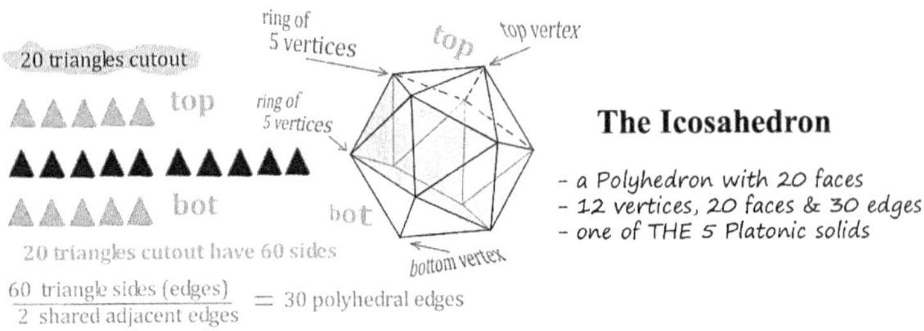

(If you remove the bottom 5 cords or edges from an icosahedron, your left with a shape of vectors that when leveled, could create the skeleton of a dome. A Class I,1v geodesic dome in fact).

41 *https://geometryofthinking.com/2024/02/01/geodesics/*

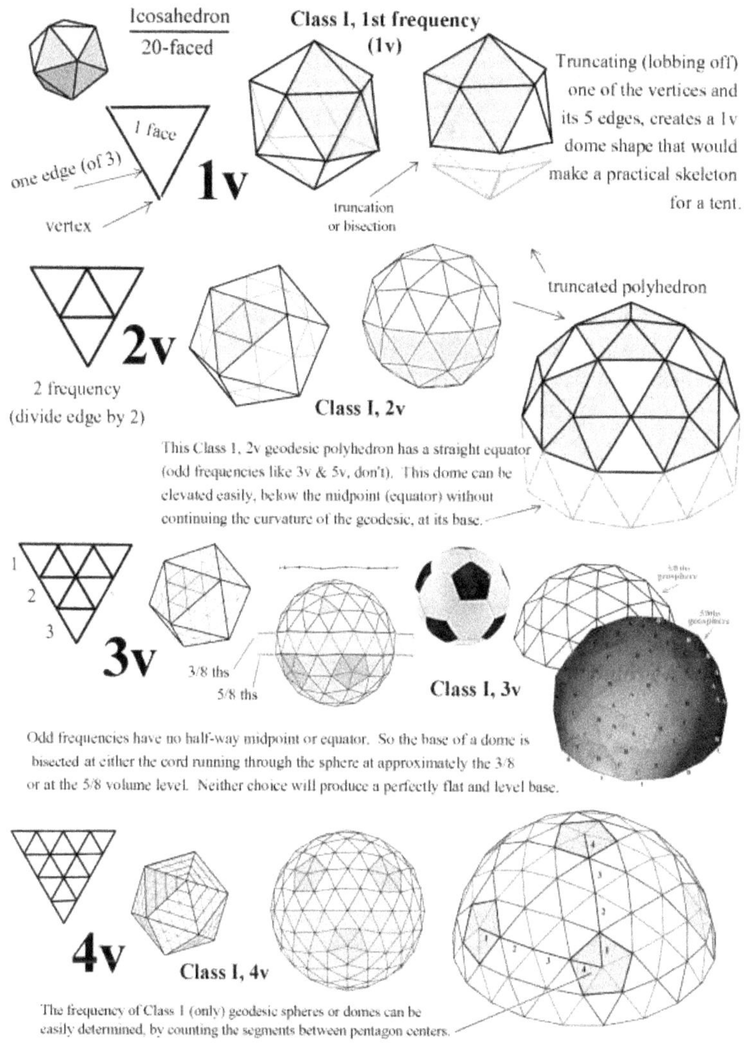

Increasing the frequency of a dome or sphere – increases the size, because more cords or struts of the same general size are being used. With the bottom removed from the icosahedron - the 1v dome has 25 cords (struts or edges) and 11 vertices. A 2v dome would have 65 cords and 26 vertices. The odd 3v dome uses 120 cords (for 3/8ths version) or 165 cords (for 5/8ths version). A 4v dome requires 250 cords. The (3/8ths) 5v dome uses 350 struts and the (5/8ths) uses 425.

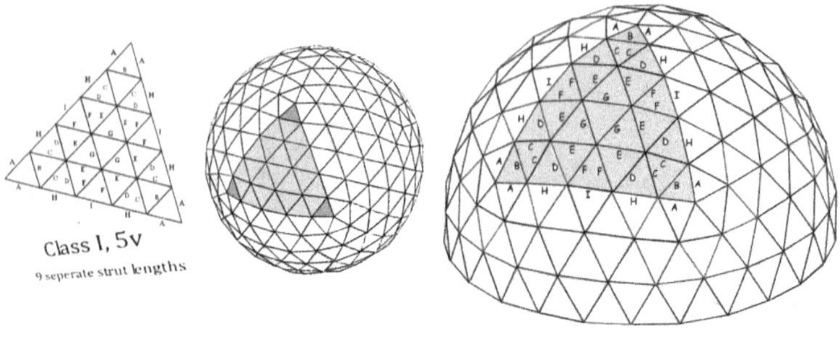

The unfortunate part is that when the frequency of the “alternate” dome (Class 1 geodesic) reaches 4v or goes beyond, the **complexity** becomes a construction concern. The 5v geodesic (above left) has an absurd 9 different strut lengths. Keeping the puzzle straight is a definite chore, especially when shuffling tools and materials while standing atop a rickety ladder or scaffold. The chopped up, old graphic below was intended to compare relative sizes when the longest cord was 5' and the frequency advances. Radius also progresses from 4'9” to 23'.

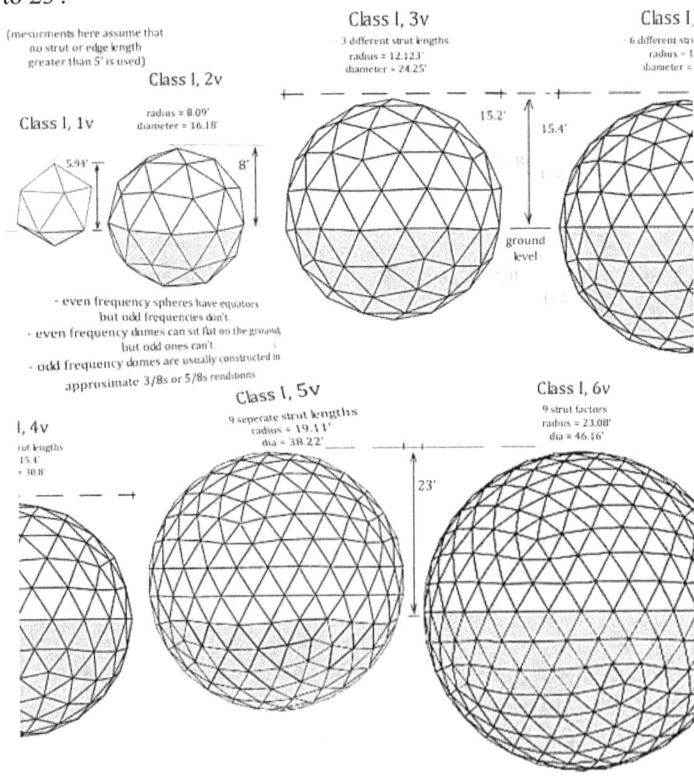

Five feet is a useful reference because it is ½ the standard (10' x ¾") length of **typical steel EMT tubing** (in the US). When struts of this length or less are used, a cautious light weight person can climb around on the structure. Longer lengths would probably bend. Materials like PVC pipe, wood, rebar (reinforcement bar) or even bamboo can be made to work adequately too. To be practical though, for whatever material used, the cords should have adequate flexural strength to support a human's weight.

Domes will need connectors to join the struts together. These will be either 4-way, 5-way or 6-way connectors. The 4-way ones will be located at the base on the ground, while the 5-ways will be at the center of the pentagons (5 joined faces looking pentagonal from a distance). Icosahedral based domes will always have (6) 5-way connectors (at the pentagons) or (12) for a sphere. That's just the nature of the namesake polyhedron, or what is left after being truncated into a dome.

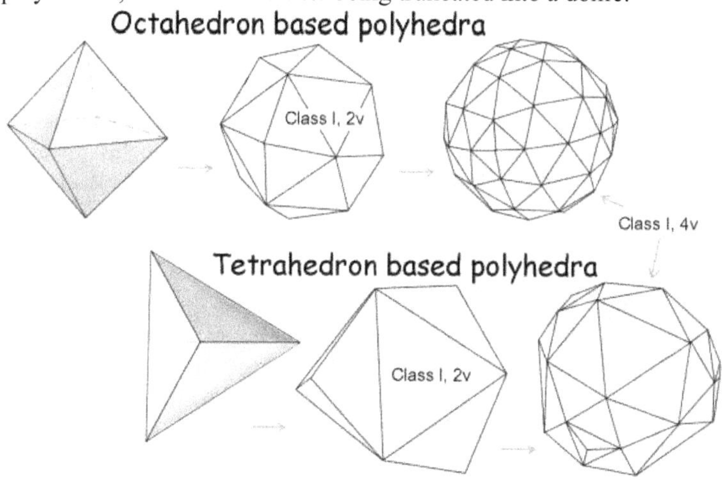

Naturally a dome can be based on another regular polyhedra, and then tessellated in many ways also. These tend to be less uniform and more complicated than the icosahedral ones. Still architects build them occasionally. But this text will focus on just a few of the simpler, smaller icosahedral domes.

It was mentioned before that a 5/8ths version of a 5v alternate dome had 425 separate struts. The even frequency 6v, requires 555 separate struts. Both 5v and 6v versions divide their strut lengths up into 9 different lengths; sometimes the difference in those lengths is very small (as in 1/8th of an inch).

Number of cord lengths as Class I frequency changes:

1v - 1	6v - 9	11v - 36	16v - 72
2v - 2	7v - 16	12v - 30	17v - 81
3v - 3	8v - 20	13v - 49	18v - 63
4v - 6	9v - 18	14v - 56	19v - 100
5v - 9	10v - 30	15v - 45	20v - 110

One can discern that the number of differing cord or strut lengths grows alarmingly, increasing the complexity of construction as the subdivision of the alternate (Class I) progresses. To combat this tendency, architects like Buckminster Fuller turned to the Class II tessellations. The triacon breakdown was the original type of subdivision in the early days of geodesics. For the larger domes today, the triacon still produces high symmetry but requires **fewer different lengths**.

U.S. Pavilion 1967 World's Fair / free to use and share filter: Flickr

The huge dome built as the U.S. pavilion for the 1967 World's Expo (in Montreal, Quebec, Canada) was a triacon, of sorts. The dome's geometry has been described as being a combination of a 32-frequency regular triacon (Class II, method 3) and a 16-frequency truncatable alternate (Class I, method 3). The dome is essentially two domes, one within the other and cross-linked for strength. During construction the builders tired of using expensive nuts and bolts, and so resorted to welding the struts together.

The project was commissioned to (Buckminster) Fuller & Sadao Inc., by a group of architects known as the Cambridge Seven. During the World's Fair an elevated railway conducted spectators through the 20 story dome. Originally the dome was covered and enclosed with transparent acrylic panels, but after 9 years these caught fire and melted away in 1976. The dome sat as a derelict until 1990, when the robust structure was turned into a museum, now called the Montreal Biosphere.

Dome calculation

Presently dome calculators abound on the Internet. Predominately these calculate dimensions for only icosahedral polyhedra with Class I tessellation. It is the simplest route to take. Most online calculators offer a way to calculate dimensions given a hypothetical dome radius, or alternatively a hypothetical strut length.

*(One of the oldest and best sites is **desertdomes.com**; which has been up and running since 2002).*

Such calculators are very useful to novices to get a feel for how things will work, or what can or cannot be achieved with differing lengths of material. However with a few formulas, the builder can construct a dome without the Internet or a calculator.

Behind the scene, years ago the distances between the vertices of various geodesic spheres were calculated using spherical geometry. (There are alternative methods to use)[42]. Those distances were converted to ratios that are unique for each geodesic. An online dome calculator just preforms simple arithmetic after a conjectural length is offered. Hiding the background geometry, the simplified Class I dome formula becomes **(radius) = (strut length) / (strut factor)**; or **(strut length) = (radius) x (strut factor)**.

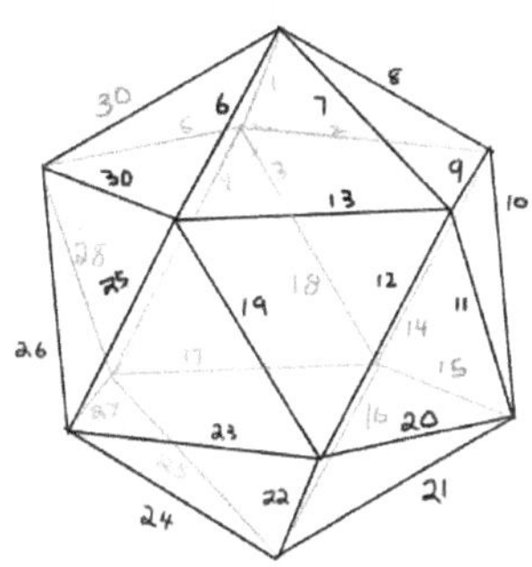

The **Class I, 1v** dome is straight forward; all the edges are equal. No calculations are needed. Such a dome would need 25 identical struts (or 30 for a sphere). The dome would require (6) 5-way connectors and (5) 4-way connectors. The dome would have 15 facets (faces); 5 less than the sphere (original polyhedra). Progressive frequencies simply divide each face of this polyhedra (the Platonic solid named the icosahedron) into smaller and smaller parts.

42 Kitrick, Christopher."A Unified Approach to Class I, II & III Geodesic Domes".9/1/1990.International Journal of Space Structures.10/10/2025

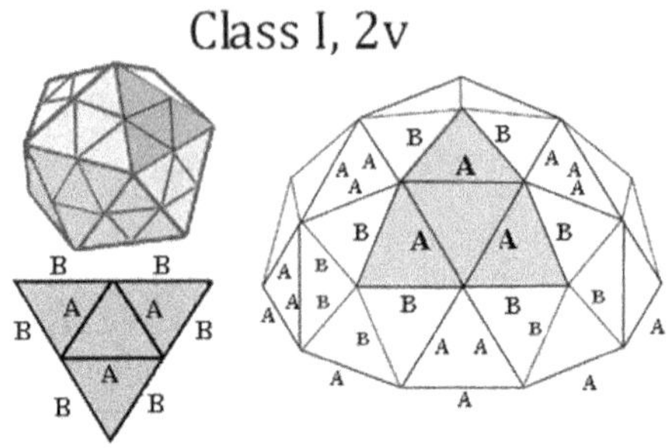

The **Class I, 2v** dome is composed of (2) struts of different lengths. A 2v sphere would need 120 struts, (half "A"s and half "B" struts). If "A" was the longer of the two, then a dome would need 35 "A" struts and 30 "B" struts. This dome would require (10) 6-way connectors, (6) 5-way connectors and (10) 4-way connectors at the base. Now the formula and strut factors come into play.

> *(Whether EMT tubing, PVC pipe or 2x4" wooden board are chosen for building material, these are all available in common 10 ft. lengths. But if the plan is to use 5ft. cords or less, then all the struts can be cut from standard 10 ft. stock, without waste. So using 5' as an example of the longest strut length possible: we plug that into an equation).*

*Assuming the A is the longer length in this 2v instance, its **strut factor** will be **0.61803**. The strut factor for B will be **0.54653**. Knowing the length of A (choosing 5' as an example) will let us determine the length of B; and then the dome's radius and therefore it's height and volume.*

*Therefore: if (radius) = (strut length) / (strut factor) then (r) = (5') / (0.61803) or **r = 8.09022'**. Now that we have a radius we can solve for the unknown length of B. If (strut length) = (radius) x (strut factor) then: (B) = (8.09022) x (0.54653) or **B = 4.4215'**.*

Since decimalized feet don't translate to a tape measure, the decimal portion needs to be converted into inches. (That's done by multiplying by 12). So 0.4215' becomes 5.058" (inches). That remaining decimal then needs to be multiplied by 16. In this case (0.058) x (16) = 0.928, which is almost 1/16th of an inch. Such a small length is not worth fussing about. Larger 1/8th" increments are.

So we have determined that when given a length of 5' for strut A, that B strut will be 4'5" and the interior height (radius) 8'. That means a diameter of 16' and a square footage of 201 sq ft (*since* $A = \Pi r^2$).

Since the Class I, 2v dome has an even frequency with an equator, it can sit flat on level ground. But it is a little cramped (assuming the same 5' maximum strut length, example). That 8' height at the center starts to loose headroom quickly as one approaches the perimeter. A builder might attempt to raise the dome's height by adding an additional ring of struts – while maintaining the 2v's spherical plan. But this would result in inward curvature at the base with loss of floor space and result with base that is no longer flat. However this dome can be raised by duplicating and adding a bottom ring using all "A"struts. That or by setting the dome upon a concrete or wooden 'pony wall' (short wall).

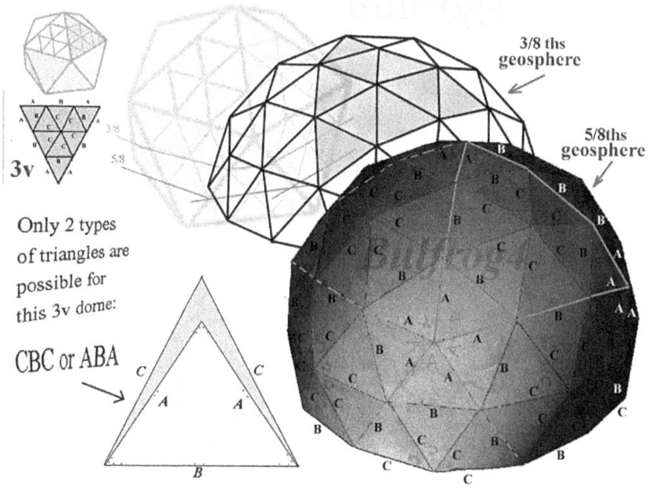

The **Class I, 3v** dome is larger than the 2v (when using similar material lengths) because more struts are being used. The shape looks a bit more spherical too. Here the builder is obliged to choose between a 3/8^{ths} or 5/8^{ths} version because there is no equator at the ½ way point. The vertexes that touch the ground will vary in levelness, a few inches different from one-another. As with the 2v, the 3v has only two types of triangles.

The 3v has 3 different strut lengths. In this example, C will be the longest, and A the shortest strut. The strut factor for **C is 0.41241**, for **B is 0.40355** and for **A is .34862**.

A complete Class I,3v sphere would have 120 C struts, 90 B struts and 60 a struts. A 5/8^{ths} dome would need 80 C struts, 55 B struts and 30 a struts. The 3/8^{ths} dome requires 50 C struts, 40 B struts and 30 A struts. The two domes require the same number of 4-way connectors (15) and 5-way connectors (6). The larger version however would require (40) 6-way connectors, while the smaller version needs (25).

Calculating the strut lengths if given 5' as an example of the longest strut, then C would = 5', B = 4' 10$^{11/16}$" and A would be 4' 2$^{11/16}$ ". The radius should be 12' 1$^{3/4}$".

To begin assembling this dome (if made from EMT galvanized pipe) one would lay 15 struts out in a big circle. Using (10)C and (5)B struts – in the pattern CCB,CCB,CCB......

** Notes on 3/4" EMT (Electrical Metal Tubing): ends can be flattened by hammer, vice or arbor press, holes are drilled ¾" in with a 3/8^{ths} bit.*

Bending both ends in about 10° helps. Watch for splitting at the welded seams in pipe, turn seams away as needed. Place any split ends towards the inside. Build "jigs" to assist speed and accuracy.

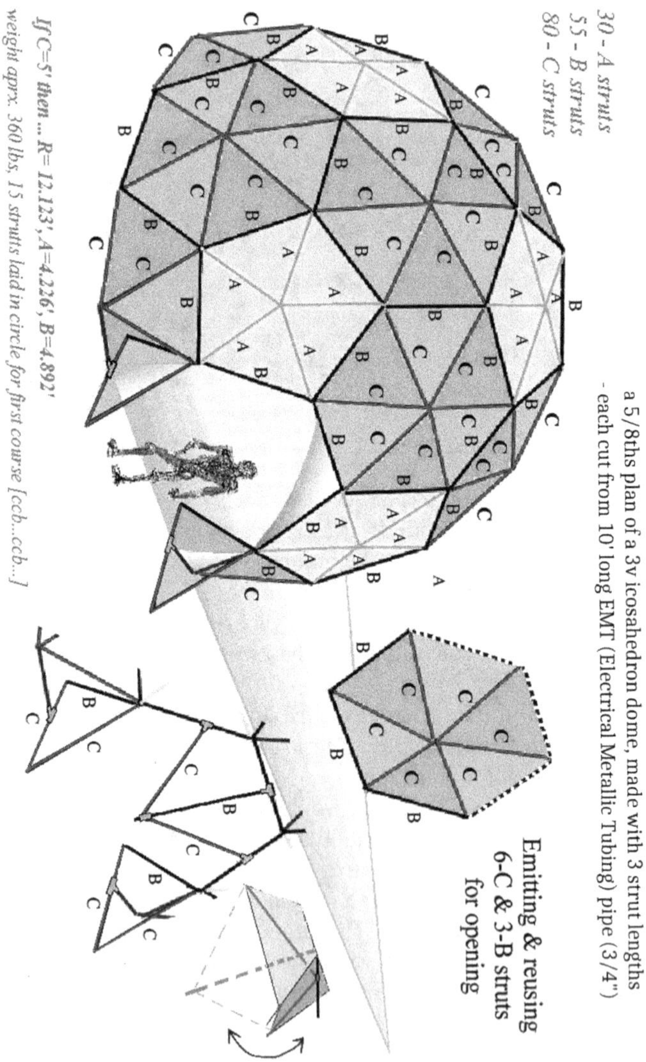

The Class I, 4v dome has 6 separate strut lengths and a flat equator for a base. With "A" being the shortest and "E" being the longest strut, the strut factors are as follows: (A's factor = A_f), A_f = **0.25318**, B_f = **0.29524**, C_f = **0.29453**, D_f = **0.31287**, E_f = **0.32492** and F_f = **0.29859**.

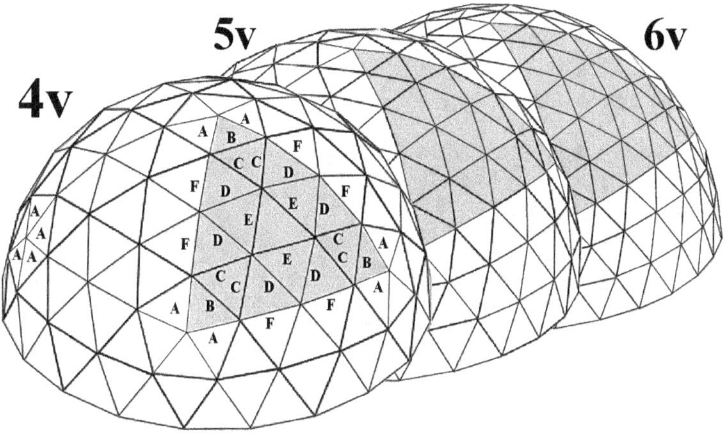

Both 5v and 6v frequency Class I domes need 9 separate strut lengths. Specifics for these and larger Class I frequencies will be skipped here. Anyone looking to build one of these to a substantive size will need a different reference, a crane, a crew of laborers and his or her head examined.

One can see the shadow of the pentagonal crown that's being lifted into place. Between 1971 and 1973 the United States Naval Construction Battalion (the Seabees) erected this dome at the Amundsen-Scott South Pole Station.

The NSF (U.S. National Science Foundation) only expected to get 10 to 15 years of service life from the structure but it lasted until 2009 before it was dismantled. There were vent holes in the crown of the dome to bleed off excess heat. Temperatures above freezing would have allowed the structure and its contents to sink deep into the snow. At least the dome blocked the fierce Antarctic winds. Designed by Temcor Inc. thedome is said to have been based upon a Class-II frequency 14 triacon breakdown.

Below triangles of OSB particle board are painted with water sealer and 4 holes are drilled in the corners with a small bit, for tie-wire to attach the board to the EMT pipes. After all faces are filled, the intention was to caulk the joints and seal the outside with a water barrier, attach a covering of chicken wire and then apply stucco or spray on gunite (shotcrete).

The inside could be completed in many ways; like attaching bubble wrap insulation covered with more OSB, or glued on polystyrene foam board insulation and wood veneer or even that insulation covered with sheetrock, plaster and tape.

Homemade chimney chase supports the stove pipe and isolates heat from combustibles.

On the next page is a "**trapezium dome**", which is not a geodesic. Neither is it as strong as a geodesic dome. But the trapezium does have some advantages.

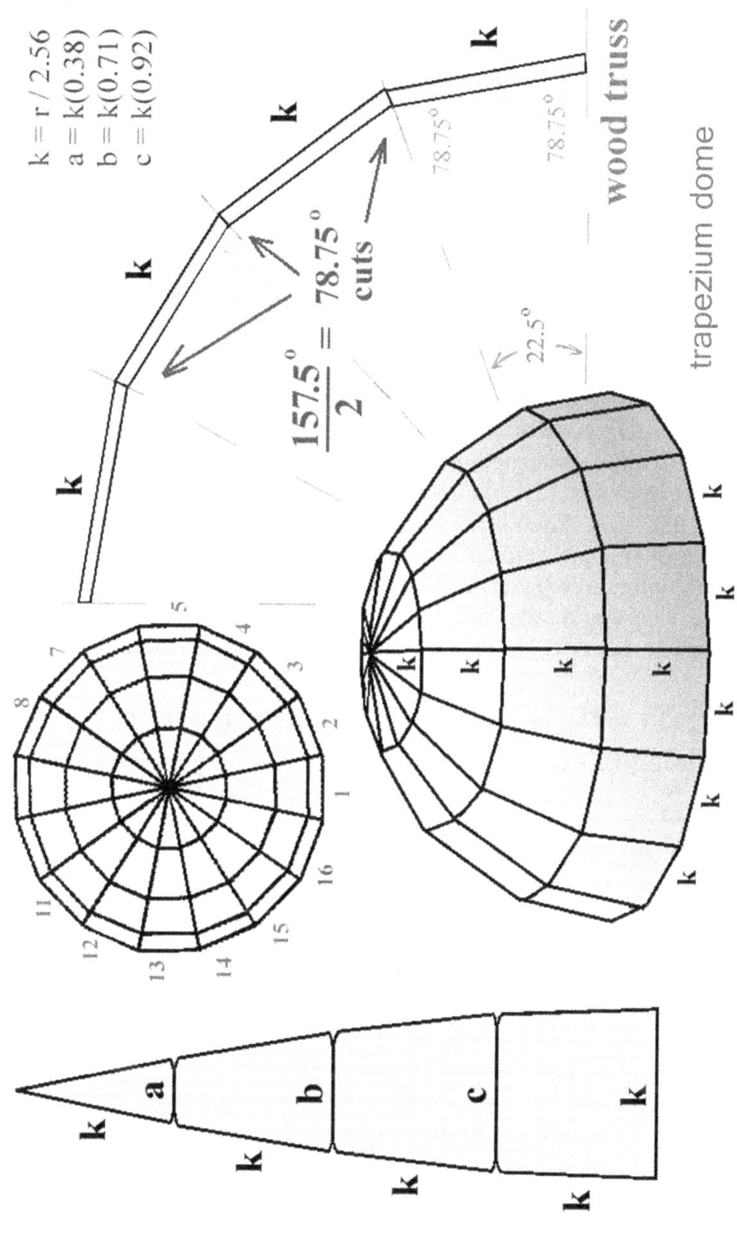

Realizing that the height of this structure should be the same length as its radius: the planner simply picks a desired diameter or a height, in order to determine the working radius. Then dividing that radius by the constant 2.56, returns the length of the cord named **"k"** in the image. Once "k" is known, the other cords are easily determined by multiplication and the plan for the dome falls into place.

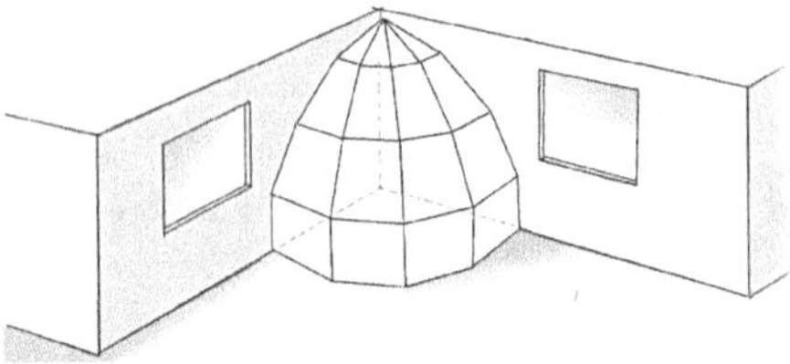
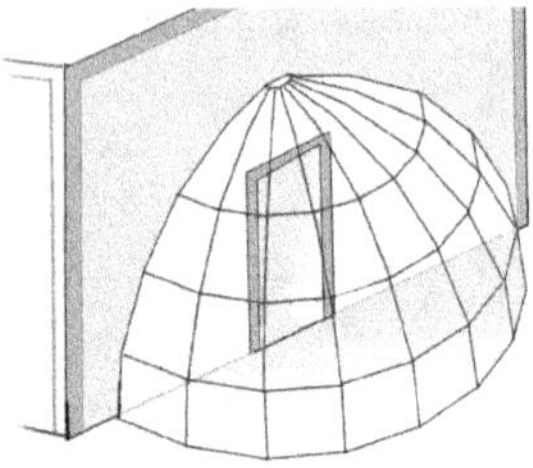

A **trapezium dome** is made of trapezoids. A trapezoid is a four sided convex quadrilateral shape that has two parallel sides. If three trapezoids and a triangle form a pie shaped peel then sixteen pie shaped peels are required to complete a circular dome (some parachutes are sewn this way). From a construction perspective the materials used for a trapezium dome would become congested at the very top. The top or crown will need some extra attention. Architects and home owners remodeling could easily adapt ¼, ½ or even ¾ manifestations of the trapezium dome to the outside of a house or other structure.

Trivia - a hodgepodge of information follows

Building Paper

"Early 20th century construction papers were made largely from softwood fibers, mechanically and chemically pulped".[43] The **sulphite process** was commonly used for low-grade papers such as newsprint, cheap magazines, and poster papers. *"Sulphite pulps were made by cooking low resin softwood chips under pressure at elevated temperatures in a solution of calcium, magnesium, or ammonium bisulfite. Spruce, balsam fir, and hemlock were typically used, and the sulphite process was the predominant pulping mode in the U.S. until about 1930".*

43 https://cool.culturalheritage.org/coolaic/sg/bpg/annual/v16/bp16-07.html

1939 / Farm Security Administration / public domain photo / Beltrami County, Minnesota

Above is an image of a well built "**tar paper shack**". The derogatory term probably arose during the Great Depression, when the majority of people were too poor to buy anything, let alone put siding on their house. While this example has something sturdy like clapboard underneath the "felt", other tar-paper shanties might have had nothing but the felt alone, covering the spaced studs in a wall.

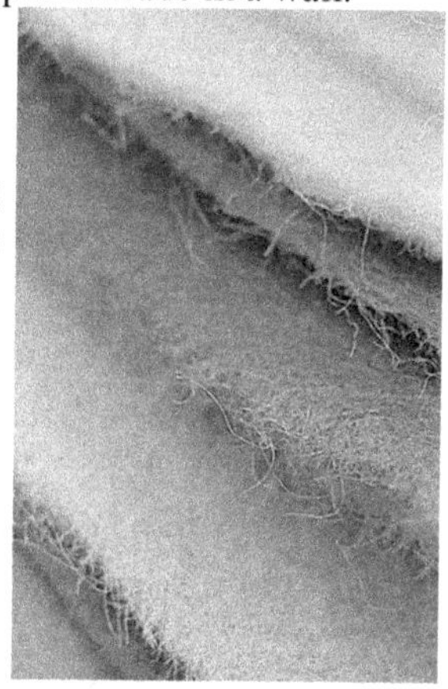

In older days "building paper" was wrapped around the house and afterword, siding installed over the wrapping. Building felt (called tar paper today) however was the heavier paper put on the roof.

*(**Felt** is a compressed, unwoven fabric of matted fibers; animal, plant or synthetic).*

Both black building paper and tar paper are impregnated with tar. Wood is mostly cellulose and so is cotton fiber. Building paper has become rare, but it can still be found. It's been replaced by synthetic house-wrap, which may or may not be progress once cost is considered.

Thicker tar paper for the roof is still labeled and sold in two forms; as 15 pound (15 lb., #15 or 15#) or as 30 lb paper. A roll of either once covered a "square". A **square** in the world of roofing, means **100 square feet**. In older days a square's worth of 15# roofing felt actually weighed 15 pounds, while the heavier, thicker felt with more asphalt and covering the same area, actually weighed 30 pounds per roll. Naturally the merchandisers found a way to cheat and change things. Not only are the newer products lighter and thinner, but the rolls now come in a variety of lengths. A modern roll of 15 pound tar paper is usually about 144 ft. long and covers about 4 squares. A new "30-pound" roll of felt today, might cover 2 squares.

Some over-hyped, new materials on the market today seem to outshine building paper and roofing felt in one way or another. But at a cost. Products like Tyvek® house-wrap are difficult to tear and are waterproof, while still allowing gas vapors from the house to breathe through and escape. New **ice and water shield** roofing underlayment is a better product for those roofs with little to no slope, or where snow and ice can dam up – causing water to back up under the overlapping roofing and therefore possibly leak. Slope serves a purpose.

But these newer materials can be unnecessarily expensive. Usually several times more expensive than the old asphalt impregnated product, oftentimes without offering a clear advantage. A home properly wrapped with building paper is still water proof from the rain and still breathes through the overlaps between the runs of paper. Tar felt on the roof when timely covered with the permanent roofing, isn't going anywhere. Without cracks and tears caused by previous winds, hot sunlight and foot traffic, fresh tar paper should preform its duty admirably; which is to conduct vapor condensation and other water away that might find its way beneath the primary roofing.

Coal-tar creosote is that oily black-brown sludge that's painted onto, soaked or pressure squeezed into telephone poles, power poles and rail road ties (sleepers). Coal-tar creosote is superior wood preservative and pesticide; a byproduct from the destructive distillation of coal, when making coke. In the past lumber used in ship hulls or just normal dimensioned construction wood was available in creosote treated form. The EPA has restricted its use, but it is still in use because it is very effective and cheap.

The first "**creosote**" (in wood tar form) was extracted by dry distillation of beechwood in 1832. Two years later (coal tar creosote) was discovered, and for a long time, was thought to be the same chemically. This was **phenol** - "Karbolsäure" (coal-oil-acid, carbolic acid).

> **Carbolic acid** (phenol) is a weak acid that once served as an important disinfectant but now is primarily used to synthesize plastics. *"Phenol and its chemical derivatives are essential for production of polycarbonates, epoxies, explosives such as picric acid, Bakelite, nylon, detergents, herbicides such as phenoxy herbicides, and numerous pharmaceutical drugs"*. Concentrated phenol is even used as an aircraft paint stripper – that removes polyurethane and epoxy coatings.

Coal tar creosote is stronger and more toxic than wood tar creosote; potentially having carcinogenic properties. Wood tar creosote still has medical applications, as an anaesthetic, antiseptic, astringent, expectorant, and laxative. (Another variety of creosote is made from both oil shale and petroleum).

Wolmanized wood

Back in 1911 a German scientist named Dr. Heinrich Wolman created a different process for preserving wood. The wood is **steamed** to open up pores in the wood cells, before it was put into a **vacuum** chamber. A preservative solution is sucked into the wood cells because of the negative pressure created by the vacuum. Then pressure is eventually increased and slowly elevated to about 150 psi. Woods like Hem & Douglass fur and Sitka spruce might have had their surfaces punctured beforehand, to better enable the preservative's penetration.

CCA (Chromated Copper Arsenate) only became an available chemical treatment after 1933 and was initially patented and sold in Europe. The green, specially processed wood became widespread by the 1970s. Wood properly **Wolmanized** is dry, its CCA becomes fixated and insoluble; safe to handle and to be around. The finished wood becomes slightly stronger, resistant to rot, fungal damage, and termites. Arsenic makes wood insoluble and leach resistant. But since 2003 concerns about the arsenic content of CCA led to a ban of its use in residential applications.

Other pressure-treated wood compounds like **ACQ** (Alkaline Copper Quaternary), **Copper Azole** and **MCA** (Micronized Copper Azole) are sold today as safer alternatives to CCA. When burned, CCA treated wood releases mild arsenic fumes into the air. Burning woods treated with the newer copper preservatives, probably releases undesirable compounds into the air also.

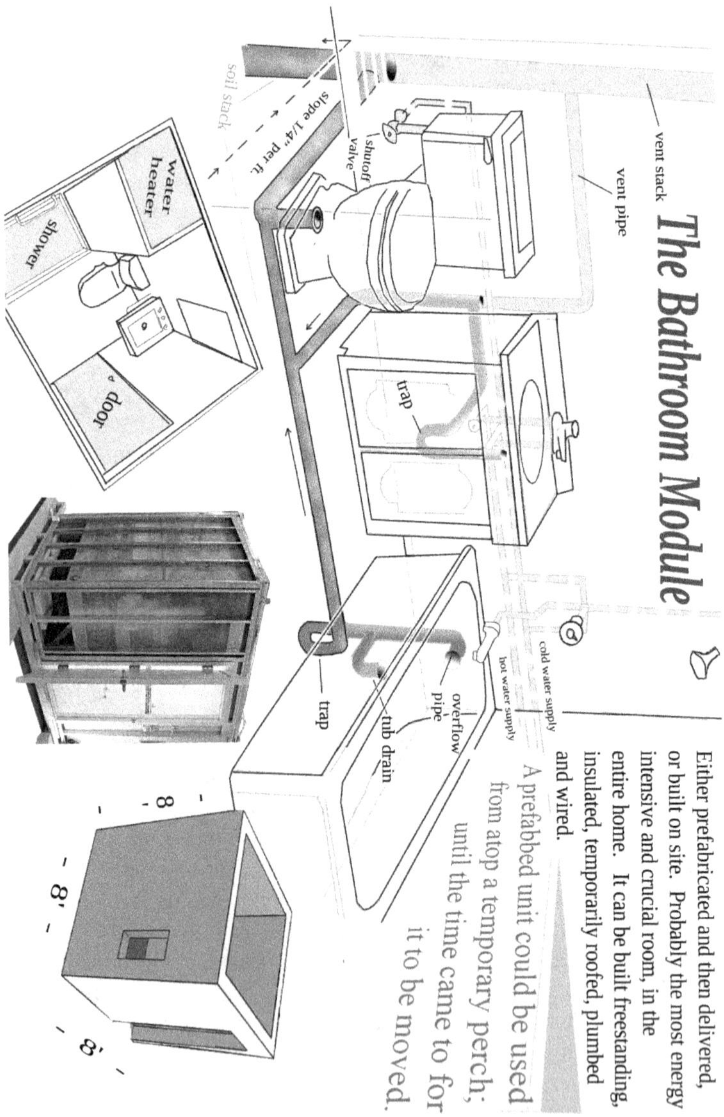

Above is a layout of a conventional bathroom plumbing wall and some miniatures of a concept for a "modular bathroom". A modular, portable, pre-manufactured bathroom that can be delivered to location, set on a foundation and connected to the utilities. A standalone unit that is usable, while the rest of the home is assembled around it.

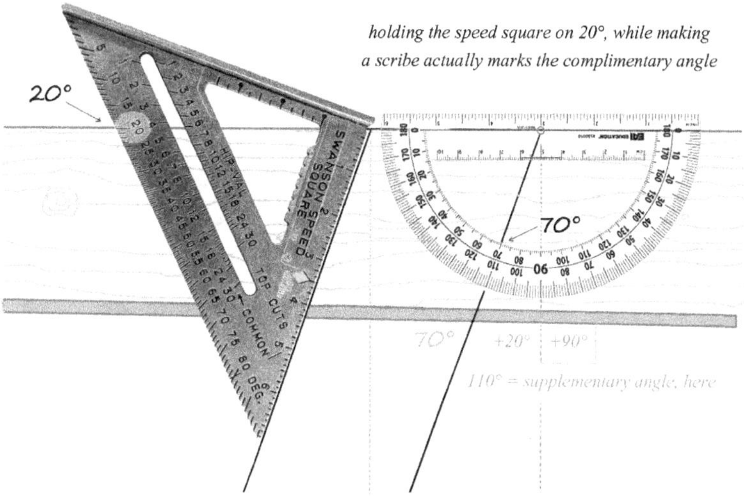

The eloquent "Swanson's Speed-Square"® was invented over a century ago now. A metric version was created, only recently.

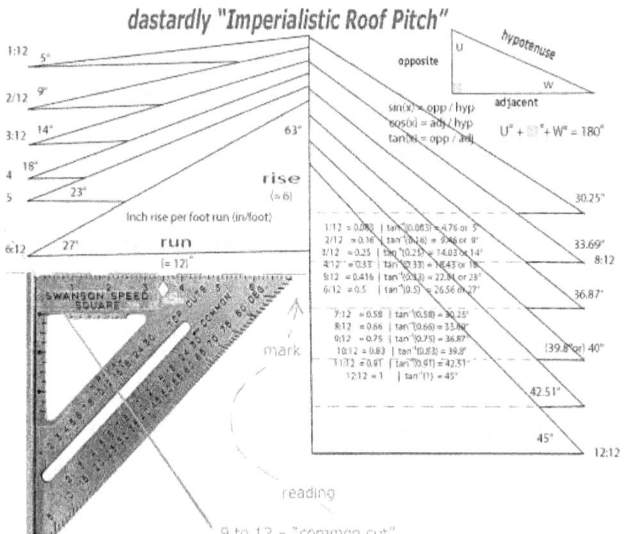

The original speed-square calculated rafter cuts using a variable ratio were 12 was constant, while hip & valley cuts used 17 as a constant. The degree scales between the Imperial and metric versions of the square are identical But on the new metric speed squares, ratios are based on scales calculated by using different constants: (10) decimeters for the common run, and (14) decimeters for the hip & valley run.

Radford's Cyclopedia of Construction: Carpentry
'Building and Architecture, Volume 8' (1909)

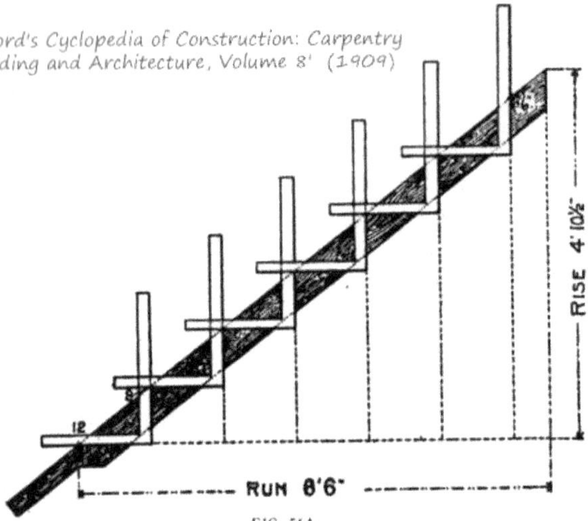

FIG. 56A

- demonstrates the "stepping off" method for laying out a stair stringer

Builder's squares of some type have probably existed since before the Pyramids were built. But over a forge in 1817, a blacksmith welded two steel rock cutting saw-blades together at a right angle – and afterword patented the first all metal "**framing square**". Later other versions would come marked with accurate increments scored along both edges and even stamped with pre-calculated common rafter lengths, according to pitch. No batteries needed.

One place where the 200 year old framing square still outshines any other tool and has no superior modern replacement, is when cutting stair jacks (runners or stringers). Staircases can challenge a carpenter's competency. Staircases are not easy, whether built of stone, metal, concrete or wood. Taken individually, the height between platforms or levels will vary and the conditions or constraints between separate projects might be unique. But all the while, the height and width of each step in a staircase should remain identical.

> * A typically standard approach to constructing a staircase is to employ a rule of thumb known as the 7-11 rule. That means calculations begin with the assumption of using 11" (inch) treads for the run, and 7" risers for elevation (rise). So for a given example where two floors are separated by a unique distance, the builder would begin by dividing that unique distance by increments of 7 inches.

Staircase

(typical indoor wood-framed type)

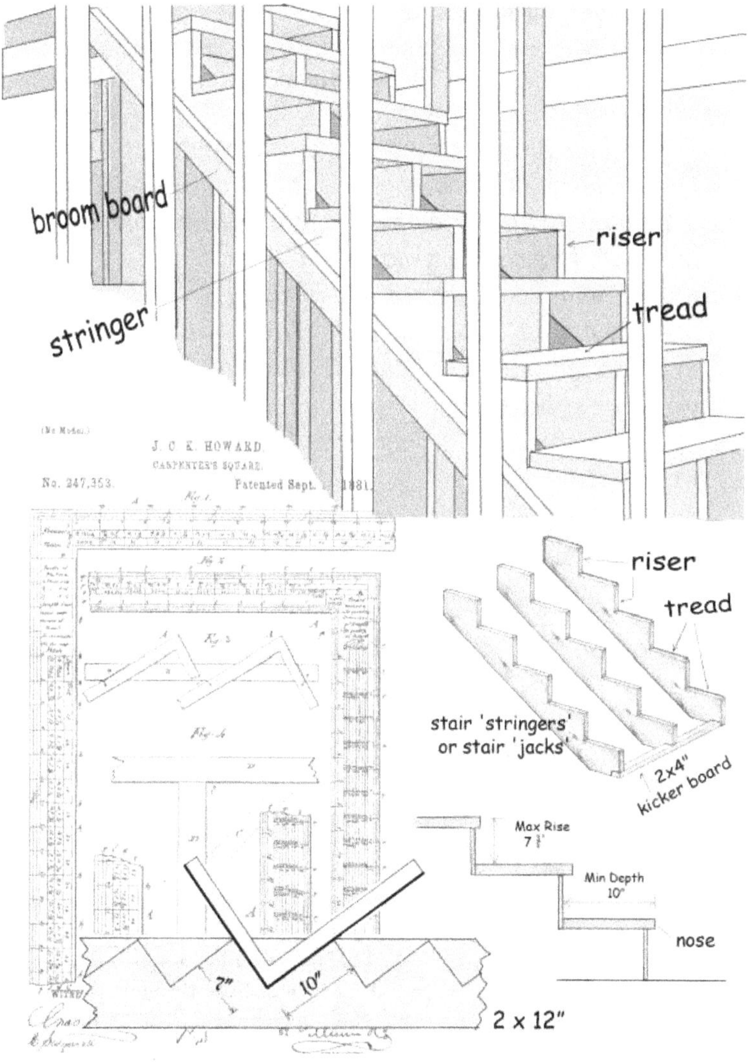

But almost never, will such a division result in a whole integer. You can't (or must not) construct a fraction of a step to make up this missing distance. Therefore the builder must mentally **shuffle riser heights and tread lengths** until he or she arrives with a solution that satisfies the situation with a **whole number** (no fractions, no decimals) of **uniform** steps. Complicating the solution to the calculation of a staircase are some rules adopted by the UBC (Uniform Building Code / now newly called the IBC or **International Building Code**).

To be brief, consider these IBC stair-mandates:

1] Minimum 36 inch clear width for stairway.
2] Minimum 6'-8" headroom height clearance for stairway.
3] Minimum 10 inch stair tread depth with nosing or Minimum 11 inches with no nosing.
4] Maximum 7-3/4 inch stair riser height.
5] Maximum 12'-7" vertical height (rise) for a flight of stairs.
6] Maximum 4-1/2 inch handrail projection into stairway width, on either side.
7] Handrail graspability: minimum 1-1/2 inch clearance from a wall.
8] Stair Handrail height to be placed a Minimum 34 inches to 38 inches

To be sure there is far more to know about stairs than this. For example, the rules listed (6, 7 and 8) concerning **handrails** are only applicable to stair flights running for four or more risers. With rule #4, some variation is allowed (any variation of risers within a flight of stairs must not be more than ⅜ inch). Rule #5 (arbitrary numbered) is to provide a **rest stop**; 151 inches (12' 7") if the height difference in floors were greater - the flight of stairs would require a level landing somewhere in-between to break up the flight of stairs. But stair landings, which fit in as height for one of the steps, are often needed for a change directions anyway, because of lack of room.

The *Oklahoma City sawhorse*

This is a superior design. It's been around for a while, I'd estimate 45 to 50 yrs. Two 'Ole Boys' that I worked for used it and they came from Oklahoma City. Its almost as simple as the more common I-beam type sawhorse, but has a few added benefits. This sawhorse is quick to build, rugged, strong and stable.

* A pair or more of these horses and a plank are handy for a primary stage of scaffolding. Simple wooden scaffold extensions or poppets for a bow lathe can be placed down through the center.

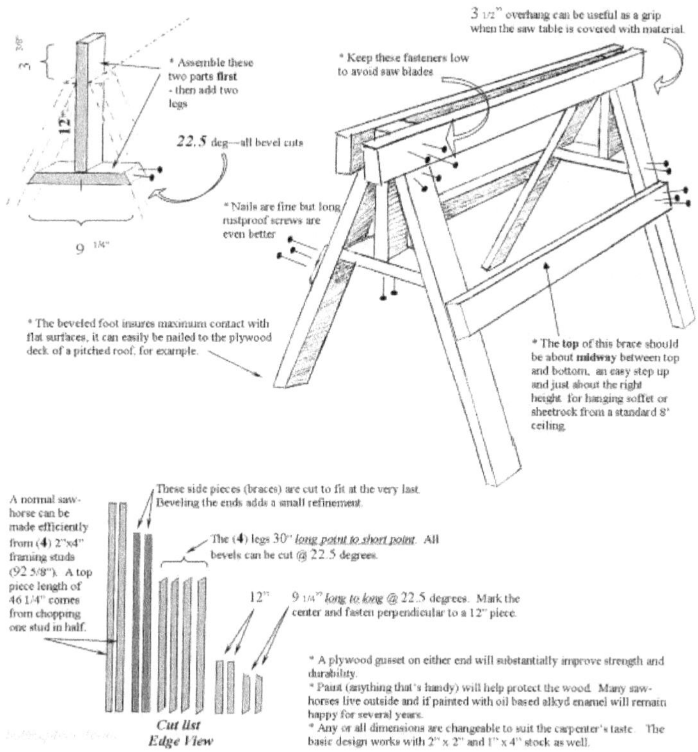

** An old publishing project, created for a web page. The next page (of wooden ladder) fitted the same purpose. Experimental, simple illustrations that could fill one whole page and occupy same with adequate explanation – needing no additional space. This image though was created with a larger "PDF" document in mind. Its resolution therefore is too condensed, to small to read when printed here on an 5.5" x 8" page.*

Anasazi Carpenters Ladder

A sturdy and quickly built tool

-- light
-- stable
-- low center of gravity

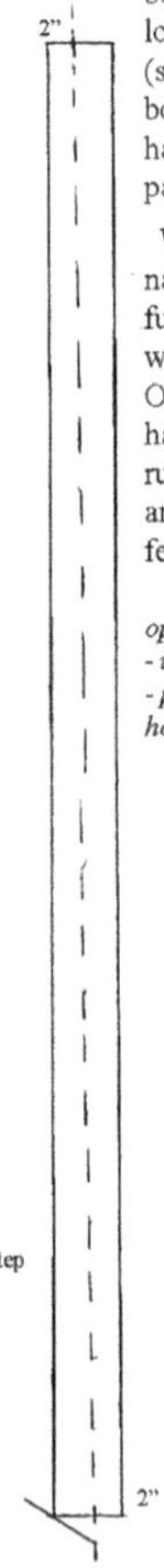

Choose a nice, straight and strong 2 x 6 that is clear of knots. Rip the board as shown, in this example a 12' long board is used. The ladder rungs (steps) can be made from 1" x 4" board, strips of 3/4" plywood or even hardwood slats taken from a forklift pallet.

When fastening the rungs ring shank nails, or screws are preferable. Be careful not to split the wood. Use about 6 well placed fasteners per rung. An OSHA approved wooden ladder must have stays (blocking) between the rungs, making the ladder a bit heavier, and may not be painted (might hide defects).

optional:
- use galvanized or zinc coated fasteners
- paint the ladder to protect the wood (if for home use)

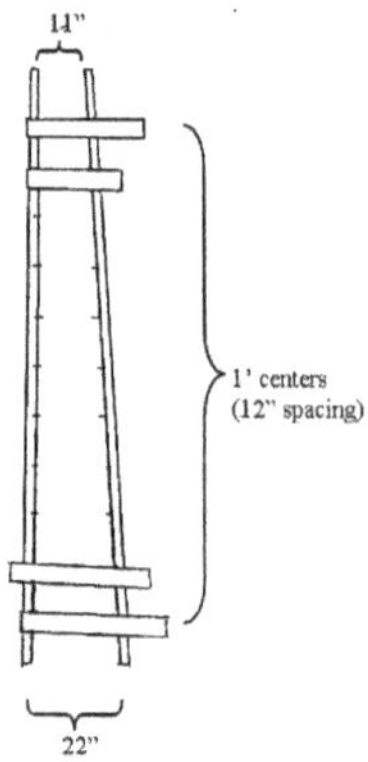

When "engineered trusses" are specified in a plan, it could mean more than one thing. First, spans are braced with struts scientifically, structured to carry a specified minimum roof load. Many configurations of engineered trusses with varying pitch and room spans exist. It is easy to observe and duplicate such trusses. However some custom plans and/or some building departments might specify that each truss is to be individually inspected by an engineer and stamped for integrity.

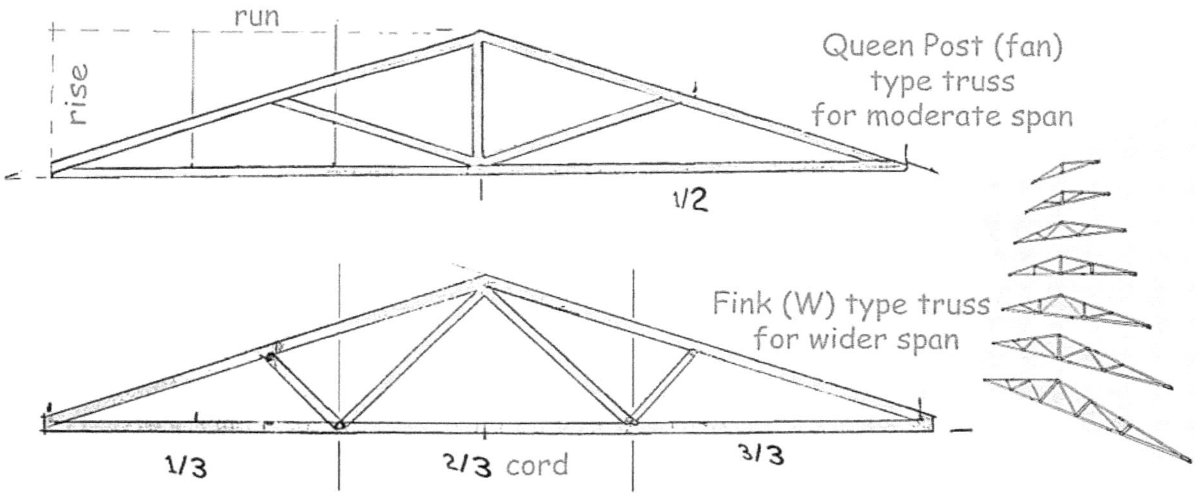

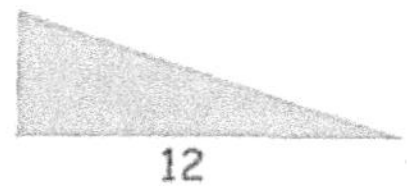

For carpentry that uses Imperial units, roof pitch is customarily described as rise (in inches) over the run (a constant, which is always 12 inches or 1 foot)..

Drawn to scale, both trusses above have the same 4:12 pitch.
As drawn: the rise = 1/3(run) or 3*(rise)= run; only one half of a truss or half a span are considered when defining pitch for a "common" roof.
So if the run is the constant 12 in, the rise then = 1/3 *(12in) = 4".

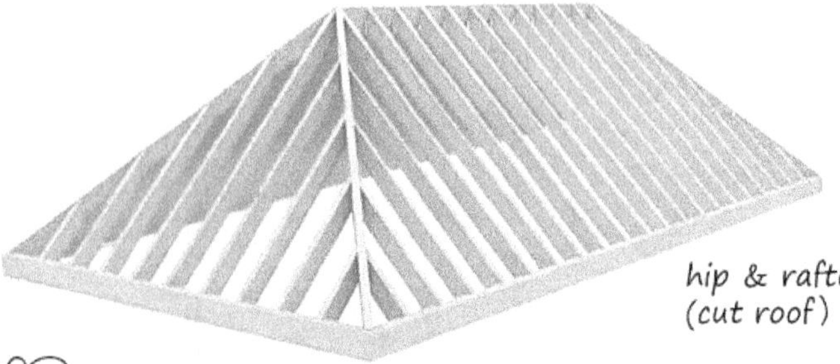

hip & rafter
(cut roof)

VS

hip style roof
made with trusses
(a delivered kit)

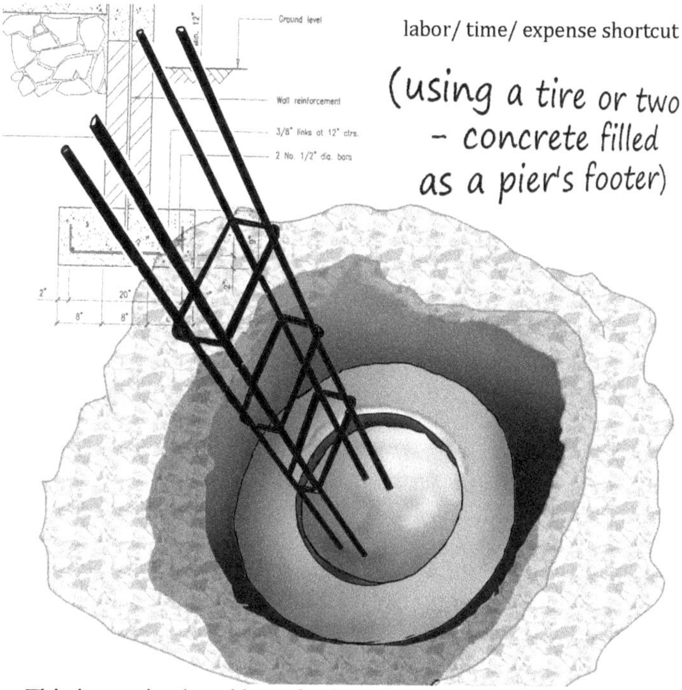

This image is placed here for humor. Sort of. In a "third world country", using an old tire as a footing form might seem perfectly logical. It is structurally sufficient and "kills two birds with one stone", so to speak. A function is preformed and trash is disposed of.

> *((Tires are one of civilization's most contrary waste products to discard. Practically indestructible, tires either take up excessive room in landfills or they are occasionally chopped up (at great expense) to be turned into sidewalk pavement)).*

But a building inspector in a despotically bureaucratic state like California might never condone letting a "private land owner" bury a tire this way. Building inspectors in some instances can be insufferable, self assured pedantocrats.

The term **private property** today is a misnomer and far cry from what it should mean, or what it once meant a century ago. Across the country, state and local entities have awarded themselves decisive authority over private land usage. This control is manifest in the forms of local zoning boards, planning departments and land management agencies. The justification is perpetually defended and written off as necessary protection for the <u>health, safety and welfare of the public</u>. Public safety is a laudable goal. But several instances occur where a planning commission's action or a building department's interpretation of code enforcement have too little to do with public safety, and smell more like a self aggrandizing, unrelenting power grab.

Then consider situations of "eminent domain" where the government simply seizes private property. Or consider annexation by a city, where a city expands its boundaries and jurisdiction, then imposes higher property taxes and changes local regulations.

The 4th amendment to the U.S. constitution protects individuals from unreasonable searches and seizures by the government. It should protect private property from warrant-less trespass by law enforcement. But it no longer does, actually. From 1791, various Supreme Courts have modified their interpretation of the amendment. In 2018 the court extended **curtilage** privacy protections to vehicles parked on private property. But then in 2026 the court was back to narrow the scope of curtilage.

> *"Curtilage refers to the area immediately surrounding a house, including any closely associated buildings and structures, but excluding open fields beyond. It is considered part of the home for legal purposes, particularly regarding privacy and protection from unreasonable searches".*

Some of the new language included in new Supreme Court interpretations of curtilage and private land includes "exigent circumstances". **Exigent circumstances** (prevention of physical harm, the destruction of evidence, or the escape of a suspect) can now allow police to enter a house without a warrant. There are numerous cases involving warrant-less entry under exigent circumstances, and it seems like a tool that could be easily exploited by ambitious cops, sheriff deputies, game wardens, FBI or ATF agents and the like.

Private property (private land in the U.S.) does not truly exist anymore in this day and age. Not when authorities can at their leisure, imagine many justifications for intrusion and trespass. Property isn't owned when it's taxed heavy and needs in perpetuity - defense against this annual extortion. A similar intrusion upon personal privilege involving personal property, occurs when local governments piously award themselves an ultimate dominion over that same property. Offices that dictate private land usage with growing zeal include state land management agencies, local zoning boards, and planning departments.

The **International Code Council** (ICC) is a non-profit organization sponsored by the building trades. A new product comes along, gets written as a suggestion into the code book – and then local governments use the varying codes, or not, as they see fit. There can be great variations in accepted codes, depending upon locality.

Planning commissions and building departments in some locations have grown into "a law unto themselves". They have considerable leverage to restrict the free will and ambitions of private property owners. They can and have used this growing power in punitive ways on some occasions (harassing builders with extra inspections and excessive or unnecessary code mandates). Ultimately, public defenders (district attorneys) are there to empower these actions.

Building codes and local ordinances can significantly increase construction costs. Permit fees are usually calculated as a percentage of the total project cost. Inspections for foundations, framing, plumbing, electrical and occupancy are scheduled and conducted by smug, clipboard toting bureaucrats wearing mirrored sunglasses and driving official looking public vehicles.

Again, building codes and zoning ordinances have commendable justifications; to a point. But today the present generation is hard pressed to find any adequate housing, let alone affordable housing. The bureaucratic red tape imposed by established officialdom, exacerbates the housing shortage by making housing ever more expensive and more difficult to acquire. By and large these institutions are hostile to the notion of someone pioneering or living off-GRID. It is a question of power and control. They want you the consumer to be plugged in and reliant upon (their) public utilities. They want a compliant public and for you to help feed the system by paying for public water delivery, sewage, trash collection and electricity. Non compliant, unauthorized practices of homesteading on private property – will be discouraged or disallowed.

As said before, there is a great discrepancy between states and local communities when it comes to building ordinances. You might expect a state like **Oregon** though to be lenient towards butterfly sniffers, tree hungers and hippies returning to nature by moving off the grid. But there: a yurt is an illegal occupancy to live in and you cannot build ponds or water catchments on your own land. In **New York** state they despise RVs and it's generally illegal to camp upon your own land! In **Texas** the authorities have codes against collecting rain water. They consider composting toilets a form of unlawful waste management and require state approved septic systems. In **Nevada** the state owns all water. You can't have any ponds and you can't use an RV as a permanent residence. You'd need a permit to live in one just part time; on your own land. A state like **Hawaii** micromanages everything. Other states vary accordingly in their chosen ordinances and vigor of code enforcement. These days, official flyovers by **snooping drones** is a new potential privacy harassment against erstwhile "private *property owners*".

*This is a **print on demand** book. Just scan the barcode on the back or enter the ISBN into a browser to order one. A copy should be printed and delivered in a few days.*

Bret Cargile

www.ingramcontent.com/pod-product-compliance
Lightning Source LLC
LaVergne TN
LVHW052214200926
848481LV00001B/104
* 9 7 9 8 2 3 4 0 8 4 2 1 7 *